Tropical Plants of Costa Rica

Tropical Plants of Costa Rica

A GUIDE TO NATIVE AND EXOTIC FLORA

Text and Illustrations by
WILLOW ZUCHOWSKI

Photographs by Turid Forsyth

A Zona Tropical Publication
FROM
COMSTOCK PUBLISHING ASSOCIATES
a division of
Cornell University Press
Ithaca and London

Copyright © 2007 by Marc Roegiers and John K. McCuen
Illustrations copyright © 2007 by Willow Zuchowski

First published as *A Guide to Tropical Plants of Costa Rica*,
2005, by Zona Tropical
Cornell edition first published 2007

All text and illustrations by Willow Zuchowski
Except where otherwise noted, photographs by Turid Forsyth

Cornell ISBN: 0-978-8014-4588-0 (cloth: alk. paper)
Cornell ISBN: 0-978-8014-7374-6 (pbk.: alk. paper)

Zona Tropical ISBN-10: 0-9705678-4-7
Zona Tropical ISBN-13: 978-0-9705678-4-0

Printed in China
10 9 8 7 6 5 4 3 2 1

Librarians: Cataloging-in-Publication Data for the Cornell edition
are available from the Library of Congress.

Editor: David Featherstone
Book design: Zona Creativa, S.A.
Designer: Gabriela Wattson

Table of Contents

Acknowledgments

The creation of this book would have been impossible without the assistance of photographer and friend Turid Forsyth, Zona Tropical publishers John McCuen and Marc Roegiers, and my husband William Haber. Turid, often struggling with the wind, composed photographs that not only depict the plant species well but also show their beauty. John made many visits to my home in Monteverde to work on the preparation of the final text. Bill was an on-the-spot source of information about plants and their natural history, in addition to being editor, photographer, and chauffeur at times.

I cannot list all of the botanists who have contributed to this book. Many based in Costa Rica at INBio (Instituto Nacional de Biodiversidad), and others working from afar, continue to describe and classify plants—their results being published in the Manual de Plantas de Costa Rica. I would especially like to thank Barry Hammel and Mike Grayum, of the Missouri Botanical Garden, who answered many of my plant identification questions. Mike made nomenclatural corrections in the manuscript and added interesting bits of information. I thank Robbin Moran of New York Botanical Garden for reviewing and editing text on ferns and John Atwood for reviewing text on orchids.

Turid and I are both grateful to neighbors in Monteverde who shared their gardens with us. Fortunately, we found a similar welcoming attitude in people as we traveled around Costa Rica to photograph and study plants.

Owners, managers, and staff of various businesses and biological stations often facilitated our photographic expeditions. These included Bruce Young, Cynthia Echeverria, Bob Matlock, and Orlando Vargas at the Organization for Tropical Studies' La Selva Biological Station, Luis Diego Gomez at Wilson Botanical Garden, the late Dora Emilia Retana at Lankester Garden, Peter and Lindy Kring at the Tropical Botanical Garden - Puerto Viejo, William Aspinall at the Arenal Observatory Lodge, Jim Wolfe and Marta Iris Salazar at the Monteverde Butterfly Garden, Gabriel Barboza of the Monteverde Orchid Garden, as well as personnel at Zoo Ave and Costa Flores.

Others who were of assistance on the many excursions Turid and I made were Gordon and Jutta Frankie, Margaret Adelman, Sarah Dowell, and Mel Baker. And we cannot forget taxi driver Guillermo Mata Valverde who, upon our demand, would brake for any plant, anywhere.

I wish to thank Francis X. Faigal, Adrian Hepworth, Mike and Patricia Fogden, William A. Haber, Richard K. LaVal, Gordon W. Frankie, Luiz Claudio Marigo, and Adrian Forsyth for assisting in photography or letting us use their photos. Mark Wainwright, who recently completed a book on Costa Rican mammals, helped by being a good listener and rallier throughout the process. David Featherstone carefully read and edited the book manuscript, and Sharon Kinsman made very helpful edits and suggestions for several chapters. The artistic and careful eye of Gabriela Wattson of Zona Creativa guided design and prepress work.

I would like to recognize the hard work and vigilance of guards and other staff at the Costa Rican refuges and national parks, many of which we visited during the preparation of this book. I thank the staff at the Tropical Science Center and Monteverde Conservation League for protecting the Monteverde Cloud Forest Preserve and Bosque Eterno de los Niños, respectively. In addition, I thank all of the other businesses and individuals who own and protect forest sites throughout Costa Rica. May we not only maintain, but expand, the fine system of parks and reserves that protects the flora and fauna of this rich country.

Introduction

Many people who visit Costa Rica's tropical forests arrive with the expectation of seeing howler monkeys, resplendent quetzals, poison-dart frogs, and giant morpho butterflies. They quickly become aware, however, of the beauty and overwhelming diversity of the plants that provide food and shelter for these animals. Ranging from miniature epiphytic orchids to towering trees, from tiny aromatic flowers to large, brilliantly-colored blossoms, and from mangrove forests lining coastal waterways to high-elevation cloud forests, the thousands of plant species found in the country attract the attention of casual visitors as well as botanists and inspire artists and horticulturalists. People throughout the world use tropical plants everyday as sources of building materials, tools, medicines, perfumes, and countless other products. They harvest food from both wild and farmed plants, and transplant them into homes and gardens as ornamentals. Tropical plants interact dynamically with atmospheric phenomena to affect climate and water cycles; and they produce oxygen necessary to sustain the life of animal species and maintain the earth's atmosphere.

A Guide to Tropical Plants of Costa Rica is an introduction to the common and conspicuous plants that both visitors to and residents of Costa Rica might observe in their travels around the country, and its organization reflects the different contexts in which people may encounter these plants. More than 430 plants are included and illustrated with a photograph and/or a pen and ink illustration. Each species account describes a plant's identifying characteristics and presents interesting facts about its natural history, chemical properties, and medicinal and other uses. Sidebars dispersed throughout the book take the narrative beyond individual species to focus on groups of plants, unusual uses for plants, and other topics of interest.

The Scope of the Book

Although Costa Rica is a small country, its wide range of habitats houses more than nine thousand native vascular plant species alone. An identification guide to all of those plants would not only require several volumes, but would be unwieldy for use in the field, which is the primary intent of this book. The author has selected for inclusion those species that are conspicuous and/or common, or that one might readily encounter following one of the more well-traveled itineraries in Costa Rica. The goal was to produce a book that would both contain enough species to be useful and be small enough to carry into the field.

All of the plants included occur in Costa Rica, but not all of them are native to the country. Many plants in this guide are exotics, meaning they are not native to Costa Rica but have been introduced from elsewhere. Throughout history, people have transplanted native plants they find useful and/or beautiful to places far from the plants' origins. A number of the tropical fruits, garden ornamentals, and large trees with showy flowers one sees throughout the country are actually native to the forests of South America, Southeast Asia, Australia, Africa, or the Pacific Islands. Thus, much of the information here is pertinent to those traveling not only in Costa Rica, but in other tropical countries also.

With this book, people traveling through Costa Rica's tropical forests, or even along highways and city streets, should be able to identify many of the plants they see and develop an understanding of the natural history of the country's native and exotic flora. The attention given to plants of the Monteverde region is greater, due both to the diversity of species found there and to its renown as a tourist destination. The author has also spent fifteen years participating in an inventory of that region's flora.

How to Use This Book

For botanists, a sensible way of arranging a plant guide is to organize chapters around plant families, or to present plant species in alphabetical order according to their scientific names. Those systems are not particularly useful to the nonbotanist, however, because the lay person simply does not tend to think of plants in terms of the family they belong to, and most people refer to plants by their common names, rather than by using Latin scientific terminology.

Authors writing for the general public have developed a number of other ways to organize plant books. These include listing species by flower color, by habitat, or by plant growth form, such as tree, shrub, vine, or herb. Each of these has advantages, but they also have disadvantages, especially when covering the diverse flora of a country such as Costa Rica.

For *A Guide to Tropical Plants of Costa Rica*, the author has incorporated several different organizing principles, building on the different ways one would encounter plants when traveling around the country. The following table provides a quick reference to the book's structure.

Chapter	General Content	Page
1. Painted Treetops	Commonly seen trees that have very showy, colorful flowers.	19
2. Other Common Trees	Trees that are seen frequently but lack large colorful flowers.	39
3. Roadside and Garden Ornaments	Plants often seen in gardens and along roadsides. Included here are both wild and cultivated species.	73
4. Fruits and Crops	Edible plants grown for commercial harvest or for home use.	161
5. Living Fences and Reforestation	Species farmers and other landowners in Costa Rica frequently plant to fence their property. Species commonly grown in tree plantations are also included here.	211
6. Special Habitats	Includes descriptions of common species that grow in distinct habitats: wet Atlantic lowlands, tropical dry forest, tropical montane cloud forest, and beach and mangrove.	235
7. Typical Tropical Groups	Groups of plants that are generally considered characteristic of the tropics: aroids, bromeliads, palms, heliconias, orchids, and ferns.	353
8. Conspicuous Grasses	An introduction to some of the many grasses commonly seen along roadsides.	459

Although this structure is intended to direct readers to information efficiently, inevitably some plants could have been placed in more than one section. Those with more background in botany may want to consult the list of species by family on page 481. It lists the plant families included in the book, and the species within those families, and provides the page number where the species account and/or photo can be found. The index contains scientific and common names of all plants in the book.

Each chapter, as well as most sections within a chapter, begins with a brief descriptive introduction. The species accounts that follow have a consistent structure to make it easy for readers to locate the information they are seeking. Each account begins with the scientific name for the plant, its common name/s, and the plant family within which botanists have classified it. Both English and Spanish common names are given, although many of the plants do not have an English common name. The categories of information in the species accounts include the following:

OTHER COMMON NAMES. Lists other English and Spanish names by which the plant is known.

DESCRIPTION. Details the physical characteristics of the plant that may aid in identification. The description for each species is a series of concise sentences that details the plant's growth form (tree, shrub, vine, etc.); its leaves; its flowers; and its fruit, including the seeds. Individual plants within a species can be variable in many ways. The leaves of a juvenile, for example, may be quite different from those of an adult plant. Environmental conditions such as sun, shade, and wind can also affect a plant's appearance. Since individual plants in a species vary considerably in the size of their parts, the descriptions often provide a range of sizes or indicate an approximate size (for example, ca. 10 cm) or a maximum size (to 10 cm). All lengths are given in metric terms. Most plant illustrations include a number that indicates the relationship between the size of the illustration and the true size of the plant it represents. The label "x 2", for example, says that the illustration is twice as large as the plant.

FLOWERING/FRUITING. Indicates the season when the plant blooms and develops fruit. Generally, the rainy or wet season is from mid-May to mid-November; the dry season lasts from mid-November to mid-May. For a more technical discussion of weather patterns in Costa Rica, see page 15.

DISTRIBUTION. Gives the geographic range of a plant worldwide as well as within Costa Rica. When a plant occurs within a well-known region of the country, such as Monteverde, La Selva, Tortuguero, Osa, or Guanacaste, that information is also provided.

RELATED SPECIES. Lists other species a plant is related to and, when a relative is also included in the book, gives the relevant page number.

COMMENTS. The information in this section covers a wide range of topics, with details drawn from scientific studies, standard reference works on the natural history of Costa Rica, and the author's personal observations. In addition to discussing the natural history of the plant, these comments include information about how people have used the plant, its importance in history and/or commerce, and its medicinal value. Note that information on medicinal use is not intended as advice on treatment of medical conditions. Readers should not ingest or apply externally any plant materials without first doing an extensive investigation of a plant's chemical properties; such information is beyond the scope of this book.

Much of the terminology botanists use may be unfamiliar, however technical terms such as those describing hairs on a plant—*sericeous, tomentose, stellate,* or *strigose*—can be extremely useful for detailing plant characteristics precisely. Although the author has limited the use of technical terms, the book does include a certain amount of botanical jargon. Its extended glossary defines these terms and includes illustrations to help readers understand them.

Botanical Classification and Nomenclature

The plant kingdom contains a tremendously wide range of organisms, from tiny mosses to enormous redwood trees. Plants fall into one of two broad categories—vascular plants and bryophytes. Vascular plants have systems of roots, stems, and leaves by which they transport the water and chemicals that sustain their life. Bryophytes, which include mosses, liverworts, and hornworts, lack these systems and must absorb moisture directly from the air or ground.

The focus of this book is on vascular plants, especially flowering plants (angiosperms), which produce flowers containing ovules that later turn into seeds. A few non-flowering plants—cycads, conifers, and a sampling of ferns—are also included.

Botanists have always struggled with how to name plants, especially in light of the number of species that exist worldwide. Common names are often descriptive and easy to remember, but they are not always unique. In different parts of the world, for example, *oregano* may refer to *Origanum vulgare* (Lamiaceae) or *Lippia graveolens* (Verbenaceae). The name *ironwood* is given to many tree species with dense wood. And in Spanish, the term *aguacatillo* (little avocado) may refer to any one of dozens of tree species that produce fruits resembling small avocados. The name *tabacón* has been given to an array of plants—from *Anthurium* species to a number of trees—that have large, tobacco-like leaves.

To avoid confusion, botanists give each species a scientific name, in Latin, known as a binomial, that is made up of two parts—a genus name and a species name. A scientific name for a plant species, such as *Origanum vulgare*, is thus unique and refers to a distinct set of plant populations.

On the other hand, the old biological concept of a species as a distinct population that can interbreed and produce fertile offspring often does not work in the plant kingdom, where breeding systems are diverse. The result is that there is substantial debate in the field about the correct way to define a species. Botanists frequently find it necessary to assign plants to taxonomic levels more specific than species—subspecies and varieties. Horticulturalists also speak in terms of cultivars and forms when they are talking about garden species that have been selected for special characteristics.

Morphological similarities between plants, especially similarities in reproductive structures, have traditionally been the basis for classification, but as scientists learn more about the chemical properties, ecology, and reproductive systems of plants, and as they gather more molecular data about them through techniques such as DNA sequencing, they can define and classify species more accurately. The result is that, as specialists gather more information and look at the larger picture of relationships between organisms, they shift plants from one category to another, lumping plants that had been considered different species together or splitting an established species and reassigning individuals into different species. Even though this leads to a better understanding of a plant's kinship and ancestry, it also requires botanists to adapt to a series of name changes. In addition there are different schools of thought

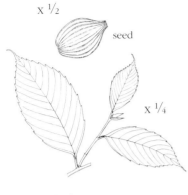

x ½

seed

x ¼

Ticodendron incognitum

in systematics that disagree about how species should be named and grouped. It is not uncommon to find that the same plant is called by one name in one source and by another name elsewhere. This is particularly true of tropical plants.

Researchers continue to discover new plant species in tropical areas, and this makes the field of tropical plant systematics exciting and dynamic. Sometimes botanists are able to describe a new genus, or, rarely, a previously unknown family. As recently as 1989, after carefully studying an unusual tree that did not fit into any known plant family, Costa Rican botanists described *Ticodendron incognitum* (illus. left) and created a new family, Ticodendraceae.

Observing and Identifying Plants

Plant identification can be relatively easy in cases where a plant has very distinctive characteristics. On the other hand, there are many look-alikes in the plant world, so one should always inspect vegetative, flower, and fruit details closely when possible.

A field botanist usually carries along a few tools to aid in plant identification. Binoculars allow you to see shapes of flowers at a distance and to see if tree leaves are compound or simple, alternate or opposite, and have toothed margins. Hand pruning shears facilitate collecting samples of plants and a pole pruner is necessary for getting branch samples of trees. A pocket magnifier (10X lens) is very useful for inspecting hairs and other minute details on plants. Binoculars held backwards act as a magnifier also. If you are unable to identify a plant in the field, pictures taken with a digital camera may be sufficient for later identification. The traditional method of preserving plant specimens for future study is by pressing and drying them, so you may want to make or buy a plant press. For short term preservation, however, you can place plant samples in a plastic bag in a refrigerator. Most botanists find that a field notebook is very useful for taking notes about location and size of the plant, presence of odor or milky latex in the twigs, and any other information that will not be evident once you are back home.

The species accounts in this book provide a template readers can use to begin to describe what they are seeing. Whether you are collecting a sample for yourself for further study or to show to a botanist, you should try to get a twig with several leaves, and preferably a flower or a fruit. When you are trying to identify a tree, watch for fallen flowers and fruits, then search the trees overhead to identify their source. You can often find some fallen leaves from the same tree to observe leaf characteristics in the hand.

Species within a plant family share a general set of flower and fruit characteristics and often have similar vegetative distinctions such as leaf arrangement or odor. So, as you become more serious about learning plants, you may want to study family characteristics. In that way, even if you are unable to identify a plant, you at least might be able to determine which family it belongs to.

When you can't find your specimen in this book, other literature and web resources listed in the bibliography may be helpful. The Museo Nacional de Costa Rica, home of the National Herbarium, and INBio (Instituto Nacional de Biodiversidad) are two important Costa Rican resources for botantists. Both institutions house collections of dried, pressed samples of most of the plant species known from the country. They also have floras (technical manuals covering the plants of whole countries or geographic regions) that you can refer to. These books contain keys, descriptions, and illustrations that can help to positively identify an elusive specimen. Then you can compare your specimen with those in the collection to see if they closely match. However, the fastest and easiest way to identify a plant is to show it, or a very good photo, to someone familiar with the flora of the area in question.

Costa Rica

It should surprise no one that, over the past few decades, Costa Rica has become a major destination for visitors from throughout the world who want to experience tropical forests and observe the many and unusual species of plants and animals that live there. It is a small country, measuring just 51,100 square kilometers (19,730 square miles), roughly half the size of the state of Virginia. Within this small area is

Map of Costa Rica indicating key landmarks.

an amazing diversity of topography, climate, and habitat types—in general, the tropics are more species-rich compared to the temperate zones—which makes it possible for travelers to encounter a wide range of organisms without going great distances.

Costa Rica is a part of the land bridge between North and South America, and this location has added to the diversity of flora and fauna found there—and in the rest of Central America—as plants and animals have expanded their ranges in response to climatic and other changes over the eons. The country's topography is rugged. Four major mountain ranges, with peaks rising to 3,819 meters (12,530 feet), plus outlying mountains in areas such as the Nicoya and Osa Peninsulas, cover more than half the country. Such a great elevation change in close proximity to the ocean on either side, combined with the influence of the northeast trade winds, has created an array of microclimates throughout the country that provide habitats for distinct populations of plant and animal species.

The trade winds' influence is increased because the Continental Divide roughly runs perpendicular to the wind direction. Moisture carried off the Caribbean Sea is discharged on the Atlantic slope or moves over the mountains in the form of a dense cloudbank. On much of the Pacific slope these clouds dissipate, and a marked dry season occurs from November until May. From mid-May to October, when the trade winds weaken, convectional thunderstorms are frequent on the Pacific side as clouds roll up the mountain slope and dump rain when the warm air cools and condenses. On the southern Pacific slope, the tall Talamanca mountains create conditions that draw the winds off the ocean during much of the dry season. The dry season there lasts only 2 to 3 months. The Atlantic slope experiences even less seasonal variation in weather than the southern Pacific slope, with rain occurring there most months of the year.

The temperature extremes within the country match the range of the topography, from freezing at the highest elevations to 35° Celsius (95° Fahrenheit) or hotter at low elevations. The annual rainfall ranges from 1.3 to 8 meters (4.26 to 26.25 feet). The various combinations of temperature and rainfall create a dozen life zones in Costa Rica, ranging from tropical dry forest and tropical rainforest to tropical subalpine rain paramo (see Holdridge 1967 for more detail). In common parlance, people rely on a less complex set of categories, and speak of dry and wet lowland forests, montane forests, and coastal habitats.

Most of the dry forest is in the northwest section of the country, which is often referred to by the name of its province, Guanacaste. The Atlantic slope and its coastal areas are wet, with rich, lowland rainforest; such forests also occur in parts of the southern Pacific region. An array of montane forests are found in the mountain ranges, or *cordilleras*, that form the backbone of the country. The beach and mangrove vegetation on the Pacific and Atlantic coasts is generally similar, although some species are confined to the Atlantic coast.

Chapter 6 of this book provides more information about these habitats and their unique vegetation types.

Conservation

The fact that tropical forests around the world are disappearing is not news. The causes of this decline are varied, ranging from exploitation for lumber to clearing forests for agriculture to climate change. Many of the tropical plants that are common today in ornamental and agricultural gardens can no longer be found in their original homes, and even though their propagation elsewhere preserves the species, their loss in the wild due to habitat destruction is a tragedy that cannot be reversed.

Costa Rica is widely praised for its exemplary system of nature reserves throughout the country, yet deforestation remains an issue. An aerial view of the northeastern part of the country reveals the decimation of large tracts of forest, and people on the ground frequently see trunks of forest giants being transported along the Braulio Carillo highway. These are dramatic reminders of how quickly centuries of growth can be turned into lumber. Not only individual species, but whole habitat types, such as premontane moist forests, are on the verge of extinction. Many of the native Costa Rican plants included in this book are very common, but others are rare species that are threatened or endangered and may soon exist only in protected reserves. Some still unknown species may never be found.

Deforested areas on Pacific slope below Monteverde.

Key to Abbreviations

CATIE: Centro Agronómico Tropical de Investigación y Enseñanza.
DGF: Dirección General Forestal (General Forestry Directorate).
INBio: Instituto Nacional de Biodiversidad.
OTS: Organization for Tropical Studies.
sp.:   a species.
spp.:   species.
TRIALS: a joint program between OTS and DGF to screen little-known tropical species for reforestation.

1. Painted Treetops

Costa Rica's most eye-catching tree species are those whose large, brilliant blossoms create splashes of color across the landscape. While all trees have flowers—or strobili in the case of conifers and other gymnosperms—the majority have flowers that go unnoticed because they are tiny or dull colored. The bright yellow, orange, pink, purple, and red flowers of the species described in this chapter make them some of the most conspicuous and easy to identify trees in the country. The majority of these showy trees are in the bean and pea family (Fabaceae) or the trumpet creeper family (Bignoniaceae).

Most of the native species in this chapter grow naturally in the seasonal dry forests of Guanacaste. The blossoms are particularly noticeable in the dry season (December through April), when many trees lose their leaves. These trees produce pollen and nectar for native solitary bees, their chief pollinators, and thus play an important ecological role in the dry forest.

A few of these native trees, such as the pink trumpet tree, are planted as ornamentals to decorate streets in San José and other cities, but most of the showy, ornamental trees you will see along city streets are nonnatives. Poró, golden shower, jacaranda, queen's crape myrtle, African tulip tree, and flamboyant are among the most common nonnative ornamentals. All of these except for the poró may continue blooming into the early wet season.

Left, *Cassia fistula*

Jacaranda mimosifolia
Family: Bignoniaceae

Jacaranda, *Jacaranda*

Other common names: Green ebony, fern tree.

Description: Deciduous tree to 15 m; smooth, gray trunk. Opposite, fernlike compound leaves, twice pinnate, ca. 30 cm long; the numerous, narrow leaflets ca. 1 cm long. Clusters of bluish lavender flowers that are bell-shaped, 5-lobed, 3–4 cm long; constricted toward the base, with unusual, glandular-pubescent staminode extending beyond stamens. Flat, disc-shaped capsule ca. 6 cm long, with flattened, somewhat wavy edge; contains delicate, winged seeds (2 cm across, including wing).

Flowering/fruiting: Flowers and fruits from February through June.

Distribution: From northwest Argentina originally, now cultivated all over the world in tropical and subtropical areas. In Costa Rica, an ornamental grown at middle elevations.

Related species: A white form ('Alba') exists. Of the 49 species of *Jacaranda*, two are native to Costa Rica: *J. copaia*, found in wet lowlands, and *J. caucana*, in the southern Pacific region. Leaflets of both are larger than those of *J. mimosifolia*.

Comments: The colors and foliage of this species create a soft, cool impression in contrast to the fiery intensity of orange, red, and yellow seen in many showy flowering trees. Planted all over the world for its shade and elegance, it is able to tolerate some cold weather. The opposite, bipinnate leaves distinguish jacaranda's foliage from ornamentals with similar (but alternate) feathery leaves. The gland-tipped hairs on the staminode in the center of the flower probably influence bees' behavior to effect better pollen placement and removal. The native species *J. caucana*

Jacaranda mimosifolia

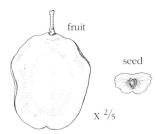

Jacaranda mimosifolia fruit with wind-dispersed seed.

contains an antitumor substance. Another, *J. copaia*, grows well in experimental reforestation plots in the Atlantic and northern zones of Costa Rica.

Spathodea campanulata African tulip tree, *Llama del bosque*
Family: Bignoniaceae

Other common names: Flame of the forest, fountain tree, *tulipán africano.*

Description: Tree from 15 to 20 m. Pinnately compound leaves to 40 cm, 7–19 entire leaflets, each 7–15 cm long. Dense terminal clusters of flowers, each ca. 10 cm long. Orange to scarlet corolla (with yellow edging) forming a scoop and then widening out into 5 broad lobes; brown buds in center of inflorescence. Erect, woody capsule, ca. 20 cm, is dehiscent, the 2 halves boat-shaped, containing many seeds, each 3 cm wide with a thin transparent wing.

Flowering/fruiting: May be found in flower in any month of the year.

Distribution: Native to tropical Africa, now planted in many tropical areas. In Costa Rica, an ornamental in parks and along streets, at low to midelevations, dry to wet habitats.

Related species: This is the only species in this genus; a yellow form (f. *aurea*) exists. This family also includes *Tabebuia*, *Jacaranda*, and *Crescentia*.

Comments: The showy flowering of this tree is not restricted to the dry season. Another common name, fountain tree, refers to how water that collects in the calyx of an unopened bud can be squeezed out squirt-gun fashion. In its native Africa, *Spathodea* has medicinal as well as magical uses.

X ¹/₂

Spathodea campanulata bud.

Spathodea campanulata

Tabebuia rosea
Family: Bignoniaceae

Pink trumpet tree, *Roble de sabana*

Description: Deciduous tree to ca. 25 m with dark gray, fissured bark. Twigs, calyx, leaves, and capsules covered with small scales. Leaves opposite, palmately compound, the five long-stemmed leaflets not uniform in size, with the largest (8–35 cm long) in the center. Dense clusters of 5–10 cm-long funnel-shaped flowers at branch tips; with 5 lobes, various shades of pink or white—the throat yellow. Slender, pendant, cylindrical capsule ca. 22–35 cm long, to 1.5 cm wide; dehiscent, with many thin-winged seeds, 2.5–4 cm broad.

Flowering/fruiting: Generally seen in flower from late December to May. Seeds mature late dry to early wet season.

Distribution: Mexico to northern South America. In Costa Rica, both slopes to ca. 1,200 m. Tolerates dry to wet conditions; common in dry forest; planted in Central Valley (e.g., Paseo Colón in San José).

Related species: Approximately 100 species in the genus *Tabebuia*; *T. impetiginosa*, with flowers a darker rose purple (or magenta), may be seen in Guanacaste. See p. 262 for the yellow-flowered *Tabebuia ochracea*.

Comments: This tree's high-quality wood, which resembles oak, gives it its Spanish name, roble (oak) de sabana. The wood is used in furniture and cabinets, tool handles, boats, yokes, interior finishing, and parquet. The tree provides shade in the wet season and a colorful crown in the dry season. This species was included in the TRIALS project (to test trees in plantation systems and in rehabilitation of disturbed land) at La Selva Biological Station. Black spiny-

Tabebuia rosea

tailed iguanas (*Ctenosaura similis*) are known to eat the flowers, which are pollinated by anthophorid bees (*Centris* spp.). Pau d'arco, the common name given to some South American species of tabebuias, is known in the herbal-medicine world as a cure for a wide variety of ailments. Lapachol, which is found in some species of *Tabebuia*, has antibiotic properties. *T. rosea* bark tea is a Costa Rican folk-medicine treatment for headache and colds.

Tecoma stans
Family: Bignoniaceae

Yellow elder, *Vainillo*

Other common names: Trumpet bush, yellow bells, *candelillo*, *sardinillo* (in Nicaragua), *tronadora* (in Mexico).

Description: Tree to 10 m, often shorter. Twigs tan and warty. Opposite, pinnately compound leaves to 25 cm with 3–11 toothed leaflets, each 5–13 cm long. Inflorescences at tips of branches; vase-shaped yellow corolla (to 5 cm long with five large lobes) very constricted at the base; thin red lines inside tube of flower; has sensitive stigma (folds up after being touched in center). Flower has delicate vanilla or candylike scent. Thin fruits, to 25 cm long, dehiscing to release many small, winged seeds.

Flowering/fruiting: Flowers generally from November to March; slender, tan papery pods follow.

Distribution: From southern Florida and Arizona to Argentina; also West Indies. In Costa Rica, occurs sea level to 1,300 m on central and north Pacific slopes. A pioneer that invades rocky areas; elsewhere occurs as ornamental.

Tecoma stans

x ⅓

Compound leaf
of *Tecoma stans.*

Related species: A number of other species in this genus, some with red flowers, occur in South America; relatives include the common ornamental Cape honeysuckle (*Tecoma capensis* or *Tecomaria capensis*), of South Africa, and trumpet creeper (*Campsis radicans*), of the United States.

Comments: The leaves of this plant resemble those of the elderberry; the flowers, however, are very different from the small white flowers of true elder (*Sambucus* spp.). Its folk use in controlling diabetes has a phytochemical basis—the leaves contain the alkaloids tecomine and tecostanine, which have hypoglycemic effects. The plant also contains an antibiotic, lapachol, as well as compounds that affect the liver. While the form of the tree is not elegant, the flowers give it ornamental value. New plants are easily started from cuttings.

Cochlospermum vitifolium
Family: Cochlospermaceae

Buttercup tree, *Poroporo*

Other common name: Silk tree.
Description: Small deciduous tree to 12 m tall with smooth gray bark and large leaf scars on branches. Alternate, palmately 3–7-lobed, long-stemmed leaf blade, to 25 cm across, usually with toothed edge. Terminal inflorescence of 5-petaled, yellow flowers, 10 cm across, each lasting just one day; many golden stamens. Five-valved, capsular fruit to 10 cm long, with velvet texture. Old dry petals, with texture of onion skin, persist on developing capsule; numerous small seeds, shaped like snail shells, are embedded in white woolly hair. Yellow-orange sap in wood.

Flowering/fruiting: Flowers in the dry season (December–April), with seed capsules forming dry season to early wet season.

Distribution: Mexico to northern South America. In Costa Rica, low elevations (to 1,000 m), generally in second growth on the Pacific slope, from the Caño Negro area south to the Osa Peninsula.

Related species: This is the only species in this family in Costa Rica; related species occur in the old world tropics; sometimes classified in the Bixaceae (annato) family.

Comments: The bowl shape of the large flowers makes them easy to distinguish from the many other yellow-flowered trees of the dry season. The tree naturally springs up along disturbed

Gordon W. Frankie

Flower of *Cochlospermum vitifolium.*

Cochlospermum vitifolium

roadsides and is a component of second growth in northwest Costa Rica, where it plays an important role in dry-forest ecology. The large, pollen-seeking bees that visit the flowers vibrate the anthers to release pollen, some of which the bees transfer to other flowers. White-tailed deer relish the leaf blades, while mice (*Liomys salvini*), as well as bruchid beetle larvae, feed on the seeds. In other parts of world the tree is used as an ornamental; a double-flower variety exists, and small trees less than a year old may bloom. Cuttings will root. The wood is soft and could be used for matchsticks, paper pulp, and boxes. The fibrous bark can be made into

x 1

Seed of *Cochlospermum vitifolium.*

string, and the woolly fibers around seeds are good for stuffing pillows. The yellow-orange sap from the wood is used in dying cotton cloth. Folk-medicine uses include treatment for jaundice.

Caesalpinia eriostachys
Family: Fabaceae
Subfamily: Caesalpinioideae

Saíno

Other common names: *Zabino, pin-tadillo* (in Nicaragua).
Description: Medium-sized tree ca. 10 m with distinctive ropy, fluted trunk. Alternate leaves, bipinnately compound, ca. 15 cm long; each section with ca. 10 pairs of 1-cm long leaflets that have a strong odor when crushed; orange-tinted new growth. Yellow flower 2.5 cm across, dark spots near center; flowers in large clusters. Elastically dehiscent legume to 12 cm long.
Flowering/fruiting: Flowers in dry season.
Distribution: Mexico to Panama, and Cuba. In Costa Rica, most abundant in dry, Pacific lowlands (to ca. 300 m). Often in fence rows on the Nicoya Peninsula and elsewhere in Guanacaste.
Related species: Gray nickernut (*Caesalpinia bonduc*, p. 338) and flamboyant (*Delonix regia*, p. 32) are in this legume subfamily (Caesalpinioideae), as are the genera *Bauhinia* and *Cassia*.

Distinctive fluted trunk
of *Caesalpinia eriostachys.*

Caesalpinia eriostachys

Comments: It is hard to miss this tree when you travel on the Nicoya Peninsula during the dry season. It is easily identified by its irregularly fluted trunk, clusters of yellow flowers, orange new leaves, and occurrence in fence lines. The trunks make attractive support posts for porches. Flowers are bee-pollinated. Recent taxonomic studies place this species in the genus *Poincianella*.

OTHER LEGUMES
WITH YELLOW FLOWERS

The legume family (Fabaceae) is a large, diverse group of plants that includes economically important crops such as beans, peas, lentils, and alfalfa. The group is split into three subfamilies (Papilionoideae, Caesalpinioideae, and Mimosoideae) that are sometimes considered individual families. Most species have a fruit that resembles a bean pod. In tropical regions, representatives of the Fabaceae range from small roadside weeds to towering trees. Most members of the Mimosoid subfamily have brushlike inflorescences of white or pinkish flowers. But many legume trees of the Papilionoid and Caesalpinioid subfamilies have brightly colored flowers. Costa Rica has many yellow-flowered legume trees. In addition to *Caesalpinia eriostachys* (p. 26) and *Cassia fistula* (p. 30), two others that you may see as you travel through the countryside are described here.

Diphysa americana, a tree that grows to 15 m, naturally occurs in the Central Valley and on the Pacific slope, but is planted elsewhere for living fence posts. It has arching, twisting branches and a deeply furrowed, gray

William A. Haber

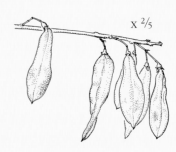

X $^2/_5$

Inflated seed pods of *Diphysa americana*.

William A. Haber

Diphysa americana (guachipelín)
Fabaceae (subfamily: Papilionoideae)

Rough bark of *Diphysa americana*.

x ⅓

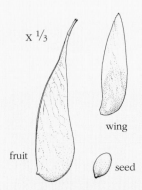

wing

fruit

seed

Form of
Schizolobium
sapling.

Schizolobium parahyba fruit
envelops a papery wing, which in
turn contains the seed. The wing aids
in wind dispersal of the seed.

trunk. The small, yellow pealike flowers are in bunches along the twigs. Its indehiscent
inflated pods (illus. opposite page) contain seeds that are less than 1 cm long. This bal-
loonlike fruit is wind-dispersed.

For a short period in the early dry season (December), this can be one of the most noticeable flowering treetops, especially in pastures at midelevations. The tree is attractive for its
ornamental qualities as well as for its utility as firewood, tool handles, and posts. The wood
is hard and long-lasting; the twisted trunks make decorative support posts in house construction. This species was formerly known as *D. robinioides*.

Schizolobium parahyba occurs on the Pacific slope, below 500 m, in moist–wet areas
or near rivers in drier regions. It is quite common on the paved road from Nicoya to
Sámara. This tree, which grows to about 30 m, has a smooth, light gray, straight, buttressed trunk. The leaves are bipinnate. Young individuals are very distinctive; they
have large, fernlike leaves (to
ca. 1.5 m) and a nonbranching
trunk (illus. above right). Large
clusters of bright yellow, 5-
petaled flowers adorn the
branch tips of adults. The 10-
cm-long fruit is flat and late-
dehiscing, with one seed (illus.
above left).

Howler monkeys have been
seen eating the flowers on trees
near Nosara Beach. The density
of the wood is variable. Since it
is a fast-growing tree, it has
potential use for paper pulp.

Schizolobium parahyba (gallinazo)
Fabaceae (subfamily: Caesalpinioideae)

Cassia fistula
Family: Fabaceae
Subfamily: Caesalpinioideae

Golden shower, *Cañafístula*

Other common names: Purging cassia, pudding pipe tree.

Description: Medium deciduous tree 5–15 m tall; trunk is cream-gray. Alternate, pinnately compound leaves to 50 cm long; leaf stem swollen at base; the leaflets are opposite (or subopposite), each ca. 15 cm long, with 4–8 pairs per leaf. Pendant flower clusters 50 cm or longer are made up of flowers with five 2.5-cm yellow petals. Fruit is a cylindrical, dark brown pod, to 50 cm long, about 2 cm in diameter; compartments within it each contain one seed. Fruit has a very-ripe-apple scent.

Flowering/fruiting: Flowers and fruits in dry season, and into wet season sporadically.

Distribution: Native to Asia; cultivated in many tropical countries. In Costa Rica, planted on roadsides, mostly low (to mid-) elevation.

Related species: Close to a total of 50 species in this genus (*Cassia*) and the now segregated genus *Senna* grow in Costa Rica—some native, some exotic. Among these are pink shower (*Cassia grandis*, p. 31), saragundí (*Senna reticulata*, p. 90), and candlestick senna (*Senna alata*, p. 92).

Comments: While all of the painted treetops are showy and eye-catching, a golden shower tree in full bloom—its bright grapelike clusters of flowers dripping from the branches—is one of the most beautiful trees seen along roadsides in lowland Costa Rican towns. The pulp around the seeds has laxative qualities due to compounds known as anthraquinones that speed up passage of material through the colon. In Nicaragua, the leaves are cooked and used as a laxative,

Cassia fistula

although its use may pose risks since it can be toxic. Researchers found that the leaf extract has hypoglycemic effects in mice. The hard, heavy durable wood is good for construction, posts, and cabinetry. The bark contains tanning agents. The pulp around seeds is used to flavor tobacco in India. This is one of a number of host plants for larvae of the large black witch moth (*Ascalapha odorata*, Noctuidae). Air pockets in each section of the seed capsule allow the pod to float on water surfaces—sometimes for as long as two years—before settling on a beach.

Cassia grandis
Family: Fabaceae
Subfamily: Caesalpinioideae

Pink shower, *Carao*

Other common names: Coral shower tree, *sandal*.

Description: Deciduous tree 10–30 m; trunk brownish gray. Alternate, pinnately compound leaves with 10–20 pair of pubescent leaflets, each to 6 cm long; new growth rusty pubescent. Large, dense, erect flower clusters along branches; flowers pink, turning orange-pink, each petal to ca. 1.2 cm long; stamens of varying lengths. Woody, indehiscent, brown-black, nearly cylindrical sausagelike pod to ca. 60 cm (and 4 cm wide), chambered, each section containing a 1.5 cm seed embedded in sweet pulp.

Flowering/fruiting: Flowers February through April; old fruits often remain on trees during flowering.

Distribution: Mexico to northern South America, and Antilles. In Costa Rica, common in Pacific lowlands, Central Valley, and Caribbean region (Limón and south); to 700 m (or higher where planted).

Related species: Another ornamental pink shower (*C. javanica*) is originally from Southeast Asia. The golden shower (*C. fistula*, see p. 30) is widely planted in Costa Rica; see also *Senna reticulata*, p. 90, and *S. alata*, p. 92.

Comments: In mid–dry season, the pink shower is one of the most conspicuous tree crowns of the Costa Rican dry forest. The flowers, which are buzz-pollinated by large bees, provide pollen but no nectar. Janzen (1983) speculates

Cassia grandis

Erythrina poeppigiana

neotropics; 13 in Costa Rica. *E. fusca*, a commonly planted native species that often accompanies *E. poeppigiana* in the Central Valley, has a pale orange flower and a leaf with a whitish underside and a more rounded tip; it is common in certain areas along the Pacific coast (e.g., on the way to Playa Herradura).

Comments: The magnificent display of a flowering poró is one of the highlights of the Central Valley in the dry season. This species belongs to a showy group of legumes known as coral trees (*Erythrina* species). While most have colorful orange to red flowers, there are two basic types: those that have a large standard (the upper petal) and are more open, as in *E. poeppigiana*, and those that are more closed and tubular (see *E. lanceolata*, p. 225). The former tend to be pollinated by perching birds such as orioles and the latter by hummingbirds; tanagers and migrant Tennessee warblers also frequent *E. poeppigiana* for nectar. Of the 13 species of *Erythrina* in Costa Rica, many are small and are commonly used as living fence posts. They have decorative, red seeds. Although the flowers and young leaves of some are edible, other parts, especially the seeds, are known for their poisonous alkaloids, which have effects on muscles similar to those of curare arrow poisons.

Lagerstroemia speciosa
Family: Lythraceae

Queen's crape myrtle

Other common name: Pride of India, *orgullo de la India.*

Description: Tree to 20 m. Opposite, entire leaves 10–15 cm long, elliptical, with pointed tips, lighter colored beneath. Flower clusters 30 cm or more long. Flowers ca. 6 cm across; 6 pink or purplish crinkly petals, basal quarter of each very narrow; many stamens. Below petals and 6 calyx lobes is a 1-cm-long, ribbed, cone- or vase-shaped receptacle. Fruit a dry capsule.

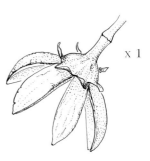

x 1

Open fruit capsule of *Lagerstroemia speciosa.*

Flowering/fruiting: Flowers late dry season into rainy season.

Distribution: Native to the Old World tropics, particularly India and China, and south to Australia. In Costa Rica, common ornamental around San José, Atenas, and Esparza.

Related species: Crape myrtle (*Lagerstroemia indica*) is smaller, often shrublike, and has smaller (to 6 cm) leaves and fewer stamens in the flowers (ca. 40 as opposed to ca. 150 in *L. speciosa*). It is often grown in the southern United States as an ornamental, with many cultivated varieties. Henna (*Lawsonia inermis*), cigar flower (*Cuphea ignea*), and purple loosestrife (*Lythrum salicaria*) are in the same family. Pomegranate (*Punica granatum*) is a close relative, sometimes placed in its own family, Punicaceae.

Comments: The flower petals resemble crape (a crinkled fabric)—thus the common name, queen's crape myrtle. Various plant parts of *Lagerstroemia speciosa* are used medicinally, and studies indicate that it has hypoglycemic activity and is potentially useful to treat diabetes. The smaller *L. indica* is used in Belize as a diuretic and externally for wounds. The durable wood of queen's crape myrtle is appropriate for boats and railroad ties.

Lagerstroemia speciosa

COLOR IN TREETOPS CAN BE DECEIVING

The blazes of color in the crown of a tree do not always come from the flowers of the tree itself. Epiphytes and hemiepiphytes often perch in high branches, where they get more sunlight, and vinelike lianas with roots in the ground climb into the tree crowns in search of the sun. As they become larger, the foliage and flowers of these plants become intermingled with that of their support tree. An observer looking up from a distance sometimes finds it hard to believe that the showy crown display is not part of the tree, but results from other plants that are covering the tree.

Species in the Bignoniaceae (trumpet creeper) and Convolvulaceae (morning glory) families frequently have a trailing habit. *Securidaca sylvestris* (photo below), another climber, is in the Polygalaceae (milkwort) family.

Plants that reside in the treetops produce an extensive array of colors. Large shrubs with yellow or lavender flowers, members of the

Securidaca sylvestris climbing in the crown of a dry forest tree.

X ⅕

X ⅖

Fruiting *Securidaca* (at Monteverde).

Asteraceae (aster) family, grow in the tops of cloud forest trees (e.g., *Sinclairia polyantha*, photo right). Canopy inhabitants in the Ericaceae (blueberry) and Bromeliaceae (bromeliad) families often produce red leaves that emerge at the same time as tender, new reddish tree leaves, fooling an observer into thinking that the tree is in peak bloom. Occasionally parasitic mistletoes that take over pasture trees will create an orange glow in the foliage.

And then there are trees that appear to be in flower but are actually in fruit. A striking example of this is a basswood relative, burío (*Heliocarpus appendiculatus*), which has dense clusters of pink-purple fruits (photo below).

Sinclairia polyantha perched in the canopy.

x 2

Fruit of *Heliocarpus appendiculatus*, enlarged to show detail.

Heliocarpus appendiculatus in fruit.

2. Other Common Trees

This chapter features trees that are common in Costa Rica. Although many lack showy blossoms and are not very conspicuous, you will frequently see them in second growth habitats on roadsides and pasture edges. Since they are grouped together here because of a lack of a single identifying characteristic, trees from a large number of plant families are included and the variety among the species described is wide.

Those species that have no value as lumber can become abundant, providing food and shelter for animals ranging from tiny ants to deer and peccaries. Some of these trees may look dull, but their fascinating natural history, folklore, and uses make them worthy of further study. Even without flashy flowers or conspicuous fruits, they still have characteristics that make them stand out. The extensive, horizontal branches of the ceiba tree, for example, create a majestic crown that dominates open landscapes.

Once you have identified a species' growth form or other special features, you may realize that it is one of the most common trees to grow along a road or path that you've traveled many times before.

Left, *Ficus goldmanii*

Anacardium excelsum
Family: Anacardiaceae

Wild cashew, *Espavel*

Description: Large tree to more than 2 m diameter and more than 40 m tall. Dark trunk with vertical fissures. Clusters of simple, alternate leaves ca. 30 cm long, rounded and wider toward tip, form dense, dark crown. Large clusters of small, 5-parted, green-white (turning pink) flowers with clove scent. Kidney-shaped green fruit, 3–4 cm, on a spiralling, fleshy stem-like receptacle (hypocarp). Crushed plant parts have a turpentine odor. Resinous sap may cause a rash.

Flowering/fruiting: Flowers January to April. Fruits late dry, early rainy season.

Distribution: Central America to Ecuador. In Costa Rica, from north to south, on both slopes but more common on Pacific, from lowlands to 900 m.

Related species: Anacardiaceae, or the poison ivy family, is a diverse set of species that includes cashew (*Anacardium occidentale*, p. 162), mango (*Mangifera indica*, p. 164), pis-tachio (*Pistacia vera*), and the Brazilian pepper tree (*Schinus terebinthifolius*).

Comments: These giants, with their dense foliage, are dominant features in many lowland *quebradas*, or stream beds, in Costa Rica. The toasted seed is edible, but the surrounding flesh has a caustic oil, cardol, that is used in making plastics, resins, insecticides, and a number of other products. Bats and monkeys eat the receptacle of the fruit. Some indigenous people crush the bark and use it to stun fish in order to trap them more easily. The fibrous wood is useful in crude construction but can cause a rash in some individuals. The local name *espavel* comes from *es para ver* (in order to see), which refers to indigenous people and explorers using this tree as a lookout point.

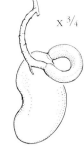

x ¾

Anacardium excelsum fruit.

Anacardium excelsum

Ceiba pentandra
Family: Bombacaceae (recently placed in Malvaceae)

Kapok tree, *Ceiba*

Ceiba pentandra flowers.

Other common name: Silk cotton tree.
Description: Giant deciduous trees to 50 m or taller; to 2–3 m across above buttress. Straight, long, grayish, cylindrical trunk with horizontal, lenticular prickly lines. Sometimes "pot-bellied" (bulging, then constricting) at base. Horizontally spreading, with flattened crown. Conical spines on branches and young trees. Alternate, palmately compound leaves with ca. 7 entire leaflets; leaflet size to ca. 20 cm. White or pink flowers, 4 cm in diameter; 5 petals and stamens. Oval or pear-shaped dehiscent seed capsule, 10–20 cm long and 5-valved, with hundreds of dark seeds in cottony fluff (kapok).
Flowering/fruiting: Flowers dry season, with fruit later in dry season; flowering irregular, not every year. Flowers when leafless.
Distribution: Native to Africa and from Mexico to Peru and Brazil; cultivated in Southeast Asia. In Costa Rica, sea level to 1,000 m on both slopes. Dry to wet regions, Guanacaste to Osa Peninsula; secondary and primary forest, often an emergent.
Related species: Ceibo (*Pseudobombax septenatum*), a related species that grows in the Pacific lowlands, has a nonspiny trunk with vertical green streaking (photo, p. 43). Other members of this family in Costa Rica include balsa (*Ochroma pyramidale*, p. 44), pochote (*Pachira quinata*, p. 216), and provision

Ceiba pentandra fruit with fluffy kapok emerging.

x ⅓

Ceiba pentandra, from Atlantic lowlands; note buttress.

tree (*Pachira aquatica*, p. 252). *Bombax ceiba* is the Southeast Asian red silk cotton tree.

Comments: Kapok trees are some of the most impressive trees seen in Costa Rica. Their profiles are distinctive, with a horizontal, compressed crown set on a straight upper trunk, and a bulge toward the base and/or some large fin-like buttresses. They may be emergents in forests or relict trees in pastures—or a roadside tree whose broad, elephantine trunk has been desecrated by a sign for a restaurant or towing service. Kapok trees grow rapidly and flower irregularly (not annually). Bats arrive by the dozens after nightfall when the flowers open, and many other visitors are attracted to old flowers. Studies at different sites in its range indicate that various insects and 33 species of birds (in Mexico) take nectar from the flowers in the morning. Nonflying mammals such as opossums, long-tailed weasels, and monkeys have also been seen visiting flowers. Hundreds to thousands of fruits form on a tree. These immense trees have been revered by various indigenous cultures, including the Maya. *Ceiba pentandra* natively occurs in both the American tropics as well as in Africa, which is unusual. All other members of the genus *Ceiba* have their origins in the neotropics, so it is probable that seeds of *C. pentandra* were carried to Africa by air currents, or perhaps whole capsules floated across the ocean. Kapok trees are fairly common in Costa Rica, probably because they are unpopular as lumber. In parts of South America, however, they are now scarce due to harvesting for plywood. The soft wood is appropriate for packing crates, forming boards, matchsticks, and paper pulp. Dugout canoes are sometimes fashioned from the trunks, and doors and tables from the plank buttresses. The tree is renowned as a source of kapok for life jackets and cushions, and for insulation. Besides being buoyant and insulating, the wax-coated

Ceiba pentandra, from Pacific lowlands; note how the trunk bulges toward the base.

Pseudobombax septenatum, a relative of *Ceiba pentandra*.

fibers surrounding the seeds are water-repellent. Medicinal applications include bark decoctions used internally (as an emetic and diuretic) and externally to treat open sores and hemorrhoids. The oily seeds are edible and contain linoleic acid; the oil can be used in soaps, margarine, and for lighting. Many children around the world have been introduced to this fascinating tree through Lynne Cherry's popular book, *The Great Kapok Tree.*

Ochroma pyramidale — Balsa tree, *Balsa*
Family: Bombacaceae (recently placed in Malvaceae)

Description: Second-growth tree to 25 m, sometimes taller, with smooth gray, sometimes buttressed trunk; mucilaginous sap. Thick branches. Long-stemmed, broad leaf blades ca. 30 cm long with heart-shaped base, entire or with 3–5 lobes; pubescent, gray-green underside. Large green-gold calyx; bell-shaped, erect, flowers ca. 12 cm tall with 5 overlapping satiny-looking, cream petals folded back. United stamens in swirled pattern, around a style that is topped by large, cream-yellow, spiral-like stigma; scent like raw pumpkin. Capsule (to 25 cm) dehiscent by 5 valves, with silvery-tan fluff surrounding many seeds (to 5 mm), looks like a short, very fuzzy, paint roller or microphone.

Flowering/fruiting: Flowers mostly in dry season, but sometimes late rainy season. Fruit is seen in dry season and later.

Distribution: Native to tropical America and now cultivated in other parts of world. In Costa Rica, found in moist and wet lowlands and near rivers in some dry regions; both slopes, from sea level to ca. 1,200 m.

Related species: Cousin to kapok (*Ceiba pentandra*, see previous species account), and pochote (*Pachira quinata*, p. 216).

Comments: This pioneer species grows in sunny open areas where there has been a fire or other disturbance. Several years ago, the author was puzzling over a sapling that appeared in her yard after an acci-

Ochroma pyramidale. From left to right: flower, developing fruit (in background), and bud.

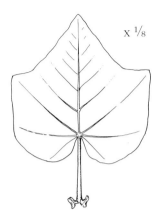

x ¹/₈

Ochroma pyramidale leaf.

dental fire; in Monteverde, at 1,450 m elevation where she lives, the tree is outside of its usual range. She deduced that it was the result of a seed from some discarded *Ochroma* kapok that had been carried up from the lowlands as a souvenir. In the warm, wet conditions of Costa Rican lowlands, a two-year-old balsa may stand 10 m tall. Though it is typical in second growth, some individuals persist in older forests. The conspicuous flowers are nocturnal—anthers surround the style during the day, but not at night, when bats visit and pollinate the flowers. Occasionally nocturnal, arboreal mammals also visit, in search of nectar. During the day,

monkeys may be seen sticking their faces into flowers, perhaps hunting for insects, but it is unlikely that these daytime visitors pollinate since the stigma is covered then. The seeds, in their kapok coat, obviously float through the air, but the kapok also may be dispersed by water. Kapok makes great filling for pillows and furniture. In contrast to most tropical lumber species, which are sought out for their high density, balsa is valued for its lightweight wood, which makes it ideal for rafts, boxes, insulation, and model airplanes. It was a component of airplanes during World War II and used to make Stim-U-Dents (remember those?). It is grown commercially in South America. Native people of tropical America have used balsa in boat-making (*balsa* means raft in Spanish). The bark fibers can also be made into twine. The various medicinal uses of balsa include preparations of root or bark as a diuretic, a flower decoction for colds and coughs, and external emollients from mucilaginous young plant parts; a decoction of flowers and bark is used as an emetic. Pittier (1978) reported that the seeds have an edible oil. *Ochroma lagopus* (whose species name means rabbit foot, the texture of which resembles that of the dehiscing capsule) is synonymous with *Ochroma pyramidale*.

Cecropia spp.
Family: Cecropiaceae

Some common names: Cecropia, trumpet tree, *guarumo*.
Description: The five Costa Rican species of *Cecropia* are very similar. These fast-growing trees are usually under 25 m tall (*C. insignis* may grow to

40 m) with a trunk that is ringed with stipule scars and hollow between the nodes; prop roots are common. Spreading, sparsely branched, open crown; often with biting ants living in branches. Large, long-stemmed, palmately

Cecropia obtusifolia

(more accurately, radially) lobed peltate leaves; whitish below in some species. Dioecious. Tiny flowers in clusters of fingerlike spikes; developing clusters surrounded by bract or spathe. Fruits small and packed densely in spikes.

Flowering/fruiting: Flowering and fruiting variable.

Distribution: *C. insignis* from Nicaragua to Ecuador; *C. obtusifolia* and *C. peltata* range from Mexico to northern South America (*C. peltata* also in Antilles); *C. angustifolia* from southern Mexico to Venezuela and Bolivia; *C. pittieri* is endemic to Cocos Island. In Costa Rica, *C. insignis* and *C. obtusifolia* in wet areas below 1,450 m on both slopes; *C. peltata* below 1,200 m, usually on Pacific slope in mostly dry climates; *C. angustifolia* in mountains 1,300–2,000+ m.

Related species: Related to *Musanga* of Africa. In Costa Rica, relatives include *Pourouma* spp. (not peltate) and *Coussapoa* spp. (all unlobed leaves, hemiepiphyte habit). Recent research places the Cecropiaceae family within the Urticaceae.

Comments: The candelabra-like branching and hand-shaped leaves make cecropias recognizable from a distance. They are often seen along roadcuts since they grow well in disturbed areas, including streamsides, landslides, and light gaps. Unlike the seeds of some other gap specialists that can endure decades in the soil, cecropia seeds are viable for perhaps a year or less and need full sunlight in order to germinate. Cecropias may grow over four meters a year. The wood is very light, and some species have been used for making crates, matchsticks, and paper pulp. The hollow trunk is easily split length-wise to make rustic gutters. In many species, the hollow stems house *Azteca* ants, which attack and bite herbivores that might try to feed on the leaves. In this elaborate case of ant-plant mutualism, the ants get glycogen from the tree via tiny egglike packets called Müllerian bodies that form on feltlike pads near the base of the leaf stems. Pollen from the short-lived male spikes, or catkins, is dispersed by wind. Bats, birds, monkeys, and probably other arboreal mammals eat the fruits. A common natural-history myth is that cecropia leaves are three-toed sloths' favorite food, but a sloth's diet is actually quite extensive. The reason they are associated with cecropias is that they are easily (and therefore, frequently) spotted in the open crowns of these trees, where they often go to sun themselves. The nymphalid butterfly *Historis odius* is frequently seen around *Cecropia obtusifolia* and *C. peltata*, the host plants for its caterpillars, which make frass chains (strings made from their own silk and excrement) to hang from, probably to avoid ant attacks. The bananaquit, a small yellow and black bird that usually feeds on nectar, at times supplements its diet with Müllerian bodies from cecropias. Other organisms associated with cecropias are cecropia petiole borers (*Scolytodes* spp.), tiny beetles that lay eggs in the stems of the fallen leaves. The ripe fruits are edible for humans. Folk remedies include decoctions of leaves for kidney problems, high blood pressure, asthma, diabetes, bronchitis, as a diuretic, and for nervous disorders. Cecropine, which has cardiotonic and diuretic effects, occurs in the bark and roots. Externally, *C. peltata* leaves, as a poultice or a decoction, are put on swellings. The sap of that species and of *C. obtusifolia* is applied to warts. Some indigenous people of the Amazon use ash from burned cecropia leaves as the alkaline component in their coca leaf preparations.

IDENTIFYING CECROPIA SPECIES IN COSTA RICA

One can easily identify the five cecropia species that occur in Costa Rica by considering a tree's location and looking at leaf and fruit characteristics. *Cecropia pittieri* is found only on Cocos Island. The other four species are listed below with some tips on how to distinguish them from one another.

Cecropia insignis
- occurs on wet Atlantic slope and wet regions of Pacific
- upper leaf surface not hairy or rough
- leaf has 10 or fewer lobes
- developing fruit spikes short to medium length (6–18 cm)
- fallen leaves tend not to curl up

Cecropia obtusifolia
- moist to wet habitat; widespread on both slopes
- leaf surface usually rough
- leaf has 10 or more long lobes
- developing fruit spikes to more than 50 cm long

Cecropia obtusifolia

Cecropia peltata
- mainly occurs on dry Pacific slope below 1,200 m
- upper leaf surface feels like sandpaper
- leaf has 11 or fewer lobes
- developing fruit spikes short (4–10 cm)

Cecropia peltata

Cecropia angustifolia (formerly *C. polyphlebia*)
- occurs in cloud forest above 1,300 m
- upper leaf surface rough
- leaf has 10 or more lobes
- developing fruit spikes short

Clethra lanata
Family: Clethraceae

Clethra, *Nance macho*

Clethra lanata

Related species: Various species in Old and New World, including sweet pepper bush (*C. alnifolia*), a North American native that is used as an ornamental. Nine species of *Clethra* occur in Costa Rica.

Comments: Clethra is one of the most common trees growing along roadbanks below the Monteverde/Santa Elena region. This species definitely has potential for erosion control; it can cling to windy, vertical slopes, helping to prevent the banks from collapsing (photo left). The sweet fragrance of the white flowers and the triangular outline of the tree give clethra ornamental qualities. The author has seen a number of individuals in Monteverde that are much taller, straighter, and less pyramidal in form than *C. lanata*, and that flower later (May–July). These are classified as *Clethra hondurensis*. Formerly the name *C. mexicana* was used for a number of species in Costa Rica, but that species is actually restricted to Mexico.

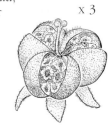

x 3

Clethra lanata fruit.

Description: To 15 m tall, often triangular in outline. Alternate, leathery leaves, to 12 cm long, bunched at twig tips; dark bluish-green above, lighter, with feltlike pubescence below; base of leaf blade turned under. Teeth on leaf margins variable—well-defined in some young leaves, hardly noticeable in old leaves. Small, white, sweetly scented flowers in arrays of fingerlike inflorescences. Small (less than 1 cm), 3-parted, capsular fruit with many winged seeds.

Flowering/fruiting: Flowers mid- to late dry season.

Distribution: Mexico to northern South America. In Costa Rica, 700 m (perhaps lower) to 2,000 m.

Clethra lanata flowers.

Enterolobium cyclocarpum
Family: Fabaceae
Subfamily: Mimosoideae

Guanacaste, *Guanacaste*

Other common name: Ear tree.
Description: Tall, spreading tree to 35 m, with rounded crown in open areas; low branching. Large (to over 2.5 m in diameter) grayish trunk smooth or with lenticels and fissures. New leaves appear toward end of dry season; leaves alternate, twice compound, with ca. 5 pairs of pinnae (but up to 15 pairs); up to 60 leaflets (ca. 1 cm long) per pinna. Some glands on rachis and one on petiole. Leaves display sleep movements (i.e., they fold at night). Small, white flowers in ball-like heads ca. 1.5 cm in diameter. Shiny, dark reddish-brown, ear-shaped pod in form of compressed, twisted spiral, ca. 12 cm in diameter; sweet pulp. Pod somewhat flattened but lumpy where the ca. 14 seeds form; seeds 1.5–2 cm long, with tan border surrounding dark brown oval on each side.

Flowering/fruiting: Flowers between January and May; fruit from the previous year is maturing then also.

Distribution: Mexico to northern South America; also the Antilles. In Costa Rica, lowlands up to 1,300 m; may be found in wet forest, but most abundant on dry Pacific slope.

Related species: The less common *E. schomburgkii*, which has pubescent twigs and smaller fruit (5–7 cm), occurs on north Atlantic slope and areas of south Pacific slope. The genus *Enterolobium* is in the mimosoid subfamily of the legumes, which includes *Mimosa pudica* (p. 92), *Inga vera* (p. 52), *Acacia* spp. (p. 269), and *Calliandra* spp. (p. 138).

Enterolobium cyclocarpum

Comments: This is one of the majestic tree species that graces the landscape of the Guanacaste region. It attracts a wide variety of animals, including small moths and beetles that visit the flowers at night, and livestock and wildlife that seek the nutritious fruits and seeds. *Amazona* parrots are often first in line, eating unripe seeds. Seeds are toxic to some rodents, but the spiny pocket mouse (*Liomys salvini*) hoards the seeds; by nicking the seed coat with its teeth, the mouse hastens germination and produces a crop of nascent seedlings that provide a less toxic meal. Today, seeds and seedlings are seen in cow dung, but Janzen (1983) suggests that, over 10,000 years ago, seeds were dispersed by large mammals (megafauna) that are now extinct. Some individual trees grow fairly rapidly, so large ones may actually be relatively young. In folk medicine, green fruits are used to treat diarrhea, and sap from the trunk is used to treat bronchitis. Saponins in *Enterolobium* species, which make the pods and bark useful as a soap substitute, are also spermicidal. The wood, often attractively streaked, makes handsome furniture; it is also used for oxcart wheels, posts, and firewood. Disposing guanacaste sawdust in streams may poison cows and fish downstream. The seeds make nice jewelry. The genus name means intestine lobe (*entero/lobium*) and the species name means circle fruit (*cyclo/carpum*).

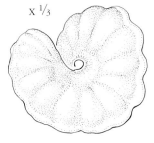

X ⅓

Enterolobium cyclocarpum fruit.

Enterolobium cyclocarpum without leaves.

Inga vera subsp. *vera*
Family: Fabaceae
Subfamily: Mimosoideae

Inga, *Guabo*

Other common name: *Guabo de río.*
Description: Tree usually less than 10 m tall; dense pubescence on new growth. Alternate, compound leaves with ca. 6 pairs of leaflets, mostly pubescent below; winged rachis. Leaflets to 17 cm long, basal ones much smaller; barrel-like nectaries between leaflets. Flowers in clusters, with 5 cm-long white stamens creating delicate, brushlike effect; actual corolla ca. 2 cm long. Curving, ridged, fuzzy legume pod to 15+ cm by 1–2 cm wide.
Flowering/fruiting: May flower any time of the year, but mostly in dry season, with subsequent fruit development.
Distribution: Mexico to northern South America; Greater Antilles. In Costa Rica, to 1,200+ m, dry to moist areas. Common on Pacific slope from north to south, also in Caño Negro region.

Related species: Ca. 300 species in the genus *Inga*, with 53 in Costa Rica; see *I. spectabilis*, p. 179.
Comments: Ingas, or *guabos* as they are known in Costa Rica, are common throughout the neotropics. All species in the genus *Inga* have once-compound leaves with an even number of leaflets and nectaries between the leaflets. Often the leaf midvein is winged—with leafy tissue spreading out on both sides. The flowers are brushlike, and the fruit contains seeds with a sweet, edible, white covering (aril). Various species are planted in coffee plantations for shade and nitrogen-fixing qualities. *Inga vera* is the most common member of this genus in the dry regions of Costa Rica, where it often grows near rivers. Though the flowers are principally bat-pollinated,

Inga vera subsp. *vera*

bees, birds, and butterflies may visit them during the afternoon; at night hawkmoths often join bats. The type of sugar in the nectar changes at night, apparently through fermentation, from sucrose to the glucose and fructose that bats prefer. A common name for some species of *Inga* is ice cream bean.

Byrsonima crassifolia
Family: Malpighiaceae

Nance, *Nance*

Other common names: Shoemaker's tree, *nancite*.
Description: Usually small tree, to 13 m; bark fissured. Simple, opposite, entire, thick leaves, more or less elliptical, to 15 cm long; new growth woolly. Terminal flower clusters 5–15 cm long, with 5-petaled (1.5 cm) yellow flowers that turn red-orange with age. Irregularly margined petals spoon-shaped, with a stem at base and a wider "bowl" at the end. Each of the 5 calyx lobes has a pair of oblong oil glands. Squat, round, yellow fruits are 1–2 cm, with 1–3 seeds in a pit or stone; each fruit has a knobby cap because of the persistent calyx.
Flowering/fruiting: Flowers as early as January and as late as September, but mostly from March to June. Fruits April to August.
Distribution: Mexico, south to Paraguay, and West Indies. In Costa Rica, typical of lowland Guanacaste rocky savannahs, but can be found to 1,400 m elevation, and from Caño Negro and Tortuguero in the north to the Osa Peninsula and Talamancas in the south. Sometimes cultivated for fruit.
Related species: Of the ca. 130 species in this genus, only three are recorded for Costa Rica. *Byrsonima lucida* (locust berry) is native to South Florida and the West Indies. This family contains ayahuasca (*Banisteriopsis caapi*), used by native people of South America as a hallucinogen in ceremonies.
Comments: The yellow, sweet-sour fruits of nance are well-known among all Costa Ricans. You may see nance being sold in plastic bags along the Inter-American Highway. They can be eaten raw, but are also made into preserves, candies, wine, and—at least in Brazil—ice cream. The fruit has a rotting-cheese odor; they are supposedly an interesting treat when left to soak in *guaro* (fermented sugar cane) for nine months. Despite its popularity as a food source, Salas (1993), in *Arboles de Nicaragua*, warns that the fruit should not be eaten because it provokes gastrointestinal disorders. The reddish-brown wood, hard and heavy, makes good firewood, and it is used for cab-

Byrsonima crassifolia flowers.

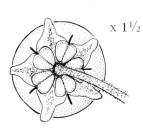

Byrsonima crassifolia
fruit, showing calyx.

Byrsonima crassifolia fruit.

inets, furniture, tool handles, and turned objects. Red dye from the bark was formerly used for coloring leather, cloth, and wooden floors. Medicinal uses include treatment for diarrhea, chest colds, and fever; in Guyana, indigenous people make a poultice out of the inner bark to speed healing of deep wounds. In Venezuela, people make fish poison from nance. Birds eat the fruit. Female *Centris* bees collect the oil produced by glands on the flower calyx; they feed this oil to their larvae and incorporate it into the walls of underground nests. Ecologically nance is of note as a fire-resistant savannah inhabitant.

Hampea appendiculata
Family: Malvaceae

Doll's eyes, *Burío*

Other common names: Hampea, *burío macho, buriogre.*
Description: To 25 m, but often half that size. Smooth, gray or tan trunk, with fissuring that looks like stitched wounds in older trees; has mucilaginous orange sap. Twigs, leaf stems, and leaves pubescent. Simple, alternate, entire leaves often broad with long petiole; leaf blade, to 20 cm long, has rusty-beige pubescence on leaf underside; little earlike flaps below point where 5 main veins join; glands on main veins on leaf underside. Dioecious; white to yellow flowers have 5 petals, with many stamens in male, a 3–4 lobed stigma in female. Brown, 3-valved capsule, 2–3 cm long, with up to 6 black seeds with fleshy, white arils (illus. opposite page).

Flowering/fruiting: Flowers in rainy season; fruits in dry season (sometimes in very late rainy season).
Distribution: Honduras to Panama. In Costa Rica, riversides, light gaps in older forest, and second-growth forest; to

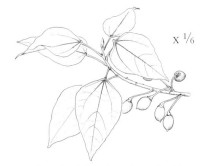

Branch of *Hampea appendiculata* with unopened fruit.

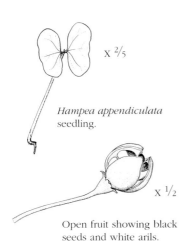

x $^2/_5$

Hampea appendiculata
seedling.

x $^1/_2$

Open fruit showing black
seeds and white arils.

Hampea appendiculata

1,700 m, moist to wet forest, throughout the country.

Related species: *H. platanifolia,* a species with lobed leaves that have pronounced cordate bases, has been collected on the Pacific slope at Manuel Antonio and Carara. Hampeas are related to cotton (*Gossypium* spp.); hibiscus (p. 140) and Turk's cap (*Malvaviscus penduliflorus,* p. 142) are also in this family.

Comments: Hampeas are easy to distinguish in second-growth forest because their distinctive golden-rusty pubescence stands out in the canopy. On the forest floor beneath trees, one often sees myriad seedlings and clusters of bright red seed bugs (Lygaeidae) eating hampea seeds. The handsome, dense, round crown and the fact that the fruits attract more than twenty species of birds, along with mammals such as white-faced capuchins, make this tree a prime candidate for restoring old pasture; the author's experience has been that young saplings do best with some shade. Butterflies and bees visit flowers. The soft wood is suitable for bed slats and forming boards; the fibrous bark can be used as twine.

Conostegia xalapensis
Family: Melastomataceae

Lengua de gato

Other common names: *María, lengua de vaca.*

Description: Generally less than 10 m tall, but to 20 m. Bark tan with vertical fissures. Simple, opposite, narrow leaves with toothy edge; leaf size variable, often ca. 10 cm, but may be much longer. The 3–5 major veins that arise from the leaf base are crossed by many perpendicular veins (illus., p. 56). A rusty-tan pubes-

Conostegia xalapensis

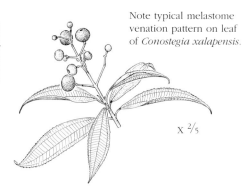

Note typical melastome venation pattern on leaf of *Conostegia xalapensis.*

x ²⁄₅

cence covers various plant parts and is especially evident on leaf undersides. Flower clusters at branch tips; 5–7 white petals, each ca. 5 mm long; yellow anthers with terminal pores. Purple-black fruit ca. 1 cm, with many seeds.

Flowering/fruiting: Seen in flower and fruit at various times of year. Fruits regularly seen in wet season, but also other months.

Distribution: Southern Mexico to northern South America. In Costa Rica, sea level to 1,500 m (occasionally to 2,000 m). Found in many regions of the country—at sea level in places along Atlantic coast; usually higher elevations on Pacific slope; common around San José and Lake Arenal; in Monteverde, to around 1,450 m, then replaced by *C. oerstediana* (photo right).

Related species: Belongs to melastome family, a large group of trees and shrubs that grows mainly in forest light gaps and other disturbed areas. One of the easiest tropical families to recognize because of the netted vein pattern in leaves. Family includes meadow beauty (*Rhexia virginica*) of the United States and the ornamental glory bush (*Tibouchina urvilleana*).

Comments: The fruits of many melastomes (as the members of this family are known), including lengua de gato, are edible. The pink or lilac coloring from fruits can be used to dye cloth. This species may be the most abundant melastome in Costa Rica; it often grows in old pastures. Pollen-seeking female bees are attracted to the fragrant flowers.

Conostegia oerstediana growing near cloud forest.

Miconia argentea
Family: Melastomataceae

María

Description: Tree to ca. 15 m with rounded crown; gray-tan flaking bark. Young stems flattened. Opposite, long-stemmed leaves, green above, with whitish or tan-flesh color below and on twigs. Leaf blade ca. 20 cm long, toothed (or not), with 5-cm petiole; 5 main veins running up from base of blade, with many cross veins; the base of the leaf is curled under, forming what may be domatia. Many small whitish flowers in large, terminal, branching clusters (to 20 cm long); flowers with 5 white petals, in the center of which are 10 yellow anthers. Purple berry under 1 cm in diameter, with many small seeds.

Flowering/fruiting: Usually flowers in dry season, with any one population being in flower for only 2-3 days. Fruits in dry and wet season.

Distribution: Southern Mexico to Panama. In Costa Rica, generally in second-growth areas; common along roads in foothills. Mostly Pacific slope, to 1,200 m, from Guanacaste south to Osa Peninsula, but also Atlantic slope.

Related species: There is a total of 97 species of *Miconia* in Costa Rica and ca. 1,000 species in the neotropics.

Comments: On a windy day, the undersides of the leaves are displayed, making these trees more conspicuous than usual. Fruits are eaten by birds and mammals. The wood is not very durable and is little used, occasionally for light construction and, in Mexico, for tool handles. In a light-gap study in Panama, the tree was found to grow up to 2.5 m per year (Brokaw 1987).

Miconia argentea

Cedrela odorata
Family: Meliaceae

Spanish cedar, *Cedro amargo*

Other common name: West Indian cedar.

Description: Medium to large, 30+ m tall, deciduous tree, to more than 1 m diameter, with furrowed gray or brown bark. Alternate, pinnately compound leaves (to more than 50 cm), with 5–12 pairs of entire leaflets (each 7–14 cm long) with asymmetrical bases. Monoecious; small (1 cm), tubular cream/green flowers in terminal, branching, pendant clusters (15–50 cm long). Woody, dehiscent, 5–valved capsules ca. 4 cm long, with many 2–3 cm winged seeds. Garlic odor in various parts of plant, including the flowers. Wood is bitter-tasting.

Flowering/fruiting: Flowers early wet season, usually in July. Fruits in dry season.

Distribution: Mexico to South America, and West Indies; also planted in Old World. In Costa Rica widespread; seen in pasture and in forest on both slopes. Generally in foothill regions, dry or wet, up to ca. 1,200 m.

Related species: In Costa Rica, both the rare *C. salvadorensis*, of low to middle elevations, and *C. tonduzii* (illus., p. 59), of higher elevations, have larger capsules than *C. odorata*. Mahogany (*Swietenia* spp., p. 259) is in this family.

Comments: This tree is best known for its wood, which has a pink-red tinge and is easy to work. It is insect and rot resistant. The resemblance of its wood to true cedar led to the common name Spanish cedar. In the 1950s, it was the main tree species being extracted from the Atlantic lowlands of Costa Rica for lumber; a large amount was exported to the United States. It is popular for cigar boxes, moth-proof chests, house construction, turned articles, guitars, doors, window frames, and furniture. Despite its classification as threatened, many medium-sized individuals are still seen in pastures and along roadsides (e.g., roads from the Inter-American Highway to Monteverde). Its popularity could result in a change from threatened to endangered status, however. A grand individual resides in Parque Morazán in San José. The moth-pollinated flowers have a distinctive garliclike odor that is especially strong under humid conditions; immature fruits and freshly cut wood also have this odor. Various plant parts, especially the bark, have folk-medicine uses.

Cedrela odorata

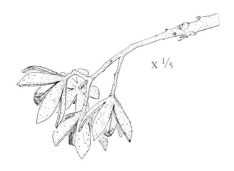

X ⅕

Fruit of *Cedrela tonduzii*, a higher elevation species.

Teas and decoctions are made to treat fever, bronchitis, indigestion and other gastrointestinal ailments, and epilepsy. Externally, bark is put in baths to relieve colds, aches, and wounds. Spanish cedar grows relatively quickly in cultivation. Interspersing the trees with other species may decrease damage by mahogany shootborer (*Hypsipyla grandella*), a pest of young *Cedrela* saplings. In Africa, farmers plant food crops among these trees, thus benefitting from a short-term harvest while the long-term tree crop is growing. This tree was formerly classified as *C. mexicana.*

Trichilia havanensis
Family: Meliaceae

Uruca, *Uruca*

Description: Tree ca. 10 m tall with dark dense foliage. Bark dark, fissured, and flaking in older trees. Alternate, compound leaves, variable in size, with 2–5 pairs of opposite leaflets, plus one at tip. Terminal leaflets to ca. 10 cm long with base usually wedge-shaped. Petiole and midvein somewhat winged. Dioecious, with small, fragrant, white to green flowers, less than 1 cm in diameter, clustered in leaf axils. Dehiscent capsules 1–1.5 cm long, shiny green, later turning brown, with 1–4 seeds with red aril.
Flowering/fruiting: Flowers generally in dry season, but some in wet season. Fruits late dry to wet season, some in other months.
Distribution: Mexico to northern South America. In Costa Rica, 600–2,600 m, from northern Guanacaste to southern regions of country; common along Pacific slope and in the Central Valley; one of the most popular street trees in San José.
Related species: In Costa Rica, there are 14 species. *Trichilia* is a genus in the mahogany family, which includes mahogany (*Swietenia* spp., p. 259) and Spanish cedar (*Cedrela odorata*, see previous species account).

Adrian Hepworth

Trichilia havanensis

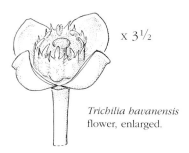

x 3½

Trichilia havanensis
flower, enlarged.

Comments: The section of San José called Uruca, now mainly an industrial zone, still has a few examples of this tree—and some saplings have been planted in recent years. Downtown San José has many streetside urucas. The flowers attract stingless bees (*Trigona* spp.). Birds disperse the seeds. People use branches to decorate church altars. In Costa Rican folk medicine, a leaf decoction is made for skin problems, and a leaf or bark decoction is taken internally for urinary tract ailments. The sap, reported to be poisonous, is used as an insecticide in El Salvador. The wood is easily worked and is good for carving.

Ficus spp.
Family: Moraceae

Some common names: Fig tree, strangler, *higuerón, chilamate, matapalo, higo, higuito.*

Description: In the Americas, figs are usually trees, many of which begin life as epiphytes on other trees and develop a strangling habit. All plant parts have milky latex. Simple, alternate, entire, often lustrous leaves. Some species are deciduous. The conical tips of twigs are made up of two, often long, pointed stipules that cover new growth; these fall off, leaving circular scars. Small male and female flowers enclosed in a globe-shaped syconium (see sidebar, p. 63) that will mature into a fig. The figs are green, yellow, or reddish, 1 to 6 cm in diameter; color and length varies from species to species.

Flowering/fruiting: Fig populations display staggered flowering, so that at any given time some individuals will be in flower.

Distribution: Old and New World, mostly tropical. In Costa Rica, sea level to high elevations, wet and dry regions; some planted for shade and ornament.

Related species: Breadfruit (*Artocarpus altilis*, p. 142), and mulberry (*Morus rubra*) are also in this family. Native relatives include a variety of forest trees

Strangling habit of *Ficus* species.

(*Brosimum alicastrum*, p. 276, and *Castilla elastica*) and treelets.

Comments: Fig trees are a common component of Costa Rican forests, and various species appear regularly in living fences or as ornamentals. The attractive foliage provides shade for people and livestock. Frugivorous birds such as guans, toucans, and trogons eat the fruit, as do bats, monkeys, and raccoon relatives. Howler monkeys eat the leaves of various species; in Africa, chimpanzees eat young leaves of *Ficus exasperata*, possibly as medicine to kill intestinal parasites (Downum et al. 1993). Many animals disperse the seeds, but parrots damage them and are therefore seed predators. Along Atlantic lowland river banks, *Ficus insipida* drops some of it fruits in the water, where they are eaten by fish. Fig species that are stranglers begin life as an epiphyte in the crotch of another tree, then produce roots that grow down to eventually anchor in the forest floor. As these roots grow, they expand in diameter, crisscross, and often fuse. The strangler eventually envelops its support tree, constricting and shading it. After the death and decay of the support tree, some stranglers end up with an interior passageway from base to crown, becoming true jungle gyms for children of all ages. These are among the largest and most spectacular trees in tropical forests, with extensive, spreading crowns and pretzel-like trunks with large buttresses. In some species, it has been shown that more than one individual will fuse together. The many-trunked, sacred banyan tree of India is a fig, *F. benghalensis*. The largest of these can cover about four acres. The common, cultivated edible fig, *F. carica*, is native to the Old World tropics. Several Old World species are common ornamentals in Costa Rica. These include the rubber plant, or Indian rubber tree (*F. elastica*), which historically was an inferior source of rubber (the best being *Hevea brasiliensis*, Euphorbiaceae family). It has very long, reddish stipules. The weeping fig, *F. benjamina*, a popular shopping mall plant, has elegant pendant branching. The bo tree, *F. religiosa*, has triangular leaves with elongated tips and long, thin petioles; one individual grows near the San José post office. The climbing fig, *Ficus pumila*, sometimes grows as a vine on buildings in San José. Edible figs (*higos*) are occasionally cultivated in Costa Rica in moist areas at midelevations. Sugared figs, not to be confused with *pasa de marañón* (dried, preserved fruit of cashew), may be found for sale in grocery stores, bakeries, and fruit stands. A number of fig species around the world are used in folk medicine. The most effective use of the latex is as a vermifuge; ficin, a proteolytic enzyme found in the latex of some species, acts on intestinal worms. Various cultures use the latex externally on ulcerating sores, herpes, fungus, and for setting bones; nevertheless, skin treated with latex and exposed to sunlight may become irritated and blistered. The leaves of some Asian and African species are edible, and a few species are used for lumber. A crude cloth can be made from the bark.

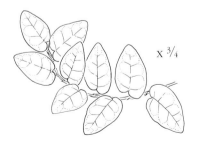

x ³/₄

Ficus pumila, a climbing fig that grows on buildings.

Ficus tuerckheimii

THE SEX LIFE OF FIGS AND FIG WASPS

Diagram of cross section of fig syconium.

All fig species have tiny male and female flowers that develop on the inside wall of a syconium, a structure that looks like, and eventually will become, a fruit that is actually many tiny fruits in one. The pollination system of figs is fascinating. A tiny female wasp (of the Agaonidae family) enters a hole in the top of the syconium by wiggling through a tunnel lined with downward-pointing, overlapping scales that only allow one-way movement. The wasp encounters two kinds of female flowers: short-styled flowers that are stalked and long-styled flowers that are not stalked. While she is pollinating the flowers, the wasp finds it easy to lay eggs in the ovules of the stalked flowers, but not in those of the unstalked flowers. After laying her eggs, she dies; and in a few weeks her offspring emerge from the stalked flowers. The wingless males mate with the winged females, chew an exit tunnel that allows the females to fly to another fig, and then die. The females pick up pollen from recently matured male flowers and then leave through the exit tunnel. They then enter syconia on a new tree, where the cycle is repeated. The pollinated syconia mature and produce figs, which are a key component of the diet of many mammals and birds. For each of the 800 or so species of figs in the world (there are about 50 species in Costa Rica), there is a matching species of pollinating wasp that appears to find its tree species by a chemical cue.

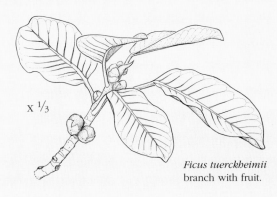

X $\frac{1}{3}$

Ficus tuerckheimii
branch with fruit.

Piper auritum
Family: Piperaceae

Root beer plant, *Hoja de estrella*

Description: Small tree to 6 m. Stems with large leaf scars and prominent, sometimes-swollen nodes; occasional prop roots. Alternate, simple, soft, pubescent leaves to 50 cm long with asymmetrical, heart-shaped base; sheathing petiole. White flower spikes, thin and arching at first, then erect to 30+ cm long. Tiny fruits packed into thicker, darker, pendant spikes. All parts of plant smell like sassafras (anise) when crushed.

Flowering/fruiting: Flowers and fruits all year.

Distribution: Mexico to Colombia, and West Indies; sometimes cultivated. In Costa Rica, sea level to 1,700+ m, on both slopes. Open areas, along streams, forest edge, and roadsides. Grows best where soil is persistently moist.

Related species: Worldwide there may be as many as 2,000 species of *Piper*, ca. 124 in Costa Rica, many of them forest shrubs. Black pepper (*Piper nigrum*, p. 191) is a cultivated member of this genus in Costa Rica. Kava (*Piper methysticum*) is an Old World species, the root of which is used to make a relaxing beverage for ceremonies in the South Pacific.

Comments: In certain second-growth areas, *Piper auritum* begins invading and spreads rapidly, forming dense shade. Stingless bees visit flowering spikes. Bats relish the fruits. The leaves, which have a distinctive, strong sassafras odor, are used to add flavor to meat and tamales. An infusion of the leaves may aid digestion, although its use is not recommend-

Piper auritum

ed due to the presence of safrole. Studies of safrole in sassafras (*Sassafras albidum*) root bark showed that it produces liver cancer in rats. Uses as a folk medicine include laying the leaves on the forehead for headache relief and on wounds to reduce swelling. A fascinating use of the leaves is as fish bait in traps set by people living along rivers on the Caribbean slopes of Panama.

Guazuma ulmifolia
Guacimo, *Guácimo*

Family: Sterculiaceae (recently placed in Malvaceae)

Other common names: Bastard cedar, West Indian elm, bay cedar.
Description: To 15+ m, with rough bark. Often branches close to the base, with many straight vertical shoots arising from main branches; star-shaped hairs on new twigs. Leafless for part of dry season. Simple, alternate, distichous leaves 6–15 cm long, more or less ovate, with tapering tip; toothed and often asymmetrical. Underside of leaves lighter colored and like soft flannel. Clusters of small, cream-yellow, honey-scented flowers with 5 petals. Oval, woody pineconelike capsular fruit, ca. 3 cm long, with pointy bumps; blackish, with sweet, caramel-like odor. Small, grayish brown seeds inside compartments.

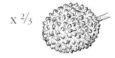

X $^2/_3$

Guazuma ulmifolia fruit.

Guazuma ulmifolia

Francis X. Faigal

Guazuma invira near Puerto Viejo de Limón.

on the Atlantic slope. Cacao (*Theobroma cacao*, p. 206) is another member of the Sterculiaceae family.

Comments: While this tree blends in with other roadside greenery in the wet season, in the dry season the absence of foliage reveals the noticeably jagged, angular branching pattern, and one realizes that this is one of the most common trees along the Inter-American Highway north of Esparza. The dark conelike fruits, also easily seen then, produce a sweet aroma along roads, where they get crushed by passing vehicles. The green fruits are edible. White-tailed deer, tapir, and peccaries, as well as horses and cows, eat the leaves and fallen fruits. The hard core and the structure surrounding the seeds appear to protect them from heavy-duty molars like those of large Pleistocene mammals—Janzen (1983) speculates that these now extinct animals may have been the original seed dispersers. Today, in Guanacaste pastures, horses with similar molars eat thousands of the fruits daily, with many seeds passing through undamaged and viable. Guacimo has multiple uses. The leaves and fruit can be ground up to make chicken feed. Its wood is used for firewood, charcoal, posts, and tool handles. Guacimo charcoal has been reported as a source of gunpowder in Central America in colonial times. The bark fibers are made into a crude string, and the mucilaginous sap is used to extract impurities in the production of crude sugar. A bark decoction treats dysentery, skin diseases, malaria, and syphilis. Other preparations are used to treat lung ailments

Flowering/fruiting: Flowering some throughout year, often greater in March and April, with fruit maturing February–March of the next year. Fruits remain on tree for many months.

Distribution: Mexico to South America, and West Indies. May be seen in Asia and western tropical Africa as an introduced species. In Costa Rica, to ca. 1,000 m, widespread on the Pacific slope. Occurs along the Inter-American Highway heading north from Esparza, and in second growth at Santa Rosa and Palo Verde. Also occurs in southwest Costa Rica.

Related species: The only other member of this genus in Costa Rica is *Guazuma invira* (photo above), found

and baldness, as well as for hemorrhoids and burns. Scientific studies have shown some activity against certain bacteria and cancers. Caffeine is present in the leaves. In the seasonally dry forests of Guanacaste province, guacimo leaves are food to weevils, various moth larvae, and the Guanacaste stick insect. The wood hosts harlequin beetle larvae, and morpho butterflies seek out fallen fruits. Easter orchid (*Encyclia cordigera*, p. 411) is a common epiphyte on this species' trunk and branches.

Apeiba tibourbou
Monkey comb, *Peine de mico*

Family: Tiliaceae (recently placed in Malvaceae)

Description: To 20 m, deciduous; fairly smooth (to slightly scaly) gray trunk; fibrous bark. Simple, alternate, pubescent leaves to 30 cm long, in a plane, with fine teeth and heart-shaped base. Petiole with long soft hairs is swollen where it connects with leaf blade. Buds and inflorescences pubescent; 4- or 5-petaled cream to yellow flowers (2–3 cm diameter) with many stamens; sepals (attached below petals) silky. Green to brown fruit shaped like squashed globe, to ca. 8 cm across, covered by dense, soft bristles coated with silver hairs. Pulp contains many small seeds (less than 5 mm).

Flowering/fruiting: Flowering is variable but mainly occurs in rainy season. Fruit may be seen at any time of year, in varying stages of maturation.

Distribution: Mexico to South America. In Costa Rica, along entire Pacific slope to 1,000 m, in disturbed forest and edge habitats.

Related species: In Costa Rica, the only other species in this genus is *A. membranacea*, a forest tree that grows on both slopes. *A. membranacea* and *A. tibourbou* are distinguished by their fruit. The black fruits of *A. membranacea* have short, stiff spikes (illus. right) in contrast to the pliable bristles of *A. tibourbou*. In Costa Rica, there are three members of the genus *Luehea* (see following species account) that are also in this family.

Comments: This is a fast-growing tree with fibrous bark that serves as makeshift string and soft lightweight wood that may be good for paper pulp. The seeds are rich in oil (30%). In folk medicine, a decoction of flowers is used as an antispasmodic, and seed oil is used to relieve rheumatism and encourage hair growth. A bark decoction is ingested to expel intestinal worms. The contents of the odd fruit, which looks like a sea urchin, is most likely eaten by mammals. In Amazonia, the fruit is boiled and opened, then inhaled as an asthma treatment.

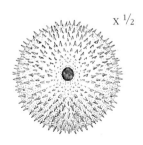

X $\frac{1}{2}$

Apeiba membranacea fruit.

Apeiba tibourbou in fruit.

Luehea alternifolia

Guácimo macho

Family: Tiliaceae (recently placed in Malvaceae)

Description: To ca. 20 m tall; deciduous; young trees with smooth gray bark. Young twigs and leaf stems covered with woolly pubescence. Leaves alternate, toothed, broad, to 20 cm long, with beige or grayish stellate pubescence beneath; 3 major veins ascending from leaf base. Flower with 5 petals, forms a saucer ca. 8 cm wide; petals large, white, narrowed at base, wrinkly along the outer edge. Five large sepals below petals, with 8–10 narrower bracts forming a starlike arrangement below; glistening rust and gold hairs on underside of sepals and bracts. Many stamens surrounding a tall style (to 2.5 cm) topped with a large, sticky stigma. Capsule 2–4.5 cm long, brown, with 4–5 valves; dehiscent at top, with many 1–cm, flat, winged, goldenbrown seeds.

William A. Haber

Luehea alternifolia

X ³/₅

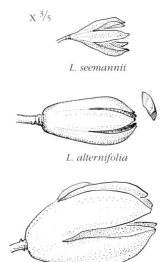

L. seemannii

L. alternifolia

L. candida

Fruits of three *Luehea* species.

Flowering/fruiting: Flowers November through January. Fruits in rainy season.

Distribution: Mexico to parts of South America, also Cuba. In Costa Rica, dry to moist forest and pastures; in Central Valley, and mid- to northern Pacific slope; to 1,200 m.

Related species: Two other species of *Luehea* occur in Costa Rica: the widespread *L. seemannii*, a large tree common in some Caribbean lowland forests, and *L. candida* of the northwest lowlands.

Comments: During the dry season, *Luehea alternifolia* has elegant white flowers that contrast with its ragged, worn leaves, which gradually fall off as the season progresses. A variety of insects, and possibly bats, visits the flowers. The wood is used in rural construction, for farm-tool handles, and as firewood. Formerly called *L. speciosa*.

Trema micrantha
Family: Ulmaceae

Trema, *Capulín*

Other common names: *Jucó, vara blanca, capulín blanco, capulín negro* (in Nicaragua).

Description: To 20+ m, with fairly smooth brown-gray trunk; fibrous inner bark. Open crown with horizontal branching. Leaves distichous, simple, alternate, toothed, with lopsided base, to 15 cm; sandpapery above, grayish below, with pubescence on veins. Separate male and female flowers, usually on same plant (monoecious), in leaf axils; flowers small green-white, 5-parted. Orange to red, 1-seeded fruits (ca. 3 mm) crowded along twigs.

Flowering/fruiting: Flowers and fruits various times of year; fruit usually abundant in rainy season.

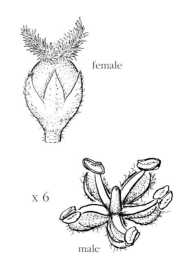

female

x 6

male

Trema micrantha flowers enlarged.

Distribution: Mexico to Argentina, and Florida and West Indies. In Costa Rica, throughout the country in disturbed, second-growth areas (e.g., old pastures, river edges); sea level–1,600 m, wet and dry sites.

Related species: Some species of *Trema* occur in the Old World. *T. integerrima* is a Costa Rican species that is typically taller than *T. micrantha*. The leaves of *T. integerrima* lack the sand-papery texture and teeth (although sometimes fine teeth are present) that characterize *T. micrantha*.

Comments: This rapidly growing (to 13.5 m in two years) tree is frequented by pigeons, thrushes, flycatchers, and a variety of North American migrant species. The fruit-laden branches are colorful since fruits of various stages of ripeness are mixed—green, yellow, and orange. The flowers are pollinated by wind or small insects. Caterpillars of hepialid moths live in the trunks of this tree in holes that they burrow and cover with a silk and sawdust plug. Country people use the fibrous bark for rope and the flexible branches for weaving baskets. The wood can be used for simple construction and firewood. *Trema* has recently been placed in the Cannabaceae family.

Adrian Hepworth

Trema micrantha

3. Roadside and Garden Ornaments

For those who appreciate plants, road trips by bus or car in Costa Rica are never boring because the array of roadside flowers and foliage changes greatly from place to place. Even within the same area, the vegetation changes during the course of the year. This chapter focuses on plants that are seen frequently and easily along the road. It includes colorful sun-loving species that grow wild as well as common yard and garden plants.

Most of the wild roadside plants in this chapter are native species, while those that grow in gardens are often nonnative. Some nonnative species, however, such as impatiens and ginger lily, have escaped to fields and roadsides and now grow alongside native species. Also, a few native species are grown so commonly in gardens that many people don't realize they also grow in the wild. Thus, you can not always tell whether a plant is native or nonnative solely by where it grows.

Most of the wild species in this chapter were included because their flowers create patches of blue, yellow, or red that make them stand out from the rest. Some of the garden species described here are distinguished by their attractive foliage, others by their flowers.

This chapter presents wild (native) species descriptions first, followed by cultivated (nonnative) species descriptions.

Left, *Bougainvillea*

Aphelandra scabra
Family: Acanthaceae

Aphelandra

Description: Shrub, often ca. 2 m tall, but may reach twice that size. Elliptical 7.5–23 cm entire, opposite leaves with edges of blades continuing down leaf stems. Leaves soft below. Candlelike, pubescent flower spikes ca. 15 cm tall, with overlapping green bracts (tinged with yellow and orange) with glands, and 4-cm tubular, 2-lipped, fuzzy, scarlet, one-day flowers.

Aphelandra scabra

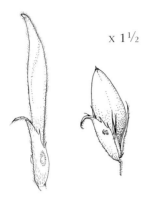

x 1 ½

Left, *Aphelandra scabra* bud with extrafloral nectaries on the bract; right, developing fruit.

Four fertile stamens and slender stigma tucked into upper lip of flower. Club-shaped, 2-cm-long capsules; 4 dark brown seeds.

Flowering/fruiting: May flower and fruit in wet or dry season.

Distribution: Southern Mexico to northern South America; also West Indies. In Costa Rica, sea level to mid-elevations (up to 1,750 m), mostly Pacific slope. In forest or disturbed areas (e.g., along rivers or roadsides).

Related species: Showier members of this genus include the zebra plant (*Aphelandra squarrosa*), fiery spike (*A. aurantiaca* 'Roezlii'), and Sinclair's aphelandra (*A. sinclairiana*); they are commonly seen in tropical gardens. Of the more than 175 species in the genus worldwide, ca. 12 occur naturally in Costa Rica, including the elegant *A. aurantiaca* (*pavoncillo* in Spanish). The Acanthaceae family also includes the golden shrimp plant (*Pachystachys lutea*, p. 109), Brazilian red cloak (*Megaskepasma erythrochlamys*, p. 108), razisea (*Razisea spicata*, p. 282) and sky vine (*Thunbergia grandiflora* p. 111).

Comments: The flowers of this species are pollinated by short-billed, territorial hummingbirds, while most Central American aphelandras are visited by long-billed, non-territorial hermit hummingbirds. The seeds are dispersed ballistically. In exchange for food from the extrafloral nectaries on the bracts, ants protect the inflorescence and developing fruit capsules. If you pick a plant and leave it in a bag for a day or so, you can see nectar from the extrafloral nectaries bead up on the bracts. This species was formerly known as *A. deppeana*.

Furcraea cabuya var. *cabuya*
Family: Agavaceae

Cabuya, *Cabuya*

Description: Enormous plants, with or without a stocky (ca. 1 m) "trunk" surrounded by skirt of old dead leaves. Rosette of many sword-shaped succulent 1.5–2 m green leaves, sometimes with a whitish coating, often with hooked reddish thorns on the edge. Each concave leaf is imprinted top and bottom with the shape of adjacent leaves—from being tightly pressed together when young. Giant asparagus-like inflorescence may

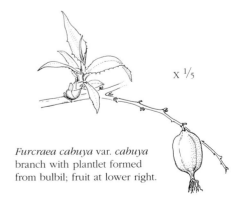

x ⅕

Furcraea cabuya var. *cabuya* branch with plantlet formed from bulbil; fruit at lower right.

Furcraea cabuya var. *cabuya*.

grow to 10 m; branches as it matures and bears 5-cm pale-green flowers and bulbils (that form plantlets). Fruits are 5+ cm bulging, dehiscent capsules with 3 chambers containing many flat seeds.

Flowering/fruiting: Generally flowers in rainy season; de-hisced fruits may persist into dry season.

Distribution: Mexico; in Central America, found in Honduras, Nicaragua, Costa Rica, and Panama. In Costa Rica, on Pacific slope on dry rocky ridges and cliffs near steep river canyons, from approximately 50 to 1,500 m elevation. Also cultivated.

Related species: Relatives include agaves, sources of pulque and tequila (*A. tequilana* and other species); century plant (*A. americana*); sisal (*Agave sisalana*); and various native and ornamental *Yucca* species (see itabo, *Yucca guatemalensis*, p. 112).

Comments: In flower and fruit, the plant is exceptionally notice-

Old fruiting stalks of *Furcraea cabuya* var. *cabuya*.

able; the developing inflorescence looks like a gigantic, out-of-control asparagus stalk. After producing the towering flower/fruit stalk, the plant dies. The side shoots, as well as the many little bulbils on the branches, will grow into new plants. Cabuya plantations near El Empalme, on the Cerro de la Muerte Highway, produce fiber that is made into rope. Two varieties of cabuya exist:

F. cabuya var. *cabuya* (leaf margin is armed with thorns) and *F. cabuya* var. *integra* (unarmed); the former makes a formidable hedge. Species of the genus *Furcraea* are sometimes confused with those of *Agave*—both have tall, branch-ing flower stalks. But, the flowers of *Furcraea* are less densely clustered, while those of *Agave* are in tightly packed heads with flat tops.

THE LARGEST FLOWER
IN COSTA RICA

The native vine *Aristolochia grandiflora* is not commonly seen in the wild, nor is it often cultivated as an ornamental, but the flower's conspicuous appearance, size, and carrion-like odor set it apart from all other plants. The enormous, foul-smelling flowers, which have maroon markings, are certainly the largest in Central America, and among the largest in the world. The flower has no petals. Three fused sepals form the odd-shaped, curving tube (to 20 cm long) that flares to 40 cm across near the mouth and has an odd thin tail to one meter long. The alternate, broad, entire, heart-shaped leaves (15+ cm long) are on long petioles. The stems become corky and ribbed with age.

Side view of *Aristolochia grandiflora* (Dutchman's pipe or matamoscas). This large member of the Aristolochiaceae family flowers throughout the year, peaking during dry season.

Front view of flower.

Cross section showing inside compartments of flower.

On the first day the flower blooms, it smells like rotting flesh, and the female parts (stigmas) of the flower located at the deepest part of the chamber are receptive. The odor and the maroon color attract carrion-loving flies (some carrying pollen from other flowers) that travel down the flower tube, which contains downward-pointing hairs and slippery papillae that prevent the flies from leaving. The deepest chamber of the flower, which is accessible, houses nectar-producing hairs and the reproductive structures. A translucent section of the wall allows light to enter the deepest part of this chamber. When the trapped flies move toward the light in search of an escape route, they end up pollinating the flower. On the second day the blossom has no odor. The anthers release pollen, the entry way widens, and the hairs in the tube wilt, allowing the flies to leave the flower with newly acquired pollen and to fly to freshly opened, malodorous flowers. The cycle starts all over again. Although the idea that these flowers are carnivorous appears in Costa Rican folklore, they just entrap the flies for a day to ensure that pollination takes place.

Butterfly larvae in the genera *Parides* and *Battus*, in the swallowtail butterfly family (Papilionidae), feed on the leaves of aristolochias, which contain alkaloids. Biologists speculate that the butterflies may store these in their body as a defense against predators.

The presence of these alkaloids in the plants may contribute to their medicinal efficacy (e.g., aristolochic acid is known to be antimicrobial), and various species of *Aristolochia* are used for treating colds, asthma, fever, and snakebite. However, some of these same compounds are toxic, and regular use or the incorrect dose can lead to coma and death. Some species induce abortion.

A. grandiflora occurs from Mexico to Panama, and in the West Indies. In temperate regions it is cultivated in greenhouses. In Costa Rica, this species grows on both slopes (to 1,000 m), usually in wet environments along forest edge or in second growth. Fourteen other species occur in Costa Rica, and about 350 worldwide. Birthwort (*A. clematitis*) of Europe and wild ginger (*Asarum canadense*) of North America belong to this family.

Ageratum spp.
Family: Asteraceae

Some common names: Ageratum, *Santa Lucía.*

Description: Herb to 1+ m, but usually shorter. Leaves usually opposite. Bluish flower heads of 40 or more florets; what appears to be a flower is actually many tiny ones packed together. Small dry fruits to ca. 1.5 mm long, sometimes with bristles along the ribs and projections on top. Though fairly similar in appearance, the species of *Ageratum* can be distinguished by pubescence, bract, and flower details, as well as the presence (or absence) of a pappus on the seed.

Flowering/fruiting: Some species can be found in flower throughout year; most flower in the dry season.

Distribution: Mexico to South America; also West Indies. Some species that occur in Costa Rica (e.g., *A. conyzoides*) have been naturalized in regions of the Old World. The genus is widespread in Costa Rica, occurring along forest edge and in pastures, in full or partial sun. From 50 to 2500+ m, several species growing mostly in the mountains.

Related species: The genus *Ageratum* contains around 44 species, with 7 or more species found in Costa Rica. It belongs to one of the largest and most diverse families of flowering plants, the aster family, which includes daisies, sunflowers, goldenrods, and artichokes.

Comments: In Spanish, the various species of *Ageratum* (and some

Ageratum sp.

Fleischmannia) are known as *Santa Lucía.* These plants are associated with early dry season since fields become covered with the lavender-blue flowers then. A Costa Rican custom is to give someone a little bouquet of these flowers in January to wish them a prosperous year. Some species are popular with certain clearwing butterflies, especially males; as they take nectar from the tiny florets, they stock up on certain chemicals (pyrrolizidine alkaloids) that serve as building blocks for the pheromones that the males will use to attract females. During mating, the females may receive some of these alkaloids, which also protect the butterflies from predators by making them taste bad.

Butterfly (*Ithomia heraldica*) taking nectar from *Ageratum* flowers.

Tithonia rotundifolia
Family: **Asteraceae**

Description: Shrubby, tough-stemmed herb 1–4 m tall. Leaves mostly alternate, variable in size, to 35 cm including long stalk; larger ones goosefootlike with 3–5 lobes, smaller leaves often toothed, but without lobes. Leaf blade decurrent on petiole. Lower leaf surface soft and pubescent, especially on veins, upper surface somewhat scratchy. Flower heads ca. 7 cm across—the disk 2–3 cm wide, surrounded by 2- to 3-cm-long ray flowers, orange-yellow (golden), each with tiny notch at end. Inner florets brown-black (= anthers) and yellow-orange (= stigmas). Flower-head bracts form a cuplike receptacle beneath florets. Slight marigold-like scent. Hairy seed 5–6 mm long with pappus (illus. left).

Flowering/fruiting: Flowers and fruits from late October through January.

Distribution: Mexico and Central America. In Costa Rica, occurs along roadsides and in fields, mostly Pacific slope, 300–1,000 m.

Related species: A number of cultivars of *T. rotundifolia* exist, some with reddish flowers, some under 1 m

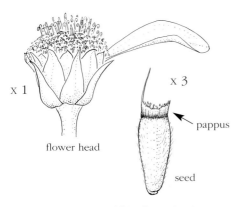

X 1

X 3

pappus

flower head

seed

A *Tithonia rotundifolia* flower head showing disc flowers and pointed bracts below (note: all but one ray flower removed to expose bracts).

Tithonia rotundifolia

tall. Costa Rica is home to two other members of this genus. *T. longiradiata* is a shrub to treelet with unlobed leaves. The Mexican or Bolivian sunflower (*T. diversifolia*) is very similar to *T. rotundifolia*, although its flower heads have a diameter that is about twice that of the latter species, and the bracts below its flowers are rounded instead of pointed.

Comments: These tall, wild sunflowers are one of many members of the aster family (Asteraceae) that flower along roadsides in Costa Rica in the latter months of the year. Its relative, the Mexican sunflower, is being investigated at CATIE as a green manure crop (Aguiar et al. 2002). Using this method of fertilization, farmers plant crops that are later plowed under to enrich the soil.

Hippobroma longiflora   Little milk-star, *Jazmín de la estrella*
Family: Campanulaceae

Other common name: *Jazmincillo.*
Description: Short herb to 40 cm, with latex. Leaves widest beyond middle, pubescent, alternate, to more than 15 cm long, with large and fine teeth; dark green above, lighter underneath. White flower with narrow tube ca. 10 cm long and 5 (2.5-cm-long) lobes; anthers fused into column; long, thin calyx lobes attached to bulging, ribbed, green floral cup. Capsular fruit to 2 cm long; tiny seeds.
Flowering/fruiting: Flowers and fruits at various times of the year, in wet or dry season.
Distribution: Recent sources state that this plant originated in the West Indies, and that it is naturalized in the neotropics, from Mexico to Peru and Brazil, and scattered in Old World tropics. In

Costa Rica, common at low elevations on Atlantic side; also seen on Pacific slope to 500 m. In moist pastures, along rivers and irrigation canals.

Related species: *Hippobroma longiflora* is the only species in its genus; in same family as lobelia (see next species account).

Comments: On the first day of flowering, which lasts 3–5 days, the flowers open at 4:30 P.M.; they produce a perfume that attracts hawkmoths, the pollinators. The plant parts contain nasty latex; in pollination studies, nectar extracted from the flowers seemed to be distasteful, though this may have been due to contamination with latex oozing from nicks in the delicate tissue. The name of the genus (*Hippobroma*) means horse poison. This plant contains isotomin, a poisonous substance that causes paralysis of the heart. The latex is very irritating to the eyes, causing extreme swelling. Mexicans consider this plant a cure for a variety of ailments, including asthma and epilepsy, and Chinese use it in treatment of snake bite and cancer. It is also an ingredient in the hallucinogenic drink *cimora* of some South American indigenous groups of the Andes. The plant is sometimes cultivated as an ornamental. *Isotoma longiflora* is a synonym.

Hippobroma longiflora

Lobelia laxiflora
Family: Campanulaceae

Lobelia, *Caragallo*

Description: Branching plant 1–2 m tall, often clumped; with latex. Stem and leaf underside sometimes finely pubescent. Alternate, simple leaves 4–10 cm long, fine teeth (with swollen purple tips) along margin. Many flowers toward top of stems, in leaf or bract axils. Yellow-tinged, red-orange flowers, 3–4 cm long, with bulging base; basically tubular, with 2 lobes and a slit above the anther tube and broader lip below; calyx green. Capsule 0.5–1 cm; tiny seeds.

Lobelia laxiflora

Related species: Four other species are reported from Costa Rica; one of those is *L. irasuensis*, with blue-violet flowers, growing at very high elevations (ca. 3,000 m). Two species common in North America are the cardinal flower (*Lobelia cardinalis*) and Indian tobacco (*Lobelia inflata*). Some very large species occur in African mountains, where they serve as nest sites for gorillas.

Comments: While many lobelias have ornamental qualities, *L. laxiflora*, perhaps because of its scraggliness, does not appear in gardens. Nevertheless, it merits a place in wildlife gardens since the flowers attract hummingbirds. A species of hummingbird flower mite, *Tropicoseius chiriquensis*, lives in the flower tubes and hitches rides on birds' bills to fresh flowers. Flowers go through a male phase first when the pollen is shed into a tube formed by fused anthers. As the style develops and elongates inside the tube, the pollen moves up and out and exposes the female stigma, which is ready to receive pollen from another flower. Devil's tobacco (*Lobelia tupa*) of South America is known for its medicinal and intoxicating properties. The genus *Lobelia* is rich in alkaloids.

Flowering/fruiting: Flowers and fruits most months of the year; flowers especially noticeable in dry season.

Distribution: Southern Arizona to Colombia. In Costa Rica, disturbed areas such as roadsides, embankments, and old pastures. From ca. 1,000 to 2,600 m. Common in the Central, Talamanca, and Tilarán ranges.

Ipomoea spp.
Family: Convolvulaceae

Some common names: Morning glory, *churristate, pudreoreja.*

Description: Most *Ipomoea* species are twining vines, without tendrils. Flowers are usually funnel-shaped, and often stamens are of different lengths. Milky latex usually present. Leaves are simple, alternate, and may be lobed. Capsular fruit 4-parted. *I. trifida* (photo opposite page), a low-growing, clambering species, has slender, pubescent stems. The ca. 5-cm-long leaf blade, usually with 3 lobes, has a cordate base and a 1.5-cm-long stem. Long stalks bear several lavender flowers, 4 cm at widest point, with darker throat. Sepals, at base of flower tube, are important characteristics in distinguishing *Ipomoea* species. In *I. trifida* the outer sepals are shorter than inner (5–7 mm vs. 1 cm for latter), hairy, with a narrowed tip. Capsule has 4, smooth, brown seeds. Closely related *I. batatas* (illus., p. 88), sweet potato, has thickened roots with sticky latex; *I. carnea*, bush morning glory, has pink flowers and grows as a shrub; *I. alba* (photo, p. 88), moonflower, a climber that may reach 30 m, has flowers to 15 cm long, white tinged green on outside, opening at night and closing before dawn. Cordate-based leaf blades, entire or 3-lobed, up to 18 cm long. Also see *I. pes-caprae* (p. 332).

William A. Haber

Ipomoea umbraticola

Ipomoea trifida

William A. Haber

Ipomoea alba

Ipomoea indica, a cultivated morning glory.

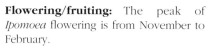

Flowering/fruiting: The peak of *Ipomoea* flowering is from November to February.

Distribution: Majority of the 500 species of *Ipomoea* in the world are in the tropics; close to 50 species in Costa Rica. The widespread *I. trifida*, found from Mexico to northern South America, grows in Costa Rican lowlands to 1,500 m, on roadsides, streamsides, and pasture edges. Common in Guanacaste, but also occurs south to Osa.

Related species: Wood rose (*Merremia tuberosa*), which has attractive seed capsules used in flower arrangements, and Mary's bean (*Merremia discoidesperma*) are also in this family.

Comments: In the latter months of the year, one can look for a variety of roadside morning glories in Costa Rica. Some trail along the ground and others scramble over shrubs or barbed-wire fences; yet others climb into tree canopies (e.g., *I. umbraticola*, p. 87). Morning glories have played various roles in the lives of humans. Their colors and forms have pleased gardeners, the thickened roots of *I. batatas* (sweet potato, or *camote*) have fed millions, and the divining qualities of the hallucinogenic seeds of *I. violacea* (= *I. tricolor*) have, for certain indigenous Mexicans, been used as a way of determining the cause and cure of diseases. The seeds of *I. violacea* (including garden varieties such as pearly gates, heavenly blue, and flying saucers) and the related *Turbina corymbosa* contain potentially harmful LSD-like alkaloids. Various species have been used medicinally; some have purgative qualities. Most morning glories open at dawn, are insect-pollinated (bees, moths, and skippers), and wilt by late morning or afternoon. The pantropical moonflower (*I. alba*) is an "evening glory" that opens at dusk and has a wonderful nocturnal perfume that draws in hawkmoths.

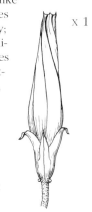

X 1

Ipomoea batatas bud.

Carludovica rotundifolia
Family: Cyclanthaceae

Panama hat palm, *Chidra*

Other common name: *Palma de sombrero.*

Description: Short-stemmed herbaceous plant to 4 m tall. Long stemmed, pleated leaf ca. 1.5 m across, divided into 4 parts, each regularly toothed— that is, with fingerlike fringes or segments of about equal lengths; juvenile leaves are usually rounded and 2-parted. Cylindrical, corncoblike spadix ca. 12 cm long has separate male and female flowers; threadlike staminodia present. Tepals of the female flowers pointed, curved, and slightly longer than the stigmas. Three spathes attached below spadix. Spadix elongates to more than 20 cm when fruiting. When mature, the outer layer of the spadix splits and curls back, starting at the top, revealing inner red-orange color (photo, p. 90); the yellowish orange 1.3-cm fruits have many small tan seeds.

Flowering/fruiting: May be found in flower and/or fruit during most of the year.

Distribution: Honduras to Panama. In Costa Rica, along the edge of and within wet forest; Caribbean region near coast, but also found in foothills and mountains (to 1,700 m in Talamancas). On the Pacific slope, it is found mostly from Carara and south to the Osa Peninsula.

Related species: There are three other species in Costa Rica: *C. drudei, C. sulcata,* and *C. palmata,* which is often cultivated.

Comments: The true Panama hat palm is *Carludovica palmata;* it's foliage looks

Carludovica rotundifolia

Carludovica rotundifolia fruit.

similar to that of *C. rotundifolia*, but the leaf, which has longer and more irregular segments or teeth, has a couple of pointed bumps on the blade close to where it joins the stem. Panama hats, or *jipijapas* as they are sometimes called, are made from the young leaf fibers of *C. palmata*. Although the hats originally came from Ecuador, they became popular during the building of the Panama Canal, thus the name. The leaves and fibrous stems of various species of *Carludovica* furnish material for thatch roofs and baskets; various parts of the plants are also used as food. The flowers are insect-pollinated; tiny weevils of the genus *Perelleschus* appear to pollinate some species (Franz and O'Brien 2001). Some developing seeds are eaten by weevil larvae. Mature fruits are eaten by various birds and mammals.

Senna reticulata

Saragundí

Family: Fabaceae
Subfamily: Caesalpinioideae

Description: Small tree to 6 m with pubescent twigs. Compound leaf ca. 35 cm, with 8–14 pair leaflets (each to 13 cm long) with medicinal odor. Inflorescences more or less erect, to more than 30 cm long including stem. Downy-velvet orange-yellow buds; yellow flowers with 5 petals, 1.5 cm long. Flat pods ca. 15 cm.

Flowering/fruiting: Flowers late rainy season through early dry season.

Distribution: Mexico to Bolivia. In Costa Rica, planted as an ornamental in a range of climates, but grows naturally

(and commonly) below 700 m in Caribbean lowlands around Tortuguero and Puerto Viejo, Sarapiquí. Also on Pacific slope (central region and south to the Osa Peninsula).

Related species: The similar *S. alata* (candlestick senna or Christmas candle, photo, p. 92), sometimes planted as an ornamental, is easy to distinguish from *S. reticulata* if it is in fruit because the pod has four longitudinal wings (the pod is flat in *S. reticulata*).

Comments: *Senna reticulata* contains the compound rhein (cassic acid),

Senna reticulata

Senna alata; note winged fruits.

which has antibiotic qualities. The plant has a wide range of medicinal uses, including preparations of leaves (and flowers) for liver and kidney ailments, venereal disease, snakebite, and rheumatism; decoctions of the root or pods are used as a purgative. A leaf infusion has been used on ringworm. Various South American peoples use fresh or burned leaves as an insect repellent. The leaves have sleep movements (folding up at night). Large bees appear to pollinate the flowers; stingless bees, often seen around the inflorescence, may be stealing pollen.

Mimosa pudica

Family: Fabaceae
Subfamily: Mimosoideae

Sensitive plant, *Dormilona*

Description: Low, spreading herb or shrub less than 1 m tall, with recurved spines. Alternate, bipinnate leaves (ca. 6 cm long) with 1–2 pair pinnae; stipules less than 1 cm long; 10–30 pair oblong leaflets per pinna, each leaflet to 1 cm long. The underside of new leaves is purplish. Conspicuous, pink, brushlike flower heads to 2 cm in diameter made up of very small calyces and corollas, and long stamens. Fruit a 1- to 2-cm-long segmented pod with bristles on margin, with several seeds (illus. opposite page).

Flowering/fruiting: Flowers and fruits principally in the dry season.

Distribution: Native to neotropics, but introduced and naturalized in many other parts of the world, including Hawaii and Australia. In Costa Rica, sea level to 2,500 m in open areas (pastures, lawns, roadsides, along rivers).

Related species: There are 22 mimosa species in Costa Rica. *M. diplotricha* (formerly *M. invisa*) and *M. polydactyla* are similar species that have more pinnae and/or leaflets than *M. pudica*. Another common roadside mimosa, *M. pigra*, is a spiny, weedy shrub of the lowlands (to 700 m). *M. xanthocentra*, generally occurring in South America, is similar to *M. pudica* and has been collected in Palo Verde and Santa Rosa National Parks. Other members of this subfamily are the Guanacaste tree (*Enterolobium cyclocarpum*, see p. 50), bull-horn acacia (*Acacia collinsii*, see p. 269), and *Inga* species (guabas).

Comments: The Spanish common name for this plant translates as sleepy head, in reference to its most striking feature. At night, or when the plant is touched, rained on, or experiences a drastic temperature change, the leaflets fold up and the stem they are on collapses downward. This may be to evade grazers or for reducing transpiration. In this action, called sleep movement, water moving in and out of certain cells results in a decrease in the turgidity of cells in the pulvinus, the thickening at the base of the leaf stem. Various preparations of the leaves and roots are purported to have diuretic, pain-relieving, and/or relaxing qualities, and the root and seeds have emetic properties. Research results indicate some anti-inflammatory, anti-spasmodic, and antiviral activity. The plant contains norepinephrine, as well as toxic compounds (various alkaloids, including mimosine). The roots of some South American species of mimosa (e.g., *M.*

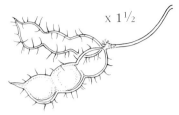

x 1½

Segmented legume of *Mimosa pudica*; sections fall off after fruit dehisces.

hostilis) have been used in the past as a hallucinogen. Folk use of the plant as a sleep inducer may have to do with the doctrine of signatures, a popular old belief that plants show signs of their potential uses (e.g., liver-shaped leaves cure liver ailments, plants that "go to sleep" bring on sleep, etc.).

Mimosa pudica

Kohleria spicata
Family: Gesneriaceae

Kohleria

Kohleria spicata

Description: Herbaceous plant to 1 m; whole plant, including flowers, pubescent; red hairs on stem; petiole and midvein often suffused with red. Leaves opposite (or sometimes in whorls of 3), with toothed margin, to 20 cm (including petiole). Uppermost leaves much shorter, even becoming bractlike where flowers are borne. Tubular, brilliant, red-orange flowers 2–3 cm long, yellowish in throat with streaks or rows of red dots on the 5 corolla lobes. Fruit a small capsule with many tiny seeds.

Flowering/fruiting: Recorded in flower from July to April.

Distribution: Mexico to Ecuador. In Costa Rica, widespread; lowlands to over 1,600 m. On steep road and stream embankments, sometimes near beaches; Guanacaste south to Osa.

Related species: Two other Costa Rican species, less-frequently seen, are *K. tubiflora*, whose flowers have a markedly narrower mouth diameter compared to that of the tube (illus. below), and *K. allenii*, which grows in the southeast region of Costa Rica. The shrub *Moussonia strigosa* (formerly *K. strigosa*) grows with *K. spicata* along the road near the Monteverde Cloud Forest Preserve. Devil's breeches (*K. bogotensis*), of Colombia, is one of a number of horticultural species.

Comments: This is one of the wild roadside plants of the African violet and gloxinia family (Gesneriaceae). Kohleria's rhizomes allow it to go into a dormant phase during particularly dry spells, sprouting anew when moisture is available. A lovely display of kohleria mixed with the blue-flowered trumpet achimenes (*Achimenes longiflora*) occurs in the latter months of the year on sunny, damp, steep banks (at ca. 1,100 m) on the road leading to Santa Elena and Monteverde. As with many of the native Gesneriads, this has potential as an ornamental, being a good candidate for use in native plant gardens in Costa Rica. It is great for attracting hummingbirds.

x 1½

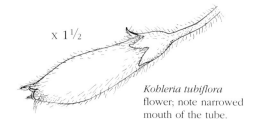

Kohleria tubiflora flower; note narrowed mouth of the tube.

Wigandia urens
Family: Hydrophyllaceae

Wigandia, *Ortiga de montaña*

Other common name: *Tabacón.*
Description: Small tree, usually ca. 3 m tall. Leaves and stem pubescent to bristly, sometimes with urticating hairs. Alternate, toothed leaves ca. 40 cm long. Five-parted, light lilac to blue, scented flowers, ca. 3 cm wide, on curled stems.

Seed capsules in bristly, branching cluster ca. 45 cm long; 1–cm–long capsule splitting along sides; many tiny seeds.
Flowering/fruiting: Flowers notable in dry season; also seen in other months. Fruits seen late dry season to early wet season.

Wigandia urens

Distribution: Central America to Ecuador. In Costa Rica, both slopes; rocky areas and on river and road banks, from southern foothills north to Nicoya Peninsula. From low to high elevations (to 3,000 m).

Related species: There are only three or four members of Hydrophyllaceae in Costa Rica. In the genus *Wigandia*, some taxonomists recognize a variety of *Wigandia urens—W. urens* var. *caracasana*—and sometimes even designate it a separate species: *W. caracasana*. The family also includes the genus *Hydrolea*, two members of which occur in Costa Rica: *Hydrolea elatior* (without spines) and *H. spinosa* (with spines), both blue-flowered semiaquatic plants that grow in wet areas such as Caño Negro and Palo Verde.

Comments: Wigandia has an affinity for the edges of gravel pits and similarly disturbed areas. The foliage and flowers are attractive; in other parts of the world, species of wigandia are planted in gardens. The controversy about its urticating qualities may be due in part to how one touches a leaf; hitting it straight on produces a burning sensation, while brushing it sideways does not. Large bees are seen visiting the flowers. An infusion of leaves is a folk remedy for rheumatism.

Calathea crotalifera
Family: Marantaceae

Rattlesnake plant, *Bijagua*

Other common name: *Cascabel.*

Description: 2–4 m herb. Broad, dark green leaves with a sheen; midvein light; swollen section in petiole below blade. Flattened golden-yellow (sometimes pinkish) flowering spikes to 40 cm; overlapping bracts with odd plastic texture. Flowers yellow, 2- to 3-cm long, with a single, petal-like stamen; other petal-like staminodes present. Dehiscent yellow capsule with blue seeds and white arils.

Flowering/fruiting: Flowers from January through October, but flowering is more profuse during the wetter months.

Distribution: Mexico to Ecuador. In Costa Rica, throughout the country in humid areas, from sea level to over 1,500 m.

Related species: More than 3 dozen species occur in Costa Rica (see *C. lutea*, next species account). Another member of this family is *Maranta arundinacea*, the source of West Indian arrowroot. Some horticultural species (e.g., the Brazilian *C. lancifolia*, also known as rattlesnake plant, and *C. zebrina*, zebra plant, as well as the native *C. warscewiczii* and *C. leucostachys*) have striking foliage with mottling above and purple below.

Comments: These flowers have a complex trigger system that picks up pollen from probing bees while simultaneously dabbing fresh pollen onto their bodies. The main pollinators are euglossine bees, although other species replace them at higher elevations. The leaves are very utilitarian. Some cooks prefer them over banana leaves for wrapping tamales. Indigenous people of the Talamancas have used them to wrap their dead prior to burial; the leaves also serve as disposable umbrellas. Calatheas in general are decorative garden plants; some have striking patterns of light and dark green on their leaves, and also often have purple underneath. Pinkish purple morphs of *C. crotalifera* may be seen on the Osa Peninsula. Formerly known as *C. insignis*.

Calathea crotalifera

Calathea lutea
Family: Marantaceae

Bijagua, Hoja blanca

Calathea lutea

Description: Large herbs ca. 3 m tall. Leaf blades ca. 1 m, whitish waxy beneath; leaf stem thickened and elbowed toward the top where it joins blade. Flowers 4 cm long, white or yellow, in cylindrical spikes to more than 20 cm long; spikes in groups of 2 or more; overlapping bracts tinged reddish-brown. Capsular fruit contains seeds with orange arils.

Flowering/fruiting: Flowers and fruits during most of the year.

Distribution: Mexico to South America. In Costa Rica, common in Atlantic lowlands in open swamps, marshes, and along rivers; also on the Pacific side, from the Osa Peninsula north to Carara.

Related species: See previous species account.

Comments: As one travels through the Atlantic lowlands of Costa Rica, one notices this plant in dense stands in wet areas along roads. The white underside of the leaf is quite noticeable, even from

Calathea lutea; note white underside of leaves.

Francis X. Faigal

a distance, as the leaves wave in the breeze. The flowers attract euglossine bees and hummingbirds. In a study that compared this species with five other Marantaceae of lowland Costa Rica, all with bird-dispersed seeds, *C. lutea* was found to be the most attractive to fruit-eating birds. The wax on the underside of the leaf can be scraped off and has qualities similar to those of carnauba wax used in furniture polish, food coatings, and cosmetics. The leaves are used in the same fashion as those of *C. crotalifera* (see previous species account). The Kuna cook the root and use it medicinally for nausea and diarrhea.

Epidendrum radicans
Family: Orchidaceae

Bandera española

Other common name: *Gallito.*
Description: Herb, usually terrestrial, to 1 m, often somewhat reclining; rarely epiphytic. Upper half of stem with distichous, clasping, purple-tinged leaves that are ca. 5 cm long; lower half of stem without leaves. Long roots, ca. 15 cm, at nodes along stem. Flower to 4 cm long, with orange to scarlet sepals and petals and orange-yellow, 3-parted, frilly lip; older flowers all scarlet. Capsule to 4 cm long; thousands of tiny seeds.
Flowering/fruiting: Flowers and fruits year-round.
Distribution: Mexico to South America. In Costa Rica, ca. 500–2,500 m (most abundant at 1,000–2,000 m) from north to south, both slopes, in open, often rocky, areas such as roadbanks.
Related species: There are over 140 species of *Epidendrum* in Costa Rica (see pp. 406-435 for photographs of other members of this remarkably diverse family). *Epidendrum radicans* is similar to *E. ibaguense*, which grows in South America.
Comments: This orchid is a common roadside plant in some regions. After pollination, usually by a butterfly, the lip of the flower darkens to a deeper orange-red. It frequently grows near the similarly colored lantana (*Lantana camara*, p. 103) and butterfly weed (*Asclepias curassavica*, p. 80). It was thought that epi-

Epidendrum radicans

dendrum flowers, which have no nectar, may receive more visits from pollinators when they grow with these other species, but see the comments on butterfly weed, p. 81. Developing parasitic wasps (genus *Eurytoma*) are sometimes found in the seed capsule; they "grow up" in the capsule, feeding on and cleaning out the contents, and disperse as adults when the capsule splits open.

Hamelia patens
Family: Rubiaceae

Firebush, *Zorrillo real*

Description: Shrub/treelet to ca. 5 m, but may reach double that size. Leaves opposite or whorled (illus. opposite page); 3–4 more-or-less elliptical blades to 15 cm, with pink-red to maroon mid-veins; 2- to 5-cm-long petiole. Stipules to 6 mm long, red-orange tubular flowers to 2.5 cm long, with 5 corolla lobes, on arching branches of inflorescences at ends of twigs. Oblong fruits ca. 1 cm long with circular scar at tip; appear in groups of dark red immature fruits mixed with purple-black mature fruits. Tiny seeds.

Flowering/fruiting: Flowers through-out year, although varies depending on region. In dry forests, for example, it tends to flower more in the wet season. **Distribution:** South Florida, West Indies, Mexico, Central America, and parts of South America. In Costa Rica, sea level to 1,600 m. Occurs in many regions; common roadside plant in some Costa Rican Caribbean areas. **Related species:** There are 7 species of *Hamelia* in Costa Rica, and most have yellow flowers. *H. patens* could be confused

Hamelia patens

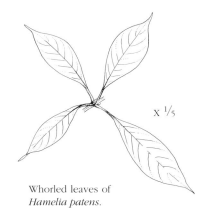

X $\frac{1}{5}$

Whorled leaves of
Hamelia patens.

with *H. rovirosae*, a less-common Caribbean lowland species that has red flowers but a shorter petiole (under 2 cm long) and visible calyx remnants (ca. 0.5 cm long) at tip of fruit.

Comments: Often found in sunny forest gaps or other disturbed habitat, firebush

is a great candidate for native-species landscaping. Its flowers attract hummingbirds and butterflies, and various birds eat the fruits, which are purported to be edible for humans as well. Hummingbirds that visit often transport tiny mites that hop off into the flower and ingest some pollen and a good deal of nectar. Although some butterflies come for nectar, others (*Heliconius* species) feed on the pollen, thus acquiring life-extending amino acids. Researchers have found that these plants flower more in wetter environments; they have also found that plants develop fruits faster when fruits are being eaten by animals at a high rate. Firebush is popular as a medicinal plant in various parts of its range. As a bath or poultice, it is applied to various skin ailments, and a leaf tea is taken for worms and to relieve fever. The plant has antibacterial and antifungal properties and is rich in alkaloids.

Russelia sarmentosa
Family: Scrophulariaceae

Wild firecracker, *Coralillo*

Description: Shrub 0.5–2 m; stem slender, with cross section consisting of 4 to 8 angles (illus. lower right). Ovate leaves opposite (or sometimes in whorls of three), usually 2.5–4 cm long with toothed margin, slightly pubescent. Red tubular flowers—less than 2 cm long, with 4–5 small lobes—in whorls at leaf axils. Dehiscent fruit capsule ca. 6 mm across, with tiny seeds.

Flowering/fruiting: Flowers and fruits year-round.

Distribution: Mexico south to Colombia; some parts of Greater Antilles. In Costa Rica, a common weedy plant occurring from sea level to 1,500 m, Guanacaste south to Osa Peninsula. Mainly along roadsides, in pastures, and

at edge of forest; also rocky areas, including near beaches and rivers.

Related species: Coral plant or firecracker (*R. equisetiformis*), a nearly leafless ornamental from Mexico that is cultivated in Costa Rica, has thin, pendant branches that resemble horsetails. Recently, *Russelia*, along with some other "scrophs" such as *Digitalis* and snapdragon

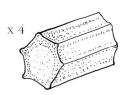

X 4

Section of angled stem
of *Russelia sarmentosa.*

Russelia sarmentosa

(*Antirrhinum majus*), have been placed in the Plantaginaceae family. *Tetranema floribundum* and *T. gamboanum* are two recently discovered Costa Rican members of this family that have tubular, red flowers (Grayum and Hammel 1996). Both species have large leaves and flowers on the ends of long stalks. Another related native, *Lamourouxia viscosa* (now in Orobanchaceae), shares the roadside habit of *R. sarmentosa*.

Comments: Wild firecracker is a variable species and has the widest range of any species in the genus *Russelia,* named in honor of Alexander Russel, an English naturalist. The plant recovers well from grazing. It could be used ornamentally, perhaps as an addition to hummingbird gardens.

Solanum wendlandii
Family: Solanaceae

Marriage vine, *Volcán*

Other common name: Potato vine.
Description: Spiny climber, ascending high into the canopy. Leaves to 20 cm, armed beneath as well as on petiole and stem with hooked prickles; entire, 3-lobed, or deeply lobed with ca. 7 segments, the lower segments sometimes appearing as small leaflets (illus. right). Branched inflorescence of many flowers; dioecious; violet-blue flowers, ca. 4 cm in diameter, but size may vary within a plant. Flower is saucer- or broad-vase-shaped, with 5 corolla lobes; 5 anthers with terminal pores are clustered in the light yellow-green center. Fruit more-or-less ovoid, ca. 5 cm in diameter, green with some longitudinal markings.
Flowering/fruiting: Flowers during most of the year.

Distribution: Once thought to be endemic to Costa Rica, collections have now been made in other parts of Central America. In Costa Rica, scattered in Tilarán Mountains and south to Talamanca Mountains, 700—1,700 m, and widely cultivated.

x ⅓

Solanum wendlandii leaf.

Related species: This is in the same family as the potato, tomato, eggplant, and bell pepper, as well as naranjilla (*Solanum quitoense*, p. 204) and angel's trumpet (*Brugmansia* spp., p. 151).

Comments: The flowers are buzz-pollinated by a variety of bees, including bumblebees (*Bombus*). To collect pollen, the bees must vibrate the anthers, which have terminal pores where the pollen billows out. Marriage vine is more often seen in cultivation than in the wild. A sometimes aggressive garden plant, it can persist at abandoned house sites, years after having been tended. Cultivated plants rarely produce fruit since most are males. Propagation is usually by cuttings. The German gardener and plant collector Hermann Wendland sent seed of this plant to Kew, the Royal Botanic Gardens in the United Kingdom, in the late 1800s, where it was grown and first described.

Solanum wendlandii

Lantana camara
Family: Verbenaceae

Lantana, *Cinco negritos*

Other common names: Red sage, yellow sage, *soterré*.

Description: Usually 1–2 m shrub, but may clamber higher. Pungent odor in all plant parts. Square stems hairy, with or without prickles. Toothed, opposite leaves to 12 cm long, lighter green and soft pubescent below, moderately scratchy above. Bracts below and within inflorescence equal in size. Dome-shaped 2–3 cm diameter flowerheads on 6-cm stalks, arising from leaf axils, made up of many 4-lobed tubular corollas ca. 1 cm long. Buds in center orange-red (or pinkish), open flowers yellow, old flowers orange to red (or pink). Clusters of shiny, metallic blue to black fruits ca. 5 mm in diameter.

Flowering/fruiting: Flowers and fruits year-round.

Distribution: West Indies, southern United States, south to northern South America. Planted as ornamental and

Lantana camara flowers.

Lantana camara

naturalized in Pacific Islands, Australia, New Zealand, South America, and parts of Asia. Throughout Costa Rica in old pastures and second-growth, or cultivated in yards, to 2,000 m.

Related species: Costa Rica has three other native lantanas, each with paler flowers than *Lantana camara*: *L. trifolia,* with elongate inflorescences of lavender flowers; *L. hirta,* with white to lilac flowers; and *L. grosseserrata,* with white flowers. Trailing lantana (*L. montevidensis*), with purplish flowers, is a South American species also used as an ornamental. Cultivars of *L. camara* vary in size and flower color. On the basis of hair details not easily seen by the naked eye, some botanists apply the name *L. urticifolia* to specimens that others would call *L. camara.*

Comments: Considered an ornamental by some, lantana is a scourge in some parts of the world (e.g., Old World tropics) where it has escaped and become a pest. The flowers, which attract butterflies and hummingbirds, begin opening on the outer edge of the inflorescence.

The old, deep-orange flowers, which persist for some time, provide landing surfaces for butterflies, deter nectar-robbing *Trigona* bees from accessing the bases of the newer, inner flowers, and augment the attractive colors in the inflorescence. A hairstreak butterfly (*Cyanophrys longula*) lays eggs in the flower head, and the caterpillars, which blend in with bracts and buds, eat the buds. In Panama, flycatchers, honeycreepers, and tanagers have been observed eating the fruit. This is one of a number of plants that has a range of internal and external folk-medicine uses but at the same time contains known toxins (in this case, lantadene A and B) and should be handled with care—literally, since it may cause dermatitis. Some domesticated animals are adversely affected by eating the plant, experiencing weakness, irritation to eyes and mucous membranes, sores in mouth, and photosensitivity (skin cracking and bleeding). Partial paralysis or damage to the internal organs, including the liver and kidneys, may occur. Humans should not eat the fruits, since unripe ones are known to cause vomiting, weakness, and respiratory and circulatory problems that may result in death. Medicinally, various preparations of the leaves are used externally for snakebite, cuts and bruises, and itchy skin. Leaf decoctions are given in cases of colds, fever, upset stomachs, and rheumatism; root decoctions are given to treat venereal disease and asthma. The leaves contain verbascoside, a substance that shows antimicrobial and antitumor properties. For an excellent article on the weed and its poisonous qualities, see Morton 1994.

Petrea volubilis
Family: Verbenaceae

Queen's wreath, *Choreque*

Other common name: Purple wreath.
Description: Scrambling shrub or liana to ca. 10 m. Opposite, stiff, sandpapery leaves ca. 15 cm long. Inflorescence to ca. 30 cm long, usually pendant. Flower, which is lilac (may be bluish or white), has 5 lanceolate calyx lobes and 5 more-rounded petals (corolla); calyx loses lilac color and persists after the corolla (in center) falls. While the fruit enclosed in the short, pubescent, cone-shaped base of the calyx develops, the calyx lobes elongate to over 2 cm long; these 5 winglike lobes aid in wind dispersal of the 1- to 2-seeded fruit (illus. below).
Flowering/fruiting: Flowers mostly in dry season.
Distribution: Lesser Antilles, Mexico, and Central America; naturalized in Old World tropics. In Costa Rica, grown as ornamental, but also native at forest edges or in overgrown fields, to 1,000 m. Central Valley and Pacific slope.
Related species: Only species reported for Costa Rica, although at least 30 more species grow in tropical America. *P. volubilis* 'Albiflora' is a white-flowered form.
Comments: A showy native that is often adopted as an ornamental in Costa Rica, queen's wreath can be grown as a

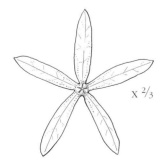

$x\ ^2/_3$

Petrea volubilis fruit with persistent winglike calyx lobes.

Petrea volubilis flower and calyces.

Petrea volubilis

bush or allowed to spread over an arbor, wisteria-style. The author has seen hummingbirds visit the flowers, although they are more likely bee-pollinated. Some researchers who are looking at plant relationships at the molecular level suggest that *Petrea* is more closely related to the Bignoniaceae than the true Verbenaceae (Grayum et al. 1997).

Stachytarpheta frantzii
Family: Verbenaceae

Porterweed

Description: Shrub to 2 m, with many branches. Leaves opposite, to 15 cm (smaller toward top), toothed on margin; soft pubescence on both sides of the leaf and on the angular branches. Leaf stem slightly winged. Inflorescence to 45 cm long; bracts with long, narrow tips; 5-lobed, purple flower with narrow tube ca. 1.5 cm long, light lilac or whitish in throat. Seed less than 0.5 cm long.

Flowering/fruiting: Flowers and fruits throughout the year.

Distribution: Central America. In Costa Rica, 200–1,300 m, mostly Pacific slope, from central region north.

Related species: Blue snakeweed (*S. jamaicensis*, p. 350), a less-pubescent plant, is common along beach and roadsides of the Caribbean and the Osa Peninsula. Some cultivated *Stachytarpheta* in Costa Rica may be *S. mutabilis* (pink snakeweed).

Comments: Stachytarpheta is a great plant for attracting short-billed humming-birds, butterflies, and moths. Although it is a native species, it is more often seen in cultivation. Easily propagated by cuttings, this is a popular garden and hedge plant in some parts of Costa Rica.

Stachytarpheta frantzii flowers.

Stachytarpheta frantzii

Hypoestes phyllostachya
Family: Acanthaceae

Polka-dot plant, *Sarampión*

Hypoestes phyllostachya

Other common names: Measles plant, freckle face.

Description: Branching herb or near-shrub to 1 m tall. New growth is downy; stem noticeably swollen above the nodes, where leaves attach. Oval or rounded, opposite leaves to 5 cm long, with much smaller leaves toward top; pale pink spots on dark green leaves. Tubular, violet, 2-lipped flowers ca. 2 cm long. Fruit a 1-cm-long capsule with 4 seeds.

Flowering/fruiting: Flowers from September through April, sometimes longer.

Distribution: Origin in Africa, but cultivated in various tropical regions; elsewhere kept as a potted plant. In Costa Rica, collected from various mountain locations (ca. 1,000–2,000 m elevation). Cultivated in yards; sometimes it spreads into partially shaded, disturbed areas.

Related species: A number of cultivars exist. This species is in the shrimp-plant family, which is very popular with horticulturalists. Native species of this family compose a significant percentage of the understory plants in Costa Rican cloud forests (see p. 282).

Comments: In parts of Costa Rica the polka-dot plant is an escaped ornamental—essentially a weed—that sometimes forms dense colonies in disturbed areas around homes, along roadsides, and in old pastures. One person's weed may be another's treasured ornament, of course, as the author discovered when she found a flower shop in Northampton, Massachusetts, selling bags of seed and small potted plants for $5.00 each. The Costa Rican name, sarampión, translates as measles.

Megaskepasma erythrochlamys
Family: Acanthaceae

Brazilian red cloak

Other common names: Sanguinea, red justicia, *pavoncillo rojo.*

Description: Semi-woody, 2- to 4-m-tall shrub with angled, slightly pubescent stems. Opposite leaves to 30 cm, entire or with some rounded teeth. Spikes 25 cm long with magenta to red bracts ca. 4 cm. White, tubular, 2-lipped flowers ca. 6 cm, with 2 stamens. Club-shaped, dehiscent capsules ca. 3 to 4 cm long; 4 seeds.

Flowering/fruiting: Flowers most profusely from September through March.

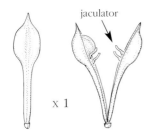

jaculator

x 1

Megaskepasma erythrochlamys seed capsule, before (left) and after dehiscence (right); note jaculators that flick seeds out of capsule.

Distribution: Venezuela; now in cultivation in many tropical countries. In Costa Rica, in yard plantings at middle elevations (common on the San José-Atenas route).

Related species: Only species in the genus. The family Acanthaceae includes many ornamentals as well as a variety of tropical forest understory plants.

Comments: This is a very showy hedge plant that is prolific. It is easily grown since cuttings root readily. As in many other Acanthaceae, the seeds are balistically dispersed (illus. above).

Megaskepasma erythrochlamys

Pachystachys lutea
Family: Acanthaceae

Golden shrimp plant, *Olotillo*

Other common name: Golden candle.

Description: Shrub 1 to 1.5 m tall. Elliptical, entire leaves to 12 cm long. Erect, terminal inflorescences, ca. 8 cm long, composed of overlapping golden-yellow bracts in 4 rows; flowers white, two-lipped, and 5-7 cm long. Fruit a dehiscent capsule.

Flowering/fruiting: Flowers during most of the year.

Distribution: Native to Peru. In Costa Rica, this species is planted in many gardens in the Central Valley.

Related species: Many relatives seen in both cultivated and wild settings. See previous and following species accounts.

Comments: The golden shrimp plant can be grown from cuttings, and is cultivated in tropical gardens as well as in containers. The true shrimp plant, *Justicia brandegeana*, which comes from Mexico, is aptly named; the overlapping pink bracts of its curved inflorescence look like segments of a shrimp's abdomen.

Pachystachys lutea

Thunbergia grandiflora
Family: Acanthaceae

Sky vine, *Emperatriz eugenia*

Other common name: Blue trumpet vine.
Description: High-climbing, ornamental vine. Opposite leaves to 20 cm with a pointed tip, and sparse, large, jagged teeth. Large, showy, 5-lobed, funnel-shaped, one-day flowers in pendant clusters. Corolla 7+ cm across, lavender-blue with yellow-white throat; constriction toward base of tube. Two large membranaceous bracts at base of flower remain after corolla falls. Capsular fruit 3 cm, dehiscent (may not fruit in Costa Rica).
Flowering/fruiting: Flowers and fruits during much of the year.
Distribution: Originally from India and now seen in tropical areas around the world. In Costa Rica, from lowlands to 1,400 m; sometimes escapes from gardens and grows on forest edges.
Related species: Several other species of *Thunbergia* are cultivated in Costa Rica, including the white flowered sweet clock vine (*T. fragrans*) and the black-eyed clock vine (*T. alata*); the latter is also popular in temperate regions of the world.
Comments: This garden vine occasionally escapes to the forest edge and climbs to the canopy. It is invasive in Hawaii and Australia. Pollination of sky vine flowers has been studied extensively in southeast Asia; a nectar guide (purple lines) in the throat attracts large bees, and a washboard path helps them climb into the flower, where pollen is deposited on their

Thunbergia grandiflora

back while they take nectar. Nectaries on the outside of the base of the corolla attract protecting ants. In Costa Rica, some flowers are robbed of nectar and pollen and damaged by *Trigona* bees. A study in a suburban area in Colombia found nectar-feeding bats visiting sky-vine flowers. The family Acanthaceae is one of the most common families in the understory of Costa Rican forests; many grow as shrubs and have red, orange, yellow, or violet flowers that are hummingbird-pollinated.

Yucca guatemalensis
Family: Agavaceae

Spineless yucca, *Itabo*

Yucca guatemalensis

Description: Small tree to ca. 8 m, single trunk or branching, thickened at base. Stiff, pointed leaves less than 1 m long. Terminal inflorescence ca. 60 cm long with branches and many whitish flowers; 6 tepals, ca. 4 cm long. Fleshy fruit capsule to 8 cm long; many small, black seeds in green to white flesh.

Flowering/fruiting: Flowers and fruits mid-dry to beginning of wet season.

Distribution: Mexico to Guatemala. In Costa Rica, cultivated in hedgerows, from lowlands to fairly high elevations.

Related species: Another popular cultivated plant in this genus is Spanish bayonet (*Y. aloifolia*), which has pendant inflorescences covered with woolly hair and purplish fruit pulp.

Comments: Country people propagate *Y. guatemalensis* by cuttings and harvest the flowers to simmer and fry with herbs and eggs. This is probably an acquired taste; a bit like asparagus, the flowers are on the bitter side. The plant is used in living fences. The leaves provide a useful fiber. *Y. elephantipes* is a synonym.

Vendor selling *Yucca guatemalensis* flowers.

Cananga odorata
Family: Annonaceae

Ylang-ylang, *Ilang-ilang*

Description: Tree to 15 m with long, pendant, upsweeping branches. Alternate, entire leaves 10 to 20 cm long, in a plane. Pendant, greenish or cream-yellow flowers, in leaf axils; with 6 petals, each 6 cm or longer; fragrant, especially at night. Each flower produces a cluster of fruits; fruit green to black, to 2.5 cm long.

Flowering/fruiting: Fruits and flowers various times of year.

Distribution: Southern India, Burma, Indonesia, Phillipines; now cultivated in tropical areas worldwide. In Costa Rica, as ornamental and escaped, lowlands to 200 m, both slopes. Common in south

Francis X. Faigal

Cananga odorata

Caribbean region and along Tortuguero canals.

Related species: Soursop (*Annona muricata*, p. 165) and custard apple (*Annona cherimola*).

Comments: The flowers yield a scented oil, ylang-ylang, which is an ingredient of Chanel No. 5 and other perfumes. In nature, the scent attracts beetle pollinators.

Allamanda cathartica   Golden trumpet, *Bejuco de San José*
Family: Apocynaceae

Allamanda cathartica

Other common name: Yellow allamanda.

Description: A liana or shrub with milky latex. Glossy 13-cm leaves with abrupt tips, in whorls of 3–4 or opposite. Bell-shaped yellow flowers 7.5 cm long, 9 cm wide, the 5 rounded (to lobed) petals overlapping; flowers in small groups toward ends of branches. Fruit a round, green, spiny capsule, 7 cm long, eventually dehiscing and releasing seeds (2.5+ cm, including encircling wing).

Flowering/fruiting: Flowers and fruits throughout the year.

Distribution: Considered to be from northeastern South America; now found in cultivation and as occasional escaped plant in tropical areas of New and Old World, including southern United States. In Costa Rica, seen as an ornamental from sea level to 1,000+ m; sometimes in abandoned pastures, or along Caribbean waterways, where it appears that it might be native.

Related species: Bush allamanda, *A. schottii* (= *A. neriifolia*), looks like a smaller version of golden trumpet. Its flowers, which are shorter than those of golden trumpet, have orange-red lines within the throat. *Allamanda blanchetii* (= *violacea*) is a species with purple flowers. Family includes rosy periwinkle (*Catharanthus roseus*, p. 116) and frangipani (*Plumeria rubra*, p. 329).

Comments: The golden trumpet is poisonous, though not as pernicious as some of its relatives (*Cascabela thevetia*, p. 115, and *Nerium oleander*, p. 117). Species name *cathartica* implies its effect—stimulating evacuation of the bowels. Pittier (1978) reports use of the leaves in an infusion as a laxative, but warns that, in large doses, it acts as an intense emetic and purgative. The

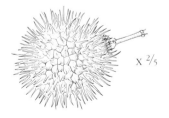

x $^2/_5$

Allamanda cathartica fruit.

leaves, latex, or bark may cause a rash. In Costa Rica and other countries, the bananaquit (*Coereba flaveola*), a bird that takes nectar, often by piercing the base of flower tubes, sometimes visits *Allamanda* flowers. The seeds have thin, winglike edges that suggest wind dispersal.

Cascabela thevetia
Family: Apocynaceae

Yellow oleander, *Chirca*

Other common name: Lucky nut.
Description: Usually a 4-m treelet, but may grow to 7+ m tall; milky latex. Alternate, spiraling, very narrow leaves to 15 cm long, glossy above. Funnel-shaped fragrant yellow flower (with tinge of green), ca. 6 cm long; 5 overlapping petals, twisted in bud; flowers single or in small groups. Green, swollen, boxlike fruit, ca. 4 cm across, turns yellow or red-purple and then black when mature; 1 brown stone with 2 seeds.
Flowering/fruiting: Flowers and fruits all year.
Distribution: As an ornamental from the southern United States to South America, but appears to grow as a native in second-growth, spiny forest of Mexico and in some inter-Andean valleys of Peru; sometimes escapes from cultivation. In Costa Rica, from 100–1,200 m.
Related species: Cultivar 'Alba' has white flowers. Two closely related, native species occur in Costa Rica: *Cascabela* (or *Thevetia*) *ovata* of the central and north Pacific lowlands and *Thevetia ahouai* of the Atlantic coastal region. Both have larger leaves than *Cascabela thevetia*. Oleander (*Nerium oleander*, p. 117) is in the same family, as are the genera *Rauvolfia* (a source of the hypertension drug reserpine), *Catharanthus*, *Allamanda*, and *Plumeria*.

Cascabela thevetia

Comments: This is an attractive tropical ornamental with narrow leaves reminiscent of yew or podocarpus. It is an extremely potent plant, as both a medicine and a poison, due to the presence of various cardiac glycosides, including cerebrin, neriifolin, theveresin, and thevetin. These are referred to as heart medicines or heart poisons. For any cardiac glycoside—even with modern medicines such as digitalis—there is a fine line between a healing dose and a toxic one. Folk remedies include application of yellow oleander to treat toothache, sores, and tumors and preparations taken internally as a diuretic and for fever. Such folk uses of the plant, however, have taken the lives of many people. Lethal qualities of yellow olean-der have been put to use as a fish poison and insecticide, as well as in murders and suicides. Death by oleander is not a pleasant way to go. After a burning sensation in the mouth, reaction to the poison includes diarrhea, vomiting, convulsions, heart failure, and then death. The seeds contain the highest concentration of poisons, but all parts of the plant are toxic; the latex is irritating to eyes and skin. A single fruit can kill an adult human, and livestock may die if they browse on the plant. For the superstitious, carrying or wearing the stone of a *Cascabela* fruit, sometimes called a lucky nut, will supposedly bring good luck (or, in Cuba, at least prevent hemorrhoids). Also known as *Thevetia peruviana.*

Catharanthus roseus
Family: Apocynaceae

Rosy periwinkle, *Conchita*

Other common names: Madagascar periwinkle, *mariposa.*
Description: Bushy plant ca. 60 cm tall; milky latex. Opposite, entire leaves to 7 cm long, shiny with light midvein. Flowers 5-parted, pink (deeper pink in center), ca. 4 cm across with narrow tube ca. 3 cm long; petals twisted in bud. Slender fruits 2–4 cm long. Acrid odor in crushed leaf.
Flowering/fruiting: Flowers all year.
Distribution: Native to Madagascar; cultivated from southern Florida and Mexico to South America and in other parts of the world, often escaping. In Costa Rica, sea level to 1,500 m (possibly higher). Most often found in yards, but naturalized in some central and south Pacific coastal areas.
Related species: Cultivar 'Albus' has white flowers. Dogbane (*Apocynum* spp.), golden trumpet (*Allamanda cathartica*, p. 114), frangipani (*Plumeria rubra*, p. 329), oleander (*Nerium oleander*, next species account), and yellow oleander (*Cascabela thevetia*, p. 115) are all in this family, which is close to, and considered by some to include, the milkweed family, Asclepiadaceae.
Comments: This common garden ornamental is famous for its alkaloids, vincristine and vinblastine. Researchers working simultaneously in the United States and Canada discovered these compounds to be useful in the treatment of Hodgkin's disease and childhood leukemia, but not in treating diabetes, which is what the plant was used for in Jamaican (and African) folk medicine. At least 70 other alkaloids are present. The plant is hallucinogenic. Ingestion may result in loss of hair, muscle deterioration, and damage to nervous system and internal organs. Despite its potential harmful side effects, var-

Catharanthus roseus

ious cultures use it medicinally, in decoctions as a gargle for sore throat, as a diuretic, to retard tumor growth, and for high blood pressure. Root decoctions have been used for toothaches, as a purgative, and to expel worms. This species was formerly called *Vinca rosea*.

Nerium oleander   Oleander, *Narciso*
Family: Apocynaceae

Other common name: Rose bay.
Description: Shrub or tree to 6+ m; with clear latex. Entire, leathery leaves, opposite or in whorls of 3, usually 10–20 cm long, 2–3 cm wide with prominent midrib. Clusters of funnel-shaped, 5-lobed flowers, ca. 3 cm long, 6+ cm in diameter; white, pink, yellow, or red, single or double, with or without scent. Long podlike fruits in pairs; seeds with tufts of white hairs.

Flowering/fruiting: Flowers November through June and possibly in other months.
Distribution: Native to the Mediterranean and parts of Asia; cultivated in many regions, including southern United States. In Costa Rica, in yards, parks, edge of highway.
Related species: This species has hundreds of cultivars, differing in flower color, foliage, and size; some with double flowers.

Nerium oleander

Comments: This is a hardy ornamental capable of withstanding salt spray, wind, and other environmental stresses, including air pollution; a few shrubs grow near the toll booth north of San José on the Inter-American Highway. All plant parts are so highly toxic that ingestion of one leaf—or even food cooked using oleander sticks as skewers—can be fatal. Oleandrin and other cardiac glycosides cause vomiting, diarrhea, abnormal heart beat and breathing, unconsciousness, and death. Contact with the plant may result in a rash, but a number of cultures use leaf baths externally for skin diseases and parasites. Although it is dangerous to ingest, it has been used as a home remedy for heart problems. In the southern United States, Mexico, and the Caribbean, the larvae of the day-flying polka-dot wasp moth (*Syntomeida epilais*) feed on the leaves.

Araucaria heterophylla
Family: Araucariaceae

Norfolk Island pine, *Araucaria*

Detail of *Araucaria heterophylla* branchlets.

Description: Tree with whorls of branches that are pleasingly symmetrical, especially when young. Can grow to 60 m. Foliage has plastic texture; the stiff, curved leaves, under 1 cm long, thickened and somewhat angular toward base, encircle the branches and branchlets. Pollen-bearing catkins ca. 5 cm long. Mature cones ca. 12 cm long.
Flowering/fruiting: This pinelike ornamental is noted more for its foliage than its flowers.

Distribution: Originally from Norfolk Island (east of Australia, in the western Pacific Ocean). Planted as an ornamental in many tropical areas of the world. In Costa Rica, cultivated in yards and parks.

Related species: Various cultivars have differences in color and compactness of foliage. Other species in this genus are *A. columnaris*, which is less pyramid-shaped and more columnlike than *A. heterophylla*, and *A. cunninghamii* (with bunches of twigs that terminate bare branches). Some species are South American, including the monkey puzzle tree (*A. araucana*) of Chile, sometimes planted in Costa Rica.

Comments: This is not a true pine, but, like true pines, it is cone-bearing and wind-pollinated. The straight trunks have been used as sailing-ship masts. The seeds of many species of *Araucaria* are edible. The Norfolk Island pine was previously called *A. excelsa*.

Araucaria heterophylla

Dypsis lutescens
Family: Arecaceae

Butterfly palm, *Eureca*

Other common names: Areca palm, bamboo palm, cane palm, *palmera múltiple*.

Description: Tree, often several meters tall, but can reach 8+ m; clustered stems. Trunk yellow-green or gray-green, ridged with old leaf scars. Sheath to ca. 1 m long, orangish and glaucous gray-green with dull yellow edging. Leaves arching, ca. 2 m long, with many narrow, tapering, orange-yellow or green leaflets. Branching flower cluster, ca. 60 m long, with white flowers. Black-purple fruit, ca. 2 cm long.

Flowering/fruiting: This ornamental palm is noted more for its foliage than its flowers.

Distribution: Origin is Madagascar and a few smaller African islands. In Costa

Dypsis lutescens

Impatiens walleriana
Family: Balsaminaceae

Impatiens, *Chinas*

Other common name: Busy Lizzie.

Description: Herbaceous plant to 1 m, though often under 50 cm; with juicy stem. Petiole variable in length, with scattered projections that are like succulent spines; toothed blades to ca. 10 cm, tapering at both ends. Bilaterally symmetrical flowers of a variety of colors (red, pink, orange, etc.), 3–4 cm in diameter; 2 of the 3 sepals green and tiny; the third, lower sepal (to 1.5 cm long)

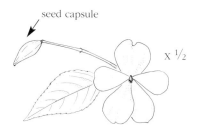

seed capsule

x ½

Flower and turgid seed capsule of *Impatiens walleriana*.

merges into an arching, 3-cm-long, narrow spur; 5 petals, bottom two longest (ca. 2 cm); purple pollen. Ribbed, bulging, green capsule with elastic dehiscence bursts open upon slightest disturbance and shoots out small seeds.

Flowering/fruiting: Flowers and fruits year-round.

Distribution: Originally from mountains of East Africa. In Costa Rica, common from lowlands to 1, 600 m, in gardens and escaped to open areas such as road- and stream-sides, sometimes even emerging from eaves troughs on houses.

Related species: Many hybrids; also more than 800 species worldwide, many in India and Africa. *I. balsamina*, another cultivated species, has deeply toothed leaves. Himalayan balsam or policeman's helmet (*I. glandulifera*) has a more cup-shaped flower than *Impatiens walleriana*. *I. turrialbana* is

Impatiens walleriana

Heliconius clysonymus visiting *Impatiens walleriana.*

a native species with red-orange flowers found at ca. 1,300–2,500 m in Tapantí, on slopes of Turrialba and Irazú volcanoes, and in parts of the Talamancas. North American jewel weed (*I. capensis*), also called touch-me-not, is in this family.

Comments: Many people are surprised to discover that the prolific impatiens, or china, is not native since it certainly seems at home in many parts of Costa Rica. The flowers attract butterflies and hummingbirds. Once you have seen a mature, turgid seed capsule burst and shoot out its seeds, the capsules are irresistible to squeeze. Another species in this genus, *Impatiens balsamina*, contains an antibiotic that acts against fungi. The juice of impatiens species of North America is applied to poison ivy eruptions and mosquito bites to relieve itching. The common name balsam is also used for various species.

Pyrostegia venusta
Family: Bignoniaceae

Flame vine, *Triquitraque*

Other common name: Firecracker vine.
Description: Liana with somewhat angular branches. Opposite, compound leaves; 2 or 3 ca. 7.5-cm-long ovate leaflets. Climbs by way of tendrils that grow from leaves; each tendril with 3 mini-tendrils at tip. Flowers in large, usually pendulous, clusters; tubular, red-orange flowers, 5–7.5 cm, with 5 lobes that curl back; protruding yellow anthers. Fruit a long thin capsule, almost 30 cm long, with winged seeds.
Flowering/fruiting: Flowers are especially abundant in dry season.
Distribution: Originally from Brazil, Paraguay, Bolivia, and northeastern Argentina. Found in cultivation throughout neotropics, sometimes naturalized. In Costa Rica, popular ornamental at middle elevations (e.g., Central Valley), often growing on walls and fences.

Pyrostegia venusta cascading down a wall in San José.

Pyrostegia venusta flowers.

Related species: Trumpet creeper (*Campsis radicans*) of the southeastern United States and many lianas native to Costa Rica, one of which is monkey comb (*Pithecoctenium crucigerum*, P. 261). Showy trees in this family include jacaranda (*J. mimosifolia*, p. 20), African tulip tree (*Spathodea campanulata*, p. 21), and pink trumpet tree (*Tabebuia rosea*, p. 22).

Comments: The typical habit of this plant is cascading down walls, scrambling along fences, and covering rooftops; it will also climb trees. The flowers attract hummingbirds. New plants are easily grown from stem cuttings.

Nopalea cochenillifera
Family: Cactaceae

Cochineal, *Tuna*

Other common names: Nopal, *caite*.
Description: Terrestrial cactus that can grow to more than 4 m. Trunk to 20+ cm diameter; larger, flat, green pads, ca. 30 cm long. Usually spineless, but if spines present, few and ca. 1 cm long. Flowers ca. 5 cm long, borne singly and often along edge of pad; lower part of flower is broad, firm, and green, upper part is rose-colored with peach tones; many yellow and pink stamens extend beyond petals. Many-seeded red fruit ca. 5 cm long.

Distribution: Origin unknown, but probably southern Mexico. Cultivated in many areas, historically for red dye, now as ornamental. In Costa Rica, seen from lowlands to ca. 1,400 m as an ornamental.

Related species: Two native nopaleas occur in Guanacaste: *N. lutea* and *N. guatemalensis*, the former treelike to 5 m, the latter only to ca. 1 m tall, both spiny.

Nopalea cochenillifera

Opuntia ficus-indica

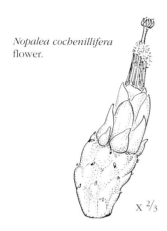

Nopalea cochenillifera
flower.

x ²/₃

Also related are the cultivated, yellow-flowered *Brasiliopuntia brasiliensis* and *Opuntia ficus-indica* (prickly pear, photo, p. 125). All of these species have

at some time been treated as members of the large genus *Opuntia*.

Comments: This cactus is the host plant of the well-known cochineal insect, *Dactylopius coccus*, a type of scale that is a source of red dye dating back to preconquest Mexico. The dye was first sent to Spain in the early 1500s. It is prepared by cooking the insects, then drying them; a not very permanent dye, it has been replaced by synthetics. The Canary Islands was a principal cultivation site that exported millions of pounds of the insect each year. The plant has also been used as animal forage and as a hedge plant. The fruit and young pads are edible. The flesh of the pads is put on burns, aching teeth, and inflamed eyes and is taken in decoction as a diuretic.

Canna x *generalis*
Family: Cannaceae

Canna lily, *Platanilla*

Description: Herbaceous plant, variable in many aspects because of the many cultivars; from less than 1 m to 2 m tall. Broad leaf blade (with variable color) merges into a sheathing base. Flowers are often red, orange, or yellow spotted with red, but may be other colors. In general, a *Canna* flower has 3 petals, one smaller than the others. Appears to have more petals because some stamens are sterile and fused into petal-like structures. Single fertile stamen also has petal-like attachment. Fruit a capsule that wears thin over time to expose and release hard, round seeds.

Flowering/fruiting: Flowers most of the year.

Distribution: Cosmopolitan garden hybrid. In Costa Rica, in gardens at various elevations.

Related species: The Peruvian *C. iridiflora*, with pendant, fuchsia flowers, is

sometimes cultivated in Costa Rica. Indian shot (*C. indica*), which is also cultivated, and sometimes escapes, has leaves 15–45 cm long, reddish flowers (3.5–5 cm), and persistent bracts. The similar *C. tuerckheimii*, native to Costa Rica, has large leaves to 70 cm long, red-orange flowers (to 7 cm), and deciduous flower bracts. Another native species, *C. glauca*, 1–2 m tall, with yellow flowers, grows in Guanacaste, at Palo Verde. *Canna flaccida* is a species of the southern United States.

Comments: Indian shot (*C. indica*) is a native Neotropical species that is also seen in cultivation. It is not as flamboyant as the canna lily, having smaller, red-orange flowers. Its round, hard seeds are used as beads; they also have potential as a substitute for lead shot. *C. indica* may actually be one of the parent species of the fancier *Canna* x *generalis*, which is the result of hybridization.

Canna x *generalis*. Arenal volcano in background.

Cheilocostus speciosus
Family: Costaceae

Crepe ginger, *Caña agria*

Cheilocostus speciosus

Description: Herbaceous plant to 3 m. Sheathing leaves, 15–30 cm with tapering tip, spiral around a corkscrew stem; underside of leaf flannel-like. Conelike inflorescence to 12 cm has spiraling, dull red bracts. What appears to be a frilly, white petal, 7–8 cm across with a delicate perfume and some yellow/orange in the throat, is actually the fusion of 5 nonfunctioning stamens (staminodes); the one pollen-bearing stamen also looks like a

petal, and it holds the style tucked between the anther sacs. True petals, which are pinkish-white, surround the aforementioned reproductive structures. Remains of calyx form crown at tip of capsular fruit that has many seeds with arils.

Flowering/fruiting: Flowers August to January, perhaps in other months.

Distribution: Originally from Southeast Asia, India, and New Guinea. In Costa Rica, usually in warm, wet or moist lowlands.

Related species: A cultivar, 'Variegatus', has leaves with white stripes. Many of the related Neotropical *Costus* (e.g., *C. woodsonii*, p. 347) have more colorful, but less frilly, flowers than the Asian *Cheilocostus*.

Comments: While not a true ginger, the crepe ginger is a member of the Costaceae, a family closely related to the ginger family, Zingiberaceae. *Cheilocostus speciosus*, and members of the genus *Costus*, are known as spiral gingers because of their corkscrew stems. Unlike the predominantly hummingbird-pollinated *Costus*, this plant is pollinated by carpenter bees of the genus *Xylocopa*. The nectar-producing glands on the bracts attract ants. The rhizome contains diosgenin, a precursor to the hormone progesterone. Detailed illustrations of the flower parts are published in Zomlefer 1994.

Cycas revoluta
Family: Cycadaceae

Cycad, *Cica*

Other common names: Sago, sago palm.
Description: Plant has an attractive rosette form; often 1–2 m tall, though may grow taller. Conspicuous rings around trunk where old leaves have fallen. Dark green, shiny leaves to 1.5 m long; the nar-

row, linear segments, which curve upward from the midvein, are ca. 15 cm long with recurved margins. Reproductive structures form at top of plant, in the center of the leaf cluster; plants have either elongate male cones (to 40 cm) or squat female cones that are broad and rounded. Seed, ca. 3 cm, has a fleshy yellowish to red coat covered with hairs.

Flowering/fruiting: This ornamental is noted more for its foliage than its flowers.

Distribution: Origin Japan, but in cultivation in many tropical regions. In Costa Rica, planted in yards and parks in moist or wet areas, low to midelevations.

Related species: Queen sago or fern palm (*C. circinalis*) is taller, with leaves to 3 m long. Cycads that are native to Costa Rica are in the genus *Zamia*, family Zamiaceae (sidebar, p. 130).

Comments: The cycads, a group of primitive plants that dates back 250 million years, were a major part of the earth's flora during the time of the dinosaurs. Although these plants look like palms, they are gymnosperms and are more closely related to pines and other cone-bearing trees. Plants are either female or male, with the pollen produced in a conelike structure; the ovules are borne along margins of bunched, modified leaves called sporophylls. As ornamentals, cycads do well in sun, or with partial shade in a drier climate. They can withstand some frost. *C. revoluta* can be grown as a potted plant; interestingly,

some bonsai specimens of this species in Japan are hundreds of years old. Even though the cycad is not a true sago palm, the name is sometimes used because the core of the trunk and other parts can be used as a source of edible starch if they are prepared properly. Long-term use, however, may be a factor in the cancer and liver problems that are seen in areas where the plant is eaten regularly. Cycasin, a poisonous compound in the plant's seeds, roots, and pith, is carcinogenic and toxic to the liver and kidneys.

Ovules on sporophylls of *Cycas revoluta*.

Cycas revoluta

THE NATIVE CYCADS

Zamia fairchildiana female cone.

The four native cycad species in Costa Rica are in the genus *Zamia*, family Zamiaceae. With ca. 50 species, ranging from the southeastern United States and Mexico to South America, *Zamia* is one of the largest cycad genera in the world.

The many parallel veins in the leaflets of zamias distinguish them from the ornamental exotic *Cycas* species (Cycadaceae), which have leaflets with noticeable midveins only. The roots of zamias, as well as all other cycads, contain nitrogen-fixing cyanobacteria that allow the plants to grow in poor soils.

Z. neurophyllidia (photo below), the most common native cycad, grows in wet Caribbean-slope forest below 1,000 m and is sometimes planted as an ornamental. The plants are ca. 1 m tall. They have short spines on the petiole and teeth along the edge of the grooved leaflets, which have a plastic texture. Orange seeds, 2 cm long, form in cylindrical cones. These plants were formerly referred to as *Z. skinneri*. *Z. fairchildiana* (photo opposite page and above right), of the south Pacific slope of Costa Rica, is usually tall—the trunk alone can exceed 1 m—and has spines on

Zamia fairchildiana

the petiole but not on the leaflet margins. Birds are known to disperse seeds of *Z. fairchildiana*. *Z. neurophyllidia* is sometimes heavily defoliated by larvae of a lycaenid butterfly, *Eumaeus minyas*.

Zamia neurophyllidia

Cyperus papyrus
Family: Cyperaceae

Papyrus, *Papiro*

Description: Water-loving sedge to 4+ m, with thick, creeping rhizome. Leafless stems thick and triangular in cross section. Dense cluster of more than 50 delicate, threadlike flowering stalks radiating from the top of stem; 4–10 bracts, to 1.5 cm wide, situated below the flowering head. Each flowering stalk, or ray, branches at its tip and bears small spikes of flowers. Seeds less than 1 cm long.

Distribution: Origin Africa; introduced to various parts of the world as an ornamental. In Costa Rica, planted from sea level to 1,100 m.

Flowering/fruiting: Recorded flowering in June and September.

Related species: Ca. 43 species of *Cyperus* occur in Costa Rica, most of them less than 1 m tall and growing in moist habitats. The few natives that reach 2–3 m lack the extremely dense heads of *C. papyrus*, which have numerous, threadlike flowering stems. Another African ornamental species, *C. involucratus*, also called papiro, is planted in Costa Rica, but usually it is ca. 1 m tall, with bracts more conspicuous than the rays.

Comments: This garden pond plant is the well-known source of a crudely made paper in Egypt, dating back more than 4,000 years ago. Strips of the wet, soft inner section (i.e., the pith) of the stem were placed overlapping one another, and another such set was placed on top perpendicular to the first. The layers were then pressed into sheets and dried. Other parts of the plant were also put to use. The outer stem fibers were used to make cordage, mats, and other woven items, and the rhizome was eaten.

Cyperus papyrus

Codiaeum variegatum
Family: Euphorbiaceae

Croton, *Croton*

Other common name: *Palo de oro.*

Description: Usually maintained as a shrub, but may grow to more than 3 m tall. Alternate, glossy leaves are quite variable in size (to more than 25 cm long), shape, and color; leaves entire or with lobes; some have a ruffly, or even stringy, appearance, and may be twisted. Blotches or streaks of white, pink, orange, yellow, bronze, or red deck the leaves—some of the colors change with age. Small, white flowers (separate male and female) are in long spikes, 20–30 cm. Three-parted, dehiscent capsules to 1 cm across.

Flowering/fruiting: The noticeable part is foliage, which is colorful year-round.

Distribution: Origin South India, Sri Lanka, and islands of Pacific; now in cultivation throughout the tropics. In Costa Rica, widespread ornamental in lowlands, but also may be seen to ca. 1,500 m.

Related species: Everything you see that looks like a variation on the theme of the croton (*Codiaeum variegatum*) is probably some cultivar of that one species. Many cultivars show variety in color, shape, waviness of leaves. Family includes poinsettia (*Euphorbia pulcherrima*, see next species account), cassava (*Manihot esculenta*, p. 175), castor bean (*Ricinus communis*, p. 177), and yos (*Sapium* spp.).

Comments: Full or morning sun produces the best color in these plants, which are useful in adding constant color in landscaping. They are seen as hedges and sometimes as potted plants. The common name for *Codiaeum variegatum*, croton, may lead one to confuse this plant with members of the genus *Croton*, which is in the same family. While *Codiaeum variegatum* is an ornamental found in Costa Rican gardens, members of the genus *Croton* are common shrubs and trees native to Costa Rica.

A broad-leafed cultivar of *Codiaeum variegatum.*

A narrow-leafed cultivar of *Codiaeum variegatum.*

Euphorbia pulcherrima
Family: Euphorbiaceae

Poinsettia, *Pastora*

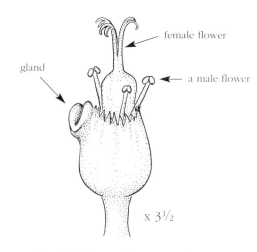

Euphorbia pulcherrima

Description: Shrub or small tree to 4 m, with copious white latex. Alternate, long-stemmed leaves with blades ca. 15 cm long, sometimes lobed; upper red leaves (bracts), which are below flowers, may be opposite or whorled. Green, red, and yellow parts in the center of the bracts make up cyathia (small clusters of flowers), each cyathium with a single female flower and several male flowers, all in a pocket with a conspicuous orange-yellow gland off to one side that looks like a pair of lips. Fruit is a 3-lobed capsule with 3 seeds.

Flowering/fruiting: Flowering peaks in November and December, but the foliage is colorful from October to March.

Distribution: Native to Mexico and parts of Central America; cultivated in many areas. In Costa Rica, common in yards from low to medium-high elevations. A "wild" population was found at ca. 900 m in the central Pacific region in 1995 by INBio collectors.

Related species: There are many poinsettia cultivars, including some with pink or cream bracts. This is a large and diverse genus with about 2,000 species worldwide. Some Old World tropical species are cactuslike. Other commonly cultivated species sometimes seen in Costa Rica are crown of thorns (*E. splendens* or *E. milii*), the white-crowned treelet pascuita (*E. leucocephala*), the purple-leaved *E. cotinifolia*, and the finger tree, (*E. tirucalli*). Some 14 species of the genus *Euphorbia* are native to Costa Rica. See previous species account for other relatives within the family.

Comments: The poinsettia is known in many countries of the world as part of Christmas season festivities. To northerners, the tropical yard plants they see in Costa Rica appear enormous compared to the horticultural potted-plant counterparts they are accustomed to. Butterflies, other insects, and hummingbirds visit the flowers. The toxic latex

female flower

gland

a male flower

x 3½

The *Euphorbia pulcherrima* cyathium is comprised of a single female flower and several male flowers.

may cause blistering of skin, and ingesting the plant results in an unpleasant irritation of mouth and gut, along with vomiting, diarrhea, and, if taken in large amounts, more severe reactions. Despite its toxicity, poinsettia latex has been used in folk medicine, externally for hair removal (depilatory) and internally as an emetic. The use of this plant in teas and baths to stimulate milk flow in new mothers may stem from the folk belief that a plant's salient characteristic—copious white latex, in this case—suggests its use.

Bauhinia variegata
Family: Fabaceae
Subfamily: Caesalpinioideae

Orchid tree, *Matrimonio*

Other common name: *Palo de orquídea.*
Description: Tree usually less than 6 m tall but may grow larger. Alternate, simple, bilobed leaves ca. 10 cm across with cloven-hoof shape; lobes rounded. Purple-pink, 10-cm-wide flower; style and 5 long stamens curving upward; flag petal with touch of yellow, red stripes, and deeper purple. Young pods flat, may be suffused with red; mature fruit ca. 20 cm long, brown and gold, split lengthwise, and spirally twisted (illus., p. 136).

Flowering/fruiting: Flowers dry season into early rainy season.

Distribution: Originally from Asia, but now found throughout tropics as an ornamental; escaped and invasive in parts of Florida. In Costa Rica, seen at low to midelevations; common in Central Valley.

Bauhinia variegata tree.

Close-up of *Bauhinia variegata* flowers.

Calliandra haematocephala

Family: Fabaceae
Subfamily: Mimosoideae

Powderpuff, *Pompón*

Calliandra haematocephala

Calliandra surinamensis

Other common name: *Pon pon rojo.*

Description: Usually large shrub, but can reach 5 m. Leaves fairly large, twice compound, although first division is only forked into 2 parts (pinnae); each pinna with 6–10 pairs of leaflets, ranging from 1 to ca. 6 cm long. Flowers in round heads ca. 7 cm in diameter; long red anther filaments the most conspicuous part of each flower; tiny calyx and corolla. Dehiscent, flat, 6- to 10-cm pod with raised edges.

Flowering/fruiting: Recorded flowering from October through February, but probably flowers in other months too.

Distribution: Originally from Bolivia. In Costa Rica, common in and around Central Valley as an ornamental.

Related species: 150–200 species in Old World and New World, with ca. 9 native in Costa Rica.

Comments: The powderpuff is most likely hummingbird-pollinated. In Hawaii, where it is called leuha haole, it is cultivated for use in leis. Another Hawaiian plant, ohia lehua (*Metrosideros polymorpha*), also has a powder-puff appearance but is in a different plant family, Myrtaceae. The South American *C. surinamensis* (photo left), which has white and pink flowers and smaller leaflets, appears in parks and yards around Costa Rica's Central Valley. *C. haematocephala* was formerly known as *C. inaequilatera*.

Arachis pintoi

Perennial peanut, *Maní forajero*

Family: Fabaceae
Subfamily: Papilionoideae (= Faboideae)

Other common names: Pinto peanut, amarillo peanut, *maní rastrero.*

Description: Trailing plant with stems rooting at nodes; usually very short, can grow to 20 cm tall. Compound leaves with 4 leaflets (ca. 3 cm long), rachis ca. 4 cm long; 2-cm-long stipules join to form sheath at base of leaf. Pealike, yellow flower, 2 cm in diameter. Fruit an underground pod with single 1 cm seed.

Flowering/fruiting: Rain stimulates flowering, though blossoms seen year-round; fruits underground.

Distribution: Native to central Brazil; planted in the United States (including Hawaii), Central America, Southeast Asia, Australia. In Costa Rica, it is planted as ground cover, from lowlands to higher elevations (1,500 m).

Related species: This species is a member of the legume family, which includes beans and peas as well as many trees. Edible peanut is *Arachis hypogaea.*

Comments: This attractive, multipurpose legume is appearing in more and more open areas of Costa Rica. It has become popular as ground cover in place of grass and grows well in full sun or with some shade. Studies in Costa Rica have found that beef and milk production increase in livestock that is pastured in a grass/*Arachis* forage combination instead of just grass. The plant does best where annual rainfall exceeds one meter; it can endure periodic inundation as well as three to four months of drought. It also keeps weeds in check and helps control erosion. All of these characteristics, and the fact that it is nitrogen-fixing, does not climb like some other legumes, and can be mowed, make it attractive in many agricultural situations. More studies must be done, however, to learn if it benefits or competes with crops. Farmers and gardeners should keep an eye on the perennial peanut because, as with many introduced species, its success at thriving in new environments could lead to its getting out of hand.

Arachis pintoi

Hibiscus rosa-sinensis
Family: Malvaceae

Hibiscus, *Clavelón*

Other common name: *Amapola.*

Description: Shrub or small tree, to 4 m. Alternate, shiny, toothed leaves to 20 cm. Five-petaled flower ca. 15 cm across; the 8- to 12-cm-long column protruding from center is topped by a starlike arrangement of 5 styles, with many stamens below. Dehiscent, 5-parted, capsular fruit with small seeds.

Hibiscus rosa-sinensis cultivar.

Adrian Hepworth

Hibiscus rosa-sinensis cultivar.

Hibiscus rosa-sinensis cultivar.

Flowering/fruiting: Flowers and fruits all year.

Distribution: Exact origin unclear; from somewhere in tropical Asia. The genus *Hibiscus* is found in tropical, subtropical, and some temperate regions of the world. In Costa Rica, it is a popular yard plant at a range of elevations, often in hedgerows.

Related species: More than 200 species, and thousands of hybrids (including red, white, and peach-colored cultivars of *H. rosa-sinensis*) within the family. Some species are native to Hawaii. *Hibiscus schizopetalus* (considered by some a variety of *H. rosa-sinensis*) is the coral, or fringed, hibiscus of East African origin (photo right); this as well as rosella (*H. sabdariffa*) are cultivated in Costa Rica. The flower calyces of the latter are a component of Red Zinger brand tea. Rose mallow (*H. moscheutos*) is native

Hibiscus schizopetalus

Malvaviscus penduliflorus

is mahoe (*H. pernambucensis*, p. 339). Turk's cap (*Malvaviscus penduliflorus*), is another popular hedge plant (photo left). Other species in this family include okra (*Abelmoschus esculentus*), marsh mallow (*Althaea officinalis*, the source of genuine marshmallow!), and the native trees clavelón de montaña (*Wercklea insignis*) and doll's eyes (*Hampea appendiculata*, p. 54).

Comments: Beautiful and easy to propagate, hibiscus is one of the most common hedgerow plants in Costa Rica. Cuttings stuck in the ground take readily. Crushed flowers yield a juice that has been used as shoe polish, mascara, and hair dye. The flowers are purportedly edible and red ones in particular are used to treat pain and heavy flow during menstruation; caution is advised, however, because antifertility qualities have been reported from extracts. Flower and leaf decoctions are used in various places to treat flu, coughs, and asthma.

to southern United States, with many cultivars. A noncultivated species that grows in Costa Rica along both coasts

Artocarpus altilis
Family: Moraceae

Breadfruit, *Fruta de pan*

Other common names: *Árbol de pan*, *castaña*.

Description: Tree to 25 m, with latex. Alternate, simple, pinnately lobed leaves to ca. 1 m long. Separate male and female flowers on same plant (monoecious); the thickened yellow male spikes to 30 cm long; the green female flowers in round clusters. Aggregate, rounded fruit, 10–30 cm across, green turning yellowish; white to yellowish flesh, with or without seeds depending on the variety.

Flowering/fruiting: Flowers and fruits throughout the year.

Distribution: Probably originating in New Guinea or Southeast Asian islands; introduced to South Pacific islands and later to West Indies. In Costa Rica, seen

throughout lowlands, but most popular on the Caribbean side, where it is grown for food or as an ornamental shade tree.

Related species: Jackfruit (*A. heterophyllus*); the family Moraceae contains figs and mulberry.

Comments: Breadfruit trees played a pivotal role in Captain Bligh's relationship to his crew on the ship HMS Bounty. The crew's discontent on the return from Tahiti was partly due to the fact that fresh, life-sustaining, water was being given to nonhuman passengers— a thousand breadfruit saplings. During the famed mutiny, the breadfruit trees, which were destined for British-held lands of the Caribbean, where they would be used as a food source for slaves, were cast to sea. It was during a later voyage in 1793 that breadfruit was finally carried to some of the Caribbean islands. It reached Costa Rica's Atlantic coast with Jamaican immigrants. There are many varieties of *Artocarpus altilis*; two occur in Costa Rica. One variety, called breadnut, has spiny fruit with seeds that, when boiled in salted water or roasted, taste somewhat like chestnuts; in Costa Rica, the common name for this variety is castaña, Spanish for chestnut. The other variety bears fruit that are smooth and seedless; this is the typical breadfruit used in Polynesia as a

Artocarpus altilis

breadlike vegetable (or as flour) when green or as dessert when mature. Breadfruit—roasted over coals, baked, boiled, or fried—is a typical part of Caribbean cooking. Caulk, for patching boats, is made from a mixture of latex and coconut oil. The latex is also used to draw worms out of the skin, and it is boiled with water to use on wounds. Decoctions of leaves or fruit are taken internally for various ailments. Quercetin and camphorol are present in the leaves. Pollination for a forest species, *A. integer*, is by gall midges that are attracted to fungus growing in male flowers (Sakai et al. 2000). Previously used names include *A. communis* and *A. incisus*.

Callistemon viminalis
Family: Myrtaceae

Weeping Bottlebrush, *Calistemón*

Description: Small tree to 10 m; branches pendant, new growth covered with silky hairs. Alternate, narrow leaves less than 1 cm wide and to 6.5 cm long. Cylindrical inflorescence to 12 cm at branch tips, with twig growth continuing beyond the flowers; although the 5 tiny petals of these flowers are inconspicuous, their long, red anther filaments, to 2.5 cm, are very showy. Fruit a persistent, woody capsule 0.5 cm in diameter, with hundreds of tiny seeds.

Flowering/fruiting: Flowers most of the year.

Distribution: Originally from the east coast of Australia, now planted in many tropical countries. In Costa Rica, common along city streets.

Related species: *C. viminalis* has various cultivars, varying in size and with inflorescences of varying shades of red. The similar *C. citrinus* (= *C. lanceolatus*) has longer, wider leaves; other species and cultivars have yellowish flowers. *Eucalyptus deglupta* (p. 229) and *Melaleuca* spp. are in this family.

Comments: Bottlebrush trees are popular street trees in cities both because they are able to tolerate a wide variety of conditions and because they flower when small. In the New World tropics, they attract hummingbirds and migrating Tennessee warblers. They are visited by birds such as honeyeaters, lorikeets, and silvereyes in their native Australia.

Callistemon viminalis

Bougainvillea spp. and cultivars
Family: Nyctaginaceae

Some common names: Bougainvillea, paper flower, *veranera*, *flor de verano* (in Panama).

Description: Shrub or liana with alternate entire leaves to ca. 10 cm; leaf shape and pubescence varies with species or cultivar. Spines above each leaf aid in climbing. Colorful papery bracts in threes, each one with a 5-lobed, narrow, tubular flower, ca. 2 cm long, attached on its inner surface (illus. right); flowers yellowish-white or similar in color to the bracts. Cultivated plants rarely produce seed. The two most common bougainvilleas in Costa Rica are probably *B. glabra* and *B.* x *buttiana* (= *B. glabra* x *B. peruviana*), a common hybrid with many cultivars.

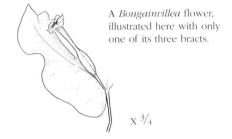

A *Bougainvillea* flower, illustrated here with only one of its three bracts.

x ³⁄₄

Bougainvillea cultivar

A red *Bougainvillea* cultivar.

A peach-colored *Bougainvillea* cultivar.

B. spectabilis may also be cultivated in Costa Rica. The different species and hybrids are difficult to distinguish using bract color alone; to sort them out, one has to look at other characteristics, such as the shape and pubescence of leaves and flowers. *B. glabra*, for example, has narrowly elliptical leaves, while those of *B.* x *buttiana* are broadly ovate.

Flowering/fruiting: Colorful year-round, but more so in dry season.

Distribution: Various species have their origin in South America, mainly Brazil; many cultivars and hybrids have been developed and exist in gardens throughout the world. In Costa Rica, bougainvilleas are very common in yards and hotel grounds throughout the country, from low to fairly high elevations.

Related species: Fourteen or more species exist in South America. The family Nyctaginaceae also includes the genera *Mirabilis*, *Neea*, and *Pisonia*. Four-o'clock (*Mirabilis jalapa*) is an orna-mental that is planted in Costa Rica and other countries. *Neea* and *Pisonia* are genera of treelets that occur in Costa Rica and other tropical forests.

Comments: The brilliance of crimson and magenta bougainvilleas on a sunny, dry-season day is such that the plants seem to shimmer. These vibrant colors are popular, but subtler shades of orange and pink are not uncommon. Hummingbirds and butterflies visit the flowers. The bracts probably play a role in the wind dispersal of seeds in naturally growing species. The intense red-purple color of bougainvillea bracts is from beta-cyanins, the group of pigments that give beet roots their red color. The common and scientific names come from Louis de Bougainville, the eighteenth-century French navigator who discovered the plant in Rio de Janeiro, Brazil. In Panama, the name *flor de verano* (summer flower) refers to its peak flowering in the dry season.

Bambusa vulgaris
Family: Poaceae
Subfamily: Bambusoideae

Common bamboo, *Bambú*

Description: Stems to 20 m, in clumps, with rhizomes. Stems yellow, usually with some vertical green striping; branching and often arching, hollow, nonspiny; internodes ca. 30 cm long. Glabrous, narrow leaf blades to 30 cm. Deciduous, stiff, tan sheaths, ca. 20 cm long, around stem; smaller sheaths on branches.

Flowering/fruiting: This ornamental is noted for its foliage; it rarely if ever flowers in Costa Rica.

Distribution: Asia; original distribution unclear because of its widespread cultivation. Introduced to many tropical areas of the world. In Costa Rica, lowlands to ca. 1,600 m.

Related species: Ca. 120 species of *Bambusa* in the world. *B. bambos* (= *B. arundinacea*), another Asian species sometimes planted in Costa Rica, has green stems and branches with thorns. Bamboo genera native to Costa Rica include *Chusquea* (*C. patens*, p. 300, and *C. subtessellata*, p. 325), *Guadua* (some cultivated), and *Rhipidocladum*.

Comments: Bamboos are a diverse subfamily of some 1,000 grasses ranging from giant 20-cm-diameter stems (*Dendrocalamus giganteus*) to 30-cm-tall miniature bamboo (*Arundinaria pygmaea*). In Costa Rica, bamboo is a popular multipurpose ornamental that acts as a windbreak and a screen, and provides shade. It is a source of poles for fencing, tubes, walls, and furniture, and it can be used for pulp in papermaking. The shoots are edible, although this does not appear to be a popular use in Costa Rica. Other possible uses for bamboo include as a high-protein forage and in making plywoodlike construction material. In recent years, a large, thorny, green-stemmed South American species, *Guadua angustifolia*, has been the focus of construction and furniture-making projects in the Atlantic zone of Costa Rica. CATIE, in Turrialba, has a large live collection of nonnative species, as does furniture builder Brian Erickson in Guapiles. In other regions of the world, the stems are made into fishing poles and are sometimes split for making mats and baskets. Various bamboo species are used extensively in China and Japan for construction, food, and in crafts, as well as for rayon production. The stems grow vertically but do not add girth, as trees can. Bamboos are known for their odd flowering cycles—all plants within a population come into flower at the same time, but usually not every year. The cycles range

Bambusa vulgaris

A stand of *Bambusa vulgaris*.

from decades long to 150 years. After a flowering bout, the plants produce an enormous crop of seeds. Rodents and birds eat a large percentage of the seeds, but not enough to deplete the total crop. The adult plants die off and a flush of new growth begins from the seeds that remain. The seed predators may increase in population due to the bumper crop of seeds, but because of the bamboo's infrequent blooming, the predators cannot maintain their high levels in a region and the next rare seed crop won't be demolished by a large seed-eater population. In the past, when natural bamboo expanses were more common in China, pandas, which feed on leaves of various species, would migrate when a bamboo species flowered and died; now, the decrease in areas to migrate to threatens the panda's survival. The green striped *B. vulgaris* is sometimes referred to as var. *vittata* or var. *striata*.

Antigonon leptopus
Family: Polygonaceae

Coral vine, *Bellísima*

Other common names: Mexican creeper, love vine, chain of love, queen's jewels, coral creeper, confederate vine.
Description: Vine with tuberous root, to more than 10 m, climbing by tendrils (at tips of inflorescences) and side shoots; alternate triangular leaves; blade glabrous to pubescent, 3–10+ cm long with heart-shaped base and tapering tip; leaf margin entire (with occasional irregular teeth) and sometimes wavy; petiole greater than 1 cm long; inflorescence ca. 15 cm long (to 25 cm); buds deep pink; the 5 pink tepals of

Antigonon leptopus

Antigonon guatemalense

Flowering/fruiting: Perhaps year-round.

Distribution: Origin probably Mexico; now grows in cultivation in southern U.S., the Caribbean, and Mexico to South America, as well as in other tropical parts of the world; in Costa Rica, in city as well as countryside yards, from lowlands to midelevations.

Related species: *A. guatemalense* (photo left) has short petioles (less than 1 cm) and is overall more pubescent (leaf, which is more broadly cordate, has carpet of grayish hairs on underside); there are also some differences in tepal size when flowers are just opening—sepals are as long as wide in *A. guatemalense* and are longer than wide in *A. leptopus*. A white cultivar ('Album') of *A. leptopus* exists.

Comments: This is a colorful vine that adapts readily to trailing on a variety of supports. Cooking the large tuberous root and inflorescences makes them edible; people of the Yucatan have used the tuber medicinally to control diarrhea. Beekeepers in some areas cultivate the vine since the plentiful, nectar-producing flowers attract bees.

the flower are largest (to 2.5 cm) when the plant is fruiting, external 3 tepals wider than internal 2; 1 cm long, 3-angled seed with persistent tepals attached.

Ixora coccinea
Family: Rubiaceae

Flame of the woods, *Flor de fuego*

Other common names: Flame flower, burning love, jungle flame, *cruz de Malta*.

Description: Usually maintained as a shrub under 2 m; dense foliage. Opposite, entire leaves to 16 cm long, sessile and often clasping the stem. Flowers tightly packed in rounded heads, pink or orange-red; 4-lobed corolla to 3 cm across, narrow tube 3.5 cm long. Maroon 2-seeded fruit, 1.5 cm long.

Flowering/fruiting: Fruits and flowers continually.

Distribution: Origin is India, but found in cultivation throughout Central and South America, West Indies, and Madagascar. In Costa Rica, sunny gardens, parks, etc., lowlands to midelevations.

Related species: Ca. 400 species in genus. Cultivars and varieties of *I. coccinea* have an assortment of flower colors (orange, pink, and yellow). *I. casei* (photo, p. 151) is a cultivated ixora sometimes confused with *I. coccinea*; it is distinguished by leaves that

Ixora coccinea

Ixora casei

have short petioles (as opposed to none) and 7-12 pairs of secondary veins (as opposed to 5-6 pairs). *I. finlaysoniana* is a planted ixora with 3-cm-long white flowers. *I. floribunda* and *I. nicaraguensis* are two native species with short white flowers.

Comments: This is a popular hedge plant. In its native habitat, the narrow-tubed flowers are probably pollinated by butterflies. The blossoms hold up well in flower arrangements. This plant is used medicinally in other parts of the world for fever, headache, bronchitis, and dysentery.

Brugmansia spp.
Family: Solanaceae

Some common names: Angel's trumpet, *reina de la noche*.
Description: Small trees, often ca. 3 m (to 5 m) tall. Leaves variable, average ca. 20 cm long, some with long petiole. Nodding or pendulous flowers with very large, funnel-shaped to flaring corollas, white or pink to peach. Taxonomy of *Brugmansia* is not that clear; characteristics such as length of corolla, number of calyx teeth, and whether flowers are nodding or pendulous, anthers are joined or not, and the calyx is inflated or not help sort out the species and hybrids. Flowers of *B. versicolor*, apricot moonflower (photo, p. 153), are 35+ cm long and constricted in the section closer to point of attachment; sheathlike calyx with one point; fresh corolla a combination of greenish-white and pink-orange, more intense in older flowers. Fruit, ca. 25 cm long on 15-cm-long stalk, looks like a giant jalapeño or cigar; green with some tiny, tan warts (illus., p. 152). Many irregularly shaped seeds ca. 1 cm, wrinkled surface. *B. suaveolens* (photo, p. 152) has

Brugmansia suaveolens

Brugmansia x *candida*

a 3–5 lobed calyx, fused anthers, and cream and dull orange funnel-shaped corolla. *B.* x *candida* flowers (photo above right) are usually white with a green tint; calyx fits tightly around corolla tube; a double flower cultivar seems quite common in Costa Rica. *B. sanguinea* flowers are much less flared than other brugmansias and are multicolored, with intense orange-red at the mouth of the tube.

Flowering/fruiting: Variable.

Distribution: Exact origin in South America not known. In Costa Rica, grown as a yard ornament, usually at mid- to high elevations.

Related species: Classification of species in this genus needs clarification; various hybrids exist.

Comments: The intensely sweet fragrance of *Brugmansia* is especially noticeable at night. Not only is the perfume strong, but the poisonous/hallucinogenic compounds in the plant are so powerful that they may induce temporary insanity. Scopolamine is the plant's chief psychoactive compound; the hallucinogenic experience begins with a violent period followed by stupor and visions. South American indigenous groups use the plant medicinally (e.g., as a treatment for rheumatism) as well as in divination to commune with ancestors. Methods of ingestion include adding seeds to chicha or making tea from the leaves and flowers; sometimes the bark is ingested. Use of *Brugmansia* is common in the Valley of Sibundoy, in the Colombian Andes. There indigenous peoples use cultivars in a variety of ways, both medicinally and as a hallucinogen. *Brugmansia* may be grown from cuttings.

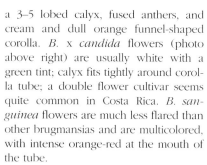

Brugmansia versicolor fruit.

X ¹⁄₅

Brugmansia versicolor has gained much popularity in the Monteverde region in recent years.

Brunfelsia grandiflora
Family: Solanaceae

Yesterday, today, and tomorrow

Other common name: *Sanjuan.*
Description: Usually ca. 1.5 m tall. Simple, entire, alternate leaves ca. 15 cm long with pointed tip. Inflorescence of many flowers that fade from violet to light purple to almost white with age; calyx ca. 1 cm long; 5-lobed corolla ca. 4 cm across with narrow tube to 4 cm long. Fruit capsule 1.7 cm long, with irregularly split calyx that does not enclose fruit. Light brown seeds 0.5 cm long.
Fruiting/flowering: Flowers for much of the year, especially from July to March.
Distribution: Originally from South America. In Costa Rica, popular yard plant in full sun at midelevations.

Related species: Two other *Brunfelsia* species that are similar and are frequently found in gardens are: *B. pauciflora*, which has a long calyx that encloses the fruit capsule, and *B. australis*, which has just a few flowers per inflorescence (whereas *B. grandiflora* has 5 to many). Lady of the night, *B. americana*, has long tubular white flowers that produce perfume at night. *Browallia americana*, a Central and South American weed, has a flower structure similar to that of *Brunfelsia grandiflora* but is much smaller. The Solanaceae, or tomato family, contains well-known food plants (tomato, potato, eggplant, bell pepper) as well as stimulants (tobacco), medicines (belladonna/atropine), and hallucinogens (henbane).

Comments: This large shrub is extremely attractive when in full bloom since the flowers nearly cover the whole plant with an array of violet, a softer rosy purple, and white. The genus name *Brunfelsia* honors Otto Brunfels, a German botanist of the early 1500s. A number of the 40 species have important medicinal uses among indigenous peoples in South America; some of these people also use species of *Brunfelsia* (*B. chiricaspi* and *B. grandiflora*) as an ingredient in hallucinogenic concoctions such as ayahuasca. Studies in the 1970s led to discovery of a substance that is not hallucinogenic but causes convulsions; it is not clear which ingredient in the *Brunfelsia* produces the perception-altering effects.

Brunfelsia grandiflora

Duranta erecta
Family: Verbenaceae

Golden dewdrop, *Duranta*

Other common names: Pigeon berry, *miguelito, once de abril.*

Description: Shrub to small tree (to 6 m), sometimes spiny; angular twigs. Leaves to 7 cm, opposite and bunched, with wedge-shaped base, entire or with teeth along upper half. Candy-scented flowers to 1 cm long on pendant arching stems; 5 rounded, lavender petals; center of flower white. Fruit ca. 1 cm, yellow-orange, in stringlike clusters.

Flowering/fruiting: Flowers and fruits year-round.

Distribution: Origin tropical America; widespread as cultivated and naturalized plant from southern Florida and Texas to Argentina; introduced to Africa, Asia, Hawaii, and Australia. In Costa Rica, appears in parks, hedges, and yards, especially at mid- to higher elevations; not native.

Related species: A white flowered variety, 'Alba', exists. *D. costaricensis* is a

Costa Rican native with pubescent, often spiny, twigs. Relatives include *Lippia graveolens* (interestingly, called orégano, and used as such, in Costa Rica), dama (*Citharexylum* spp.), lantana (*Lantana camara*, p. 103), and teak (*Tectona grandis*, p. 231).

Comments: This ornamental attracts a wide variety of bees, moths, and butterflies. More than 90 species of bees and lepidoptera visit the flowers (W. A. Haber, pers. comm.), which have a sweet scent in the late afternoon and evening. Hummingbirds also visit the flowers. Various parts of the plant are poisonous, containing alkaloids and saponins. Although the fruits are abundant, in Monteverde the author has noted just a single instance of a bird eating the fruit (a young prong-billed

x ¹/₂

Duranta erecta
flowers and buds.

Duranta erecta

barbet, *Semnornis frantzii*). Ingesting the fruit results in fever, nausea, vomiting, convulsions—and sometimes death. Used medicinally in Guatemala and Mexico to reduce fever and as a stimulant. The juice of the fruit kills mosquito larvae. The plant is easily grown from cuttings. *D. repens* is a synonym.

Alpinia purpurata
Family: Zingiberaceae

Red ginger, *Antorcha*

Description: Usually under 2 m tall, but may reach 4 m. Large, sheathing leaves, to 75 cm, alternating up stem, which is topped by an inflorescence, 30–50 cm long, made up of spiraling, red bracts and white flowers. The latter often less noticeable than the plantlets that form in the bracts. Only one stamen is fertile; petal-like, sterile stamens are joined and form lip of flower. Fruit is a capsule.
Flowering/fruiting: Showy bracts year-round.

Distribution: Originally from South Pacific islands and the Malay Peninsula. In Costa Rica, garden ornament in many areas, especially common in Atlantic lowlands.
Related species: Very similar plants with pink bracts or larger flower heads are most likely cultivars of this species. Several other species used horticulturally, including shell ginger (*Alpinia zerumbet*). Ginger (*Zingiber officinale*, p. 208), cardamom (*Elettaria cardamomum*), turmeric (*Curcuma longa*), and

Alpinia purpurata

Alpinia purpurata cultivar.

white ginger (*Hedychium coronarium*, see following species account) are also related.

Comments: The red bracts add color to Hawaiian leis, and the inflorescences are popular in tropical flower arrangements. The scientific name *Alpinia* honors an early Italian botanist, Prospero Alpino. Rhizomes of various Asian species serve as flavoring or medicine.

Hedychium coronarium
Family: Zingiberaceae

White ginger, *Lirio blanco*

Other common names: Ginger lily, butterfly lily, *heliotropo*.

Description: In clumps, usually 1–1.5 m (to 3 m) tall. Flower spikes to 20 cm at end of leafy stems. Leaves to 50 cm by 10 cm, arranged alternately in a plane along the long stems. White flowers with narrow tube to 9 cm long; as in other gingers, one stamen fertile, the 4 others modified into petal-like structures (staminodes), 2 of which are fused to form the 5-cm-wide, 2-lobed, white lip; anther wrapped around hollow style; 3 true petals very narrow. Fruit an orange capsule with many red seeds and red-orange arils. Rhizome looks like ginger root.

Flowering/fruiting: Flowers most of the year.

Distribution: Originally from Himalayan region of Asia. In Costa Rica, a common ornamental in gardens; escapes, especially into wet areas, such as roadside ditches and river banks; sea level–2,000 m.

Francis X. Faigal

Hedychium coronarium

Related species: The genus *Hedychium* includes other garden species, with white, red, or yellow flowers, including kahili ginger (*H. gardnerianum*), which has a tall spike of yellow flowers. Other ornamentals include torch ginger (*Etlingera elatior*, photo opposite page), *Kaempferia rotunda* (photo below), and red ginger (*Alpinia purpurata*, see previous species account). *Renealmia cernua* (photo right) is one of 14 native ginger species in Costa Rica.

Comments: The heady, nocturnal fragrance of these flowers attracts hawkmoths seeking nectar. Hairs on the anthers secrete a glue that coats the pollen as it is released, helping it stick to the wings of a hovering moth when it visits the flower and hits the reproductive structure. Another name, garland flower, perhaps arises from the use of these flowers in Hawaiian leis. The leaves have been

A native species, *Renealmia cernua*, in fruit.

A typical leafy stem of the Zingiberaceae.

used in the Caribbean region in tea for fever, and in the Brazilian Amazon for relief of abdominal pain. Various indigenous people of the Amazon use root preparations externally or internally to relieve pain.

Kaempferia rotunda often flowers when leafless.

Etlingera elatior

4. Fruits and Crops

Tropical fruits are a feast for the senses. They come in all shapes and sizes. Many are colorful, and some have bizarre surface textures or flesh with intriguing flavors and aromas. A few fruits, such as coffee, cacao, and bananas—along with sugar, which comes from a tropical grass—find their way into almost every kitchen in North America and Europe. Others are common in Costa Rica, but are rarely seen beyond the tropics. Some fruits have certain qualities that lead to their widespread use in cosmetics, meat tenderizers, oils, and medicines.

Many plants that originated in one tropical region have been transported widely and are now cultivated in far-flung corners of the world. The coffee plant, for example, was first discovered and used in Africa, but it is now a major crop in Latin America. And cassava, native to the Americas, is now more commonly grown in the Old World tropics.

Humans have always taken advantage of natural mutations in plants that yield more useful crops, but frequently they have used hybridization and other methods to create more productive plants yielding sweeter and bigger fruits. One result of this process is that, over the centuries, the original wild ancestors of certain plants have been ignored. In some cases, habitat destruction has led to the extinction of wild ancestor species.

Now, agricultural researchers generally recognize the importance of trying to find and save the original habitats of economically important food plants, because the genetic material from ancestors can be used to improve disease resistance in today's crops. CATIE (Centro Agronómico Tropical de Investigación y Enseñanza) in Turrialba, Costa Rica, is one of the world's major research and conservation centers for tropical food plants.

Some lesser-known relatives of edible tropical fruits may have food-crop potential. In parts of South America, for example, local people harvest dozens of wild species of plants to sell in markets. Some of these fruits are becoming so popular that destructive harvesting is threatening their survival (Vasquez and Gentry 1989). It is thus critical that both agriculturists and conservation biologists investigate not only the possible cultivation of these plants but also techniques for collecting them on a sustainable basis.

You will see many of the fruits included in this chapter in the markets of Costa Rica. In addition to fruits eaten out of hand, the author has also included a few beverage and spice plants.

Anacardium occidentale
Family: Anacardiaceae

Cashew, *Marañón*

Description: Shrubby tree usually less than 12 m tall. Leathery leaves simple, alternate, oval or obovate to ca. 15 cm long, clustered toward branch tips; new growth reddish. Terminal inflorescences of many small, 5-parted flowers, white-green with pink stripes, turning pink-maroon with age; have spicy cinnamon-like scent. True fruit is the 3-cm-long kidney-shaped gray knob; what appears to be a fruit (the 10-cm-long, fleshy, yellow-orange-red part) is actually the fruit stem (pedicel); inner flesh of pedicel is yellow and astringent. The true fruit develops very quickly compared to the pedicel.

Flowering/fruiting: Flowers and fruits in dry season, with some fruiting in early wet season also.

Distribution: Origin thought to be in the region of Venezuela and Brazil, although native range may be more extensive. Introduced to Asia and Africa in 1500s. In Costa Rica, found planted on both slopes, low to midelevations; more suited for Pacific-slope climate.

Related species: Costa Rica has one other species in this genus (*Anacardium excelsum*, p. 40). Other species, with edible fruit parts, occur in South America. Relatives include mango (*Mangifera indica*, p. 164), red mombin

Anacardium occidentale

(*Spondias purpurea*, p. 213), pistachio, and poison ivy.

Comments: Besides the delectable nuts that end up in cans on grocery-store shelves, cashew trees furnish an array of other products. In Costa Rica, there is some small-scale processing of the nuts, but the cashew "apple," called *marañon*, is the main product. The tasty marañon, which is actually the swollen stem of the true fruit (the curved gray part that encloses the nut), is astringent and rich in vitamin

Anacardium occidentale fruits for sale in market.

C. It is made into juice or eaten fresh, dried (*pasa de marañon*), or stewed, as in a syrup recipe that includes jocote, mango, brown crude sugar (*dulce*), and clove or cinnamon. One of the most widespread medicinal uses of cashew is in alleviating diarrhea; both wine made from the fruit and bark and leaf decoctions are taken for this purpose. A bark decoction is used as a contraceptive in Amazonia. Anacardic acid, which is found in the covering around the nut, shows antibiotic activity. Liquid from the cashew nutshell is sometimes carefully used on warts. Don't try to crack the gray knob of a fresh cashew because the flesh surrounding the nut is full of cardol, a caustic oil that can cause blistering irritation to the skin and burning of the eyes. The principal photographer for this book, who is especially sensitive to poison ivy, once did a watercolor painting of the cashew, including a cross section of the freshly cut nut in its gray coating. Due to contact with the oils, her face swelled up and she felt a burning sensation for two weeks. Heat destroys the irritants, so roasting a nut in its gray shell until it turns black makes it safe to peel and eat. The fumes from roasting can also be irritating, however. Nutshell oil has many industrial uses, such as in brake linings, varnishes, inks, linoleum, paints, and electrical insulation; it can also protect timber from termite damage. In places where the oil is extracted from the nut husks through a heat process, the nuts are removed mechanically; but in some parts of the world (e.g., India), much of the work is still done by hand. The women who do the work usually keep a combination of oil (castor or linseed) and lime or ash on their hands to protect their skin from the caustic liquid. A gum, similar to gum arabic, that flows from the cut trunk has insect repellent qualities and is also useful in book binding. The wood is sometimes used for construction and firewood. Since cashew trees can grow in poor, sandy soil, they have potential in recovering damaged land. The flowers are insect-pollinated. Birds and mammals (including bats) carry off and eat the colorful "apple" while discarding the gray knob and its seed; other animals (parrots, tapir, bats) seem not to be bothered by biting through the caustic covering to get at the nut. A recent article (Rickson and Rickson 1998) reports on the presence of extrafloral nectaries on various parts of the plant (leaves, inflorescence, developing nut); these attract ants, as well as wasps and spiders, that may ward off insect pests. This ant-plant relationship could play a role in the biocontrol of insect pests on cashew trees.

Mangifera indica
Family: Anacardiaceae

Mango, *Mango*

Other common name: Manga.

Description: Tree to 30 m with large crown of dense foliage; rough gray-brown bark; aromatic, resinous sap. Simple, alternate, entire, lanceolate leaves to 25+ cm long, dark green on top, lime green below, with ca. 6-cm-long petiole swollen at base; reddish new growth. Small, 5-parted, green-yellow to pinkish flowers (some perfect, some male) in large, pyramid-shaped clusters at branch tips. Fruit to 20 cm, kidney-shaped or more rounded, sometimes S-shaped, often on long stem. Immature fruit is green, sometimes tinged purple; mature fruit are usually a mixture of yellow, orange, and red; deep yellow or orange flesh, somewhat stringy, sweet and aromatic. One seed enclosed in large, bony pit or stone.

Flowering/fruiting: Flowers from dry to wet season; fruits from dry season into wet season, especially March through July.

Distribution: Native to parts of India, Burma, and China; planted and naturalized in many tropical and subtropical parts of the world. In Costa Rica, planted in much of the country, doing best in warm areas with a distinct dry season.

Related species: There are 40–60 species in the genus *Mangifera*, none native to the New World; red mombin (*Spondias purpurea*, p. 213), cashew (*Anacardium occidentale*, previous species account), and wild cashew (*A. excelsum*, p. 40) are all in this family.

Comments: Mango, a superb Asian fruit with many uses, has been adopted by diverse cultures, including those of the Americas. Cultivars from other Latin American countries arrived in Costa Rica in the early 1800s. Besides being the basis for chutneys and sauces, the fruit is delicious in ice cream, jams, and pies, and it can be dried for later use. It has a high level of vitamin A. Small green fruits eaten with lemon and salt are popular among Costa Ricans; mature fruits that are large and not stringy are called mangas. Many home medicines are derived from mango. Roasted, powdered seed kernels are used as a home remedy for getting rid of intestinal worms. A leaf decoction is used for treating coughs and diarrhea, and as a contraceptive or abortifacient. The bark, which is astringent, is used in decoctions for diarrhea, fever,

Mangifera indica

asthma, and gonorrhea. The plant contains mangiferin and isomangiferin, which have shown antiviral activity on the herpes simplex virus in laboratory tests (HerbalGram 1989; Zheng and Lu 1989). Although the resin is applied to various skin afflictions, it can be irritating, producing a reaction similar to that of the mango's cousin, poison ivy (*Rhus radicans*)—a rash with swelling and itching and eye irritation. Even being in the vicinity of a tree in flower can lead to breathing discomfort and rash on the face of those who are particularly sensitive. When eating the fruit, one should try to avoid the peel since it is also resinous. The ground seed can be made into flour, and its oil has potential as an alternative to cocoa butter in products such as chocolate. Mango trees provide shade and greenery in certain cities in Costa Rica (e.g., Alajuela's central park and the road leading into Liberia). The trees also furnish lumber for furniture, chopping blocks, turned articles, flooring, and crates. The flowers are visited by an array of insects; seed dispersal is probably by bats in the mango's native habitat. In Costa Rica, a number of mammals eat the fruit. Many mango seeds sprout when they are tossed out into a field or compost heap. Some fine cultivars have been developed from such accidents, but the best mangos (tasty, but not too fibrous) are usually from grafted trees; not all cultivars produce fruit every year. There may be as many as 1,000 cultivars in India, the major producer of mangos, where the fruit has a long history, having appeared in symbolism and religion for thousands of years. Preservation of forests that contain wild *M. indica* and its close relatives is important for future crop improvement (see introduction to this chapter, p. 161).

Annona muricata
Family: Annonaceae

Soursop, *Guanábana*

Description: Tree to 8 m. Alternate, entire, shiny, oblong-elliptical leaves yellow-green above, lighter below, ca. 12 cm long, held in a plane. Aromatic flowers green-yellow with 6 sturdy petals, outer ones ca. 3 cm long, inner smaller; many stamens compacted into a disc. Some flowers originate, as do the fruits, from thick branches and trunk. Aggregate, green, ovoid fruit, commonly more than 20 cm long, with fleshy, curved spines; white pulp inside, juicy, with wet mop texture and acidic-sweet taste; many shiny, brown to black seeds ca. 3 cm long.

Flowering/fruiting: Flowers and fruits during most months; highest fruiting from January through July.

Distribution: Exact origin unknown, though probably West Indies or Brazil; has been cultivated in many parts of tropical America. In Costa Rica, cultivated in lowlands on both slopes, doing best below 1,000 m; prefers moist climate, but can withstand some dry spells.

Related species: More than 100 species in genus, including the smaller and sweeter custard apple, *A. cherimola*. There are ca. 10 native species with a variety of fruit sizes and textures; *A. montana* is somewhat like soursop, but with smaller fruits that have straight, rather than curved, spines; there are more than a dozen native genera of Annonaceae in Costa Rica, many of which are found in the forest. Ylang-ylang (*Cananga odorata*, p. 113), a highly fragrant, cultivated tree, is also in this family.

Annona muricata

Comments: As tropical fruits go, the large size (to 40 cm) and odd reptilian texture of the soursop most likely catches the eye of a foreign visitor more often than many other fruit. Soursop is very popular in fruit drinks, conserves, yogurt, and ice cream; it is a good source of vitamins C, B1, and B2. The Annonaceae family has many representatives in the neotropics, and many have interesting natural history aspects relating to beetle pollination. The female part of the soursop flower is made up of many carpels; if a cluster of these does not get pollinated, the fruit develops unevenly and ends up looking curved and lopsided. The tree possesses many alkaloids, including muricine and muricinine, both of which have insecticidal activity. The seeds are toxic and they (and sometimes the leaves) are used to get rid of external parasites such as head lice or to kill other insects. Rubbing leaves on your skin is purported to repel chiggers and ticks. The bark and seeds have been used as fish poison. Leaf tea has been used for dysentery and internal parasites, and various plant parts are used in teas for diabetes, flu, fever, and other ailments. People in the Caribbean Islands mix leaves with other species to make teas, some of which are taken medicinally to calm the nerves. Externally leaves are placed on wounds as a poultice. When planting soursop, one needs to plan ahead since seeds taken out of a fruit may take several months to germinate. The genus *Annona* has a lot of potential for future human use because many of the wild species have interesting edible fruits.

Colocasia esculenta
Xanthosoma sagittifolium
Family: Araceae

Taro, *Malanga*
Tannia, *Tiquisque*

Note: Although they are from different genera, these two species are described here together because they are quite similar.

Description: *Colocasia esculenta* and *Xanthosoma sagittifolium* both have more-or-less arrow-shaped leaf blades on long stems clumped at the top of edible underground corms. They can be distinguished easily by looking at how the leaf is connected to its stem—in *Xanthosoma*, the stem is attached along the edge at the base of the blade, whereas in *Colocasia*, it is peltate (attached within the area of the blade) and the point of attachment appears as a purple blotch on upper side of the blade (illus. right). Plants of both species

Colocasia esculenta

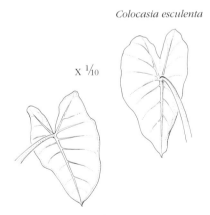

X ¹⁄₁₀

Xanthosoma sagittifolium

The manner in which the leaf stalk attaches to blade is a distinguishing character.

Colocasia esculenta

Colocasia esculenta corm.

Xanthosoma sagittifolium

are less than 2 m tall; height varies depending on the cultivar and environmental factors. *Xanthosoma sagittifolium* has a collective vein along margin of blade. In both species, a spathe envelops the inflorescence (spadix), which is composed of female flowers in the lower half and male flowers in the upper half. In *Colocasia*, there is a sterile extension of the spadix. *Colocasia* has reddish sap and *Xanthosoma* milky sap; *Xanthosoma* sometimes has a whitish cast and purple shading in the stems and veins.

Flowering/fruiting: *Colocasia esculenta* recorded flowering August and October; *Xanthosoma sagittifolium* recorded flowering February, and September through November.

Distribution: *Colocasia esculenta* is from Southeast Asia, possibly India, while *Xanthosoma sagittifolium* is from somewhere in the New World tropics. In Costa Rica, both species in cultivation or as escapes, sea level–1,500 m, in fairly wet areas.

Related species: Several wild xanthosomas grow at various elevations in Costa Rica. *X. undipes* is a giant "elephant ear" that is common from 1,200–1,800 m; *X. mexicanum* is much smaller and pubescent, and *X. wendlandii* has an attractively dissected leaf and purple-blotched stem. Approximately eight species of *Colocasia* are found in Asia. Another member of Araceae planted in Costa Rica is *Alocasia macrorrhizos*, an Asian species that has shiny arrow-shaped leaves and reddish sap; gives off a cyanide odor when bruised.

Comments: These two similar food plants are very common in Costa Rica and are seen in many pastures and country yards and serve as food as well as ornamentals. As members of the Araceae family, their plant tissues contain calcium-oxalate crystals and other toxins that cause swelling and burning of the mouth and throat if eaten raw, as well as diarrhea and vomiting. The juice of the plants may irritate skin and eyes. Nonetheless, their corms have been important starch sources for thousands of years. Cooking decreases the amount of acridity. Sometimes the nutritious young leaves are cooked, with baking soda, as a potherb. Taro (*Colocasia esculenta*) has its origins in the Old World, while tannia or tiquisque (*Xanthosoma sagittifolium*) is a New World plant; both have traveled many miles in different directions to end up as food sources in diverse cultures. Hawaiians make poi out of taro and have dozens of cultivars. The common names given to these plants around the world are many, and there is some confusion as to what refers to what. In Costa Rica, for example, malanga, ñampi, dachín, and chamol are common names for *Colocasia esculenta*, and the name chamol is sometimes used in place of tiquisque for *Xanthosoma sagittifolium*. In some other countries, malanga refers to *Xanthosoma*. In addition, other species names may be applied to *Xanthosoma sagittifolium* (e.g., *violaceum*) because its classification is unclear. Both *Colocasia* and *Xanthosoma* are planted vegetatively by using offshoots of the corms or pieces of corms with buds. *Xanthosoma* and some cultivars of *Colocasia* can be stored for several months. Some farmers on the Atlantic slope of Costa Rica grow these crops for export.

Carica papaya
Family: Caricaceae

Papaya, *Papaya*

Other common name: Pawpaw.

Description: A soft-wooded tree; main distinguishing characteristics are unbranched trunk (3–6+ m tall) with conspicuous scars left by fallen leaves, leaves bunched at top of stem, and large, pendulous fruits coming right off the trunk. Leaves large and palmately lobed, with 7 or more primary veins, and with petioles up to 1 m long; milky latex found in leaves, stem, and unripe fruit. Flowers may differ from plant to plant because some individuals are male and some female; hermaphroditic flowers may be present, and male plants can change sex; flowers whitish or yellowish, with 5 petals, female flower with large ovary and single or few flowers in a group, male flowers tubular and in long clusters. Fruits are melonlike, yellow-orange when ripe, and typically the size of a football; many small, slippery, black seeds in cavity in middle of yellow to red-orange flesh.

Flowering/fruiting: Flowers and fruits may be seen any time of year.

Distribution: Native to New World tropics, but not clear if it originated in Mexico, or Central or South America; possibly it came from the Andes, where a number of wild species are found. About six of the 22 species of papaya (*Carica* and *Vasconcellea*) are cultivated, and they have become one of the most popular tropical fruits in the world. In Costa Rica, *C. papaya* is popularly cultivated throughout the country, not only in large commercial patches but as yard plants, doing best in humid areas below 200 m; occasionally escapes into second growth forest.

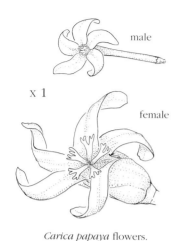

Carica papaya flowers.

Related species: Wild species of papaya (considered by some taxonomists to be in the genus *Vasconcellea*), growing in gaps and along streams in forests in Costa Rica, have 5 primary veins in the lobed leaves and small orange fruits on the lower part of the trunk. Similar genus, *Jacaratia*, which also occurs in Costa Rica, has palmately compound leaves, branching trunks, and may be spiny (e.g., *J. spinosa*).

Comments: Papaya fruit is eaten raw or sometimes cooked, and in some Costa Rican households the cooked leaves and/or immature fruits are added to *picadillo* or soup; a fresh slice with a squeeze of lime juice or a *papaya en leche*, a blender drink with milk, are two of the best ways of eating the fruit. The fruit is 80% water and 10% sugars, with lots of vitamin A and C; the redder the flesh, the more vitamin A. Papain, a proteolytic enzyme, is another important product of the plant. While known mostly as a meat tenderizer, papain prevents shrinkage in wool and silk and appears in medicines for digestive ailments. Its meat-tenderizing ability can be seen by wrapping meat in papaya leaves for a few hours. In parts of the world where papayas are grown specifically for this enzyme, shallow slits are made in the green fruits while they are still on the tree. The latex drips down onto cheesecloth-covered trays, where it dries and is collected. Folk uses include applying the latex from the green fruit on warts and fungal skin diseases and taking it internally to get rid of worms. Decoctions of green fruit are taken for high blood pressure. Some individuals are allergic to papain. Studies of various parts of the plant show antibacterial, antiyeast, and antifungal properties. Compounds affecting the heart are present as well; carposide, a glycoside, and carpaine, an alkaloid, occur in the leaves. Chymopapain, another enzyme from papaya, is an ingredient of enzymatic contact lens cleaners and is used in injections for treating slipped disks in the back. A typical Costa Rican papaya is large, but the small (ca. 15 cm long), pear-shaped cultivar is increasingly popular. Wind, thrips, and small flies have been implicated as pollinators, but the most likely pollinators are moths since the flowers produce the typical scent that attracts hawkmoths at night. Male flowers offer nectar; the female flowers appear to offer no reward, but trick the moth into pollinating by having showy petal-like stigma lobes. While probing for nectar with its tongue, the moth deposits pollen that it picked up from a male flower. Hermaphroditic flowers, when present, are self-fertile. A plant will produce from 30 to 150 fruits a year for three years, after which productivity declines notably. Fruits attract both mammals and birds; wild species get pecked at by birds.

Carica papaya

WHAT GIVES EDAM CHEESE ITS COLOR?

Annatto (*Bixa orellana*) is a popular dye plant that spans the cultures of industrialized countries and Amazonia. The yellow to red coloring obtained from the seed coating has been an important body paint, fabric coloring, and insect repellent to various native peoples of the neotropics, where it has been in cultivation for centuries. In the industrial world, bixin, present in the red-orange covering of annatto seeds, is a safe food coloring that not only lends color to cheese, butter, chocolate, and lipstick, but provides consumers with a dose of vitamin A. It is also an ingredient in some oils, paints, and varnishes. In Costa Rica, where annatto is called achiote, it gives rice dishes a yellow-orange tint and imparts extra appeal to the cheddar, gouda, and edam cheeses of the Monteverde Cheese Factory. Country people say that if you feed the seeds to your chickens, their egg yolks will be yellower.

This natural dye's popularity and value has fluctuated over the years. Before the emergence of synthetic colorings such as Red Dye No. 3, annatto was a valuable cash crop and was exported from South America to the United States and Europe. For about twenty-five years, from the mid-1940s to 1970, manufacturers of Red Dye No. 3 had their heyday, but that declined with the discovery that the dye may be cancer-causing. It was finally banned in 1990, and there is now a resurgence of interest in bixin.

Annatto can be grown in pastures since the leaves contain calcium oxalate crystals that deter cows from browsing on them. The agricul-

Edam cheese with achiote paste.

Bixa orellana (annatto)
Bixaceae

tural institute CATIE, in Costa Rica, is one of the few places in the world doing research on improving annatto as a crop.

The tree, which may be seen in cultivation or, occasionally, growing wild, is shrubby, has ovate to heart-shaped, papery leaves and delicately scented pink or white flowers ca. 5 cm across. The dull reddish brown to bright scarlet fruit capsule is variable in size and shape, ranging from globose to ovoid, generally less than 5 cm in diameter and length, and usually, but not always, covered with soft spines. It splits open to reveal ca. 50 small seeds with a red-orange coating (aril).

The wood is soft and not very useful, but the fibrous bark yields a crude rope. Folk remedies use various parts of the plant for diarrhea and dysentery, fever, epilepsy, and gonorrhea. Other uses include bathing sores and rashes, and as an anti-inflammatory, expectorant, and aphrodisiac. Annatto is also used to control dandruff and to stimulate hair growth. Studies indicate activity against some bacteria.

The flowers attract carpenter (*Xylocopa*), orchid (*Eulaema*), and stingless bees (*Trigona*), all of which collect the pollen. Leaf-footed bugs (family Coreidae) eat immature seeds in the showy capsules, and larvae of a seed beetle (family Bruchidae) feed on them when they mature. Some populations of *Bixa* have extrafloral nectaries on the stem and just below the flower (and fruit). These secrete nectar that attracts ants that, in turn, protect the plant from munching insects; a variety of different ant species patrol the seed capsules round the clock. This relationship makes for a much more successful seed crop. The spines on the capsules may also deter some larger animals from eating the seeds.

Sechium edule
Family: Cucurbitaceae

Vegetable pear, *Chayote*

Description: Vine climbing to 10+ m; tuberous root; channeled stem and branching tendrils. Alternate, long-stemmed leaves to 20+ cm, heart-shaped to 5-lobed, with short, light-colored teeth on margin, rough above. Monoecious; green to yellow, 5-parted flowers, 1.5–3 cm across, male flowers more numerous than female. Fruit spiny or smooth, dark to light green, to 20 cm, pear-shaped with subtle grooves; puckered at both ends; single, flat seed ca. 4 cm long.

Flowering/fruiting: Flowers and fruits year-round.

Distribution: Origin is probably Mexico. Now in cultivation from southern United States to parts of South America, West Indies, and the tropics and subtropics of Old World. In Costa Rica, grown in many regions, probably doing best at midelevations.

Related species: *Sechium tacaco,* formerly *Frantzia* or *Polakowskia tacaco,* a 5- to 6-cm oval edible squash with spines at one end, is endemic to Costa Rica and cultivated; it occasionally escapes. Melons, squashes, loofah (*Luffa* spp.), and gourds are all in the Cucurbitaceae, or cucumber, family.

Sechium edule

Comments: This is the perfect plant for the nongardener. If you don't eat it, a chayote will begin to grow while sitting on your kitchen counter. The seed sprout emerges from the wider end of the fruit, its tendrils reaching out, clinging and climbing onto whatever is in reach. Transferred to the edge of a compost heap or other fertile area, the chayote will take off and need little care. Bees and wasps pollinate the flowers. Costa Rican folklore suggests that a chayote fruit with two shoots emerging develops into a more productive plant. The bland fruit, which is not highly nutritious, is fried, stuffed, and baked and is added to stews or *picadillo*. With sugar and the appropriate spices, it can be concocted into a mock apple pie. The seed, the tender leaves, and the large roots may be cooked and eaten, too; the shoots have some minerals, and vitamin A and B. Skin irritation, or even numbness in the hands, sometimes occurs when peeling raw fruits; in some regions, a poultice of fruit pulp is put on irritated skin and wounds. The fruit and root are used as diuretics, and leaf decoctions or juice of the fruit for lowering blood pressure.

A wild *Sechium* species.

Fiber from the stems is woven into sacks or made into twine. The name comes from the Nahuatl *chayotl* or *chayotli*. The most commonly planted squash in Costa Rica, thousands of tons of chayote are exported to the United States each year.

Manihot esculenta
Family: Euphorbiaceae

Cassava, *Yuca*

Other common name: Manioc.
Description: Shrub or treelet to 3+ m with milky latex; many varieties, so characteristics vary. Cone- or spindle-shaped tubers, to 1 m long. Alternate, deeply lobed leaves, ca. 34 cm across, dark green above, light below; with ca. 7 lobes; long leaf stem to 30+ cm, sometimes red-tinged. Flowers, ca. 1 cm long, have 5 sepals but lack petals; monoecious inflorescence with yellowish male and red-green female flowers. The three-parted capsular fruit, 1.5 cm

long, with 6 ridges, shoots out 1 cm blotched, grayish seeds.
Flowering/fruiting: Generally flowers July through February.
Distribution: Origin unknown, probably Brazil; now in cultivation in tropics worldwide. In Costa Rica, sea level–1,500 m, throughout the country.
Related species: There are perhaps up to 200 varieties of *M. esculenta*. The genus *Manihot* includes ca. 100 species ranging from southwestern United States to Argentina. Four species total in

Tamarindus indica

Tamarind, *Tamarindo*

Family: Fabaceae
Subfamily: Caesalpinioideae

Tamarindus indica

Description: Medium-sized tree to ca. 20 m tall; rough bark. Alternate, pinnately compound leaves ca. 10 cm long, with ca. 12 pairs of leaflets 1–2 cm long. Red buds; light yellow flowers; 3 upper petals 1 cm long with red streaking, two lower ones reduced (rudimentary). Leathery reddish- to tan-brown fruit, 5–15 cm long by 2 cm wide, bulging where 1 cm seeds form; acidic, sticky, brown pulp surrounds seeds.

Flowering/fruiting: Flowers and fruits in various months of the year.

Distribution: Most probable origin is tropical Africa, now found planted and escaped in tropics and subtropics worldwide. In Costa Rica, commonly seen near Pacific coast beaches, plus planted elsewhere; grows at sea level to 1,000 m.

Related species: This is the only species in the genus. A native legume in the mimosoid subfamily, *Inga spectabilis* (photo below right), is also cultivated for the pulp surrounding its seeds.

Comments: There is nothing like a refreshing tamarindo *refresco* to quench your thirst on a hot day at the beach, but go light on it because it has a laxative effect. As a matter of fact, it is taken as a home remedy for constipation. The drink is made by soaking the fruit, then rubbing the pulp off the seeds and adding sugar to offset the natural acids (tartaric, citric, malic, and others). Tamarindo is easily found in markets in Costa Rica; a handful of pulp is wadded up into a piece of plastic wrap; it may have some crude brown sugar (*tapa de dulce*) mixed in. The pulp has high levels of vitamins C and B and also contains calcium and iron. Besides tamarindo drinks, the pulp lends itself to candies, syrups, and jams; in other parts of the world, it is an ingredient in chutneys and sauces (including Worcestershire sauce). Although the seeds are not commonly eaten, they may be made edible by removing the seed coat and boiling or roasting them. Powdered seeds mixed with gum arabic make a decent homemade glue. The trees provide shade, firewood, and lumber; the wood is hard and durable and is good for furniture, tool handles, and beams. Medicinally, the pulp is also diuretic and helps combat bacteria that cause urinary tract infections. Studies of extracts from various plant parts indicate activity against bacteria, viruses, and fungi. Leaves are used in decoctions for diabetes and colds; an infusion of leaves serves as a gargle for sore throat and as a bath for wounds and rashes. In its native habitat, the seeds are probably mammal-dispersed; Bagaces is the tamarindo capital of Costa Rica, with many large trees along the town's streets. *Tamarindo* (or *tamarindo de montaña*) is also a name for a native species of a separate genus, *Dialium guianense*.

Inga spectabilis (guaba) is another legume cultivated for the pulp surrounding its seeds.

Persea americana
Family: Lauraceae

Avocado, *Aguacate*

Other common names: Alligator pear, *palta*.

Description: Tree to 30 m with a dense crown. Alternate, simple, entire leaves ca. 25 cm long (including long petiole), dark green above, gray-green below; with a distinct odor when crushed. Inflorescence of small, yellow-green, scented flowers, each ca. 1 cm in diameter. Pear-shaped, oval, or globose fruit variable in size depending on variety; some more than 15 cm long; yellow and green flesh with a single large seed; fruit surface smooth or bumpy, green to dark purple-black.

Flowering/fruiting: Flowers usually in dry season, to May. Fruits through much of the year, with peak May through July.

Distribution: Mexico to Colombia, and planted in United States, Asia, and Africa. In much of Costa Rica, from sea level to 2,500 m, in forest and in cultivation.

Related species: Ca. 200 species of *Persea* worldwide and 12 species total in Costa Rica. In some regions of Costa Rica (e.g., Monteverde), what is considered *P. americana* may be more than one species, and hybrids may exist. *P. schiedeana* is a large-fruited, wild avocado of the cloud forest with broad leaves that are rusty pubescent below. Ca. 130 species of the avocado family (Lauraceae) in Costa Rica (see *Ocotea* spp., p. 298).

Comments: Street vendors along streets in San José, calling out *"pura mantequilla"* are not selling butter, but a fruit that has creamy-textured flesh with up to 30% oil (rich in oleic acid). The avocado, mashed into guacamole, added to soups, or simply eaten with a little lemon and salt, is a good source of vitamins A, B, and E. Archeological studies in Mexican caves indicate its use by people as early as 10,000 B.C. The common name in

Persea americana

Spanish, *aguacate*, is derived from an Aztec word that signifies *testicle tree*, probably referring to the dangling fruits. It appears that the avocado was taken from the forest and domesticated in various Mexican and Central American sites. There are three varieties—the West Indian, Guatemalan, and Mexican—from which many cultivars have been developed. In the United States, avocados did not gain popularity until the 1900s. California, Florida, and Mexico are major commercial growing areas. Avocado cultivation requires having a mix of what are referred to as type A and type B trees. Flowers of type A trees are receptive to pollen in the morning but do not shed pollen until the next afternoon; type B flowers are receptive to pollen in the afternoon and release pollen the next morning. Bees and a variety of other insects visit the flowers. Uses of avocado in home medicine are many—decoctions of leaves are used for menstrual problems, diarrhea, colds, and fevers; seed decoctions are made to treat diarrhea and to induce abortion, or used externally on wounds or as a hair wash to eliminate gray. Various Amazonian groups take a seed decoction as a contraceptive. Research findings show antibiotic properties and steroids in the seed. Experiments with animals indicate that the leaves contain substances that cause relaxing of smooth muscles and stimulation of the uterus. The fruit is said to be an aphrodisiac. Oil from the fruit is an ingredient in soaps and other cosmetics. Unripe fruit, seeds, bark, and leaves can poison livestock, fish, and other organisms. The wood is sometimes used for construction, furniture, ox yokes, or turnery. While the large fruits of forest avocados are eaten by a variety of mammals, many other Lauraceae species produce small fruits favored by birds (see *Ocotea* spp., p. 298). Wild populations of *P. americana* and *P. schiedeana* are important sources of genetic material for future crop improvement (e.g., resistance to diseases).

Musa acuminata
Family: Musaceae

Banana, *Banano*

Description: Juicy herb, 2–7 m tall. Paddle-shaped leaf blades ca. 2.5 m long, to 0.5 m wide, spirally arranged, erect to arching, green to purple-tinged; sap produces brown stains. True stem is underground; "trunk" is made up of rolled, overlapping leaf-stem sheaths; suckers (*hijos*) arise from base. Long, pendant inflorescence of red-purple bracts covering clusters of female, perfect, and male flowers (latter toward tip) with musty scent. Large clusters of green to yellow, crescent-shaped, seedless fruits, size of which varies from ca. 8 cm to 25 cm, depending on the cultivar; soft, white to yellow pulp. Commercial cultivars may have hundreds of fruits in one bunch.

Flowering/fruiting: Flowers and fruits year-round.
Distribution: Origins of today's banana are thought to be in India, Southeast Asia, New Guinea, Australia, and the Philippines; bananas are grown throughout Costa Rica, up to ca. 1,500 m; plantations are in steamy lowlands along Caribbean coast and in the southwest.
Related species: There are hundreds of banana forms and cultivars worldwide. Ca. 40 wild species of *Musa* exist, including Manila hemp, *M. textilis*, of the Philippines. Plantains (*Musa* x *paradisiaca*), which are starchy and usually cooked, come from hybridizing *M. acuminata* and *M. balbisiana*. Ornamentals with erect

Musa acuminata

inflorescences and colorful bracts include *M. coccinea* and *M. velutina* (photo opposite page); bird of paradise (*Strelitzia reginae*, p. 404) and heliconias (p. 394) are related, but in separate families.

Comments: As one of the commonest tropical fruits in our households, we don't usually think about the bizarre aspects of bananas—how they grow, their shape, and the fact that they do not have seeds.

What we call a banana tree is not a tree at all, but a giant herb with an underground stem (or corm) that sends up multiple shoots. Each one of these pseudostems produces one cluster of bananas and then dies back. The fruits develop by parthenocarpy (without pollination and without seeds). Since the fruits lack seeds, propagation is done vegetatively by planting suckers, usually with a part of the corm. Researchers turn to wild species or to varieties of bananas that produce seeds (or viable pollen) to achieve new, improved banana hybrids. Much work is also being carried out using tissue culture and other biotechnological techniques, including genome sequencing, to develop new hybrids. Bananas and plantains (*Musa* x *paradisiaca*) are major staples for many people around the world; while bananas are usually eaten raw, starchy plantains are usually cooked before eating. Prior to the development of seedless and pulpier fruit, humans and livestock probably ate various other parts (the corm, inner stalk, etc.) of the banana. The male flower clusters are edible, and the giant leaves have been used as food wrappers and thatch. Certain species and cultivars make good fiber and paper pulp. The fruits, besides being eaten, are made into alcoholic beverages (beer in Central Africa, *chicha* in Latin America). Some indigenous people of Costa Rica make banana vinegar and use dried plantain to make flour. Medicinal uses vary around the globe. Plant sap and leaves are used externally for skin problems, and the sap is taken internally for dysentery, epilepsy, hysteria, and as a diuretic. The peel and pulp of ripe bananas have antifungal and antibacterial properties and contain dopamine and serotonin. Nutritionally, bananas are rich in potassium and vitamins B6 and C; plantains are high in vitamin A. Green bananas are sometimes cooked and added to soups in Costa Rica, and plantain chips are a favorite snack food. Sweet, ripe fruits have innumerable uses in desserts; excess

bananas can be dried for later use. Even though bananas are not native to Costa Rica, they are attractive to many animals. Mammals, including bats, are attracted to ripening fruits and morpho and caligo butterflies feed on rotting fruit; caligo caterpillars feed on the leaves. Nectar-feeding bats and hummingbirds visit the flowers. Bats are the most likely pollinator for the ancestors of the typically cultivated bananas; *Musa* flowers that attract bats produce nectar that is more sugary and has some protein; they tend to be purple-bracted, scented, pendant, and open at night. Wild banana species with pink bracts are usually bird-pollinated and have flowers open during the day. *M. salaccensis*, which has an erect inflorescence, has sunbirds and tree shrews as pollinators; other examples of erect inflorescences include the ornamentals pink velvet banana, *M. velutina*, and scarlet banana, *M. coccinea*, the latter of which has brighter red bracts; the former has a 6-cm, pink, fuzzy fruit that splits open and reveals seeds embedded in white pulp. *M. velutina* can be invasive.

Musa velutina

BANANAS IN COSTA RICA

This fruit was perhaps first cultivated in India as early as 500 B.C., and it was transported soon thereafter to Africa. Bananas were introduced in about 1510 to the Canary Islands, and from there carried by explorers to the West Indies and South America. It is not known precisely when and how bananas entered Costa Rica, although it is interesting that a few indigenous peoples in the southern part of the country maintain a sizeable variety of banana and plantain cultivars.

The first commercial plantations in Costa Rica were established in the 1870s; the first shipments of the fruit to New York were made in 1880, via boat from Limón. Minor Keith, who built the San José-Limon railway, was a prime mover behind those initial commercial ventures. Banana exporting in the country grew into a big business involving the United and the Standard Fruit Companies. Challenges such as fungal diseases, degraded soil, price fluctuations, and disgruntled workers have contributed to the shift of the plantations from one side of Costa Rica to the other. In recent decades, revenue from banana exports has contributed significantly to the national economy.

Banana growing can be an environmentally unfriendly activity because of a number of different practices. Much of the deforestation that Costa Rica's Caribbean slope experienced came as a result of growers clearing land for banana plantations. In addition, commercial farmers regularly apply pesticides and add fertilizers to the soil. The blue plastic bags that they place on fruit clusters to protect them from insect and bird damage, and to hasten the fruit's development, litter the plantations or wash into rivers and the ocean.

Some independent farmers, as well as large companies such as Chiquita Brands, participate in the Rainforest Alliance's Eco OK program, which certifies growers of tropical crops who follow specified guidelines relating to pesticide use, treatment of workers, waste management, and other factors. On small farms, it is not uncommon to see bananas mixed with other crops, but large commercial plantations are endless monocultures, a situation that can lead to devastation should a pathogen appear on the scene. A great deal of time and money thus goes into research on combating diseases and developing disease-resistant cultivars. Conservation of both wild banana species and the hundreds of cultivars scattered around the globe is extremely important in order to sustain future breeding efforts.

Harvested fruits of *Musa acuminata*.

Psidium guajava
Family: Myrtaceae

Guava, *Guayaba*

Description: Small to medium-sized tree to 10 m; twisted, musclelike, smooth trunk and branches with light bronze green and pinkish gray bark that chips off; angled twigs. Opposite, rather stiff leaves ca. 10 cm long, with prominent veins and short pubescence; minute translucent dots in leaf, seen most clearly in new growth; older leaves dark above, light below. White, lightly perfumed flowers single, or a few, in leaf axils; 4–5 petals (each 1.6 cm long), 4 green-white sepals; center of flower brushlike with many stamens. Sweet-smelling, yellow fruit 4–8 cm, round (sometimes pear-shaped), with dark scar on top; pink or white inside, with many small, hard seeds.

Flowering/fruiting: Flowers and fruits intermittently throughout the year

Distribution: Tropical America; exact origin unclear since this tree is widespread in the neotropics and often associated with people. Now occurs from Florida, the Antilles, and Mexico to South America, sea level to 1,700 m; also naturalized in parts of Old World where introduced for cultivation. In Costa Rica, seen in most parts of the country, planted or in disturbances; very abundant in some old pastures; generally low to midelevations.

Related species: More than 100 species in the genus. Cas (*P. friedrichsthalianum*), native to Costa Rica and popular in fruit drinks, has similar fruit with light-colored flesh and acidic flavor; has fewer, less prominent leaf veins (8 pairs versus 12+) and is nearly glabrous. Güísaro (*P. guineense*), a smaller species, has unpleasantly acidic fruits. A half dozen other species grow in the forests of Costa Rica. Strawberry guava (*P. cattleianum*), of Brazil, is another cultivated favorite.

Psidium guajava

Comments: In certain regions of Costa Rica, having a guava tree (or two or three) on your plot of land is not a choice—they just appear; and since they bear an edible fruit, are rather handsome trees, and provide firewood, you might as well allow some to grow. Also, fermenting fruits on the ground attract butterflies, including beautiful morphos. To a novice, the many ripe, yellow guavas found in abundance under trees are a real temptation, but it turns out that these very ripe, tasty fruits are full of fly larvae (i.e., maggots). In Costa Rica, guava fruits are picked from the trees that grow in old pastures and are made into jam, jelly, paste, and filling for pastries and candies. Elsewhere in the world (India, South Africa, California, and Florida), more selection

Pink pulp of *Psidium guajava* fruit.

eaten and dispersed in pastures by livestock. Findings at archaeological sites in Peru and Mexico indicate that humans have been using guava for thousands of years. Portuguese and Spanish traders introduced the fruit to other parts of the world. The wood is hard and dense and is good for tool handles and firewood. Tannins in the bark make it suitable for tanning hides. Guava has a number of medicinal uses, the most popular being for diarrhea and dysentery. People in various cultures make decoctions of leaves, flower buds, bark, or unripe fruits for this purpose. The leaves have antibiotic properties and contain a flavonoid, quercetin, that blocks release of acetylcholine, which is involved in intestinal contractions. The Chinese have a method of making medicinal pellets by feeding guava leaves to a large stick insect (*Heteropteryx dilatata*) and collecting the pellets of feces—drop about 10 pellets into warm water and drink! (Lutterodt 1989; McCaleb 1989). External uses include applying a bark decoction on wounds and dried leaf powder on ringworm.

and grafting has been done, and seedless varieties are cultivated. Guava juice is appearing in the United States market, sometimes mixed with passion fruit, but in Costa Rica, the acidic cas (*P. friedrichsthalianum*) is a more popular fruit drink. Fresh guava is a good source of vitamin C and also contains vitamin A, iron, calcium, and phosphorous. Fruits are attractive to squirrels, monkeys, and some birds, and they are

Syzygium malaccense
Family: Myrtaceae

Malay apple, *Manzana de agua*

Description: Large tree to ca. 20 m with patchy tan and gray bark and dense, dark green foliage. Thick, opposite leaves ca. 20 cm (but sometimes up to 35 cm long), with pointed tips. Pink-red brushlike flowers clustered along branches; short petals, but many long stamens. Somewhat pear-shaped, pink to red fruit to 8 cm, with white flesh.

Flowering/fruiting: Fruits from June to August.

Distribution: Originates in Malay Peninsula. In Costa Rica, may be grown in a variety of sites, but does best on Atlantic slope.

Syzygium malaccense fruits.

Syzygium malaccense flowers.

Related species: Other cultivated members of this family include cloves (*S. aromaticum*), rose apple (*S. jambos*), Surinam cherry or pitanga (*Eugenia uniflora*), and guava (*Psidium guajava*, p. 185).

Comments: This is one of several popular fruits that are sold in bags along the Inter-American Highway. The flesh, which is somewhat applelike but more watery and not as dense, has a pleasant, refreshing flavor. Even though the fruit is mostly eaten raw, it can also be made into preserves. Rose apple (*S. jambos*), another Old World species, is grown in windbreaks and for its round, yellowish, rose-scented fruits.

Averrhoa carambola
Family: Oxalidaceae

Carambola, *Carambola*

Other common name: Star fruit.

Description: Small to medium, low-branching tree, 5–10 m, with dense foliage. Alternate, compound leaves with ca. 6 pairs of opposite (sometimes alternate), entire leaflets, plus terminal one, ranging in size from 1.5 cm to more than 8 cm, each one asymmetrical. Small, pinkish-purple 5-petaled flowers. The yellow 5-ribbed fruit is 13 cm long

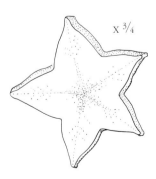

x ³⁄₄

Cross section of *Averrhoa carambola* fruit.

Averrhoa carambola fruits.

and star-shaped in cross section; the many fruits often weigh branches down.
Flowering/fruiting: Fruits develop at various times of the year.
Distribution: Native to parts of Indonesia. Now cultivated in many parts of world where climate is warm and humid, including southern Florida. In Costa Rica, grown from sea level to midelevations.
Related species: *Averrhoa bilimbi* (tiriguro or mimbro) is more vegetable-like, being added to curries, jams, and *picadillo*. Sour-tasting wood sorrels

(*Oxalis* spp.), common weeds of temperate areas, are also in this family.

Comments: It is sometimes a surprise to northerners to learn that star fruit is in the wood sorrel family, which is represented in temperate areas by small yellow- or purple-flowered herbs. Oxalic acid imparts a sour taste in both sorrel and its tree-sized cousin. Some varieties of carambola are a bit sweeter (or less sour) than others. The fruit is a good source of vitamins A and C and also contains iron. It is frequently used in Costa Rica as a *refresco natural*, a natural fruit drink. It also makes a good marmalade, and occasionally you will see its characteristic star-shaped slices as a garnish in restaurants such as Tin Jo

Averrhoa carambola flowers.

in San José. The fruit can be used as a stain remover (on cloth and skin) and as a brass polish. The leaves have sleep movements similar to those of the sensitive plant (*Mimosa pudica*, p. 92). The flowers are insect-pollinated.

Passiflora ligularis
Family: Passifloraceae

Sweet passion fruit, *Granadilla*

Other common name: Sweet granadilla.
Description: Climbing vine with tendrils and woody base. Alternate, entire leaves to more than 15 cm long, with heart-shaped base, on long stalks with several pairs of threadlike glands; leafy stipules ca. 3 cm long. Flowers, to 10 cm in diameter, with pale green-white sepals and petals; the flower has a complex structure with a central column where 5 stamens and 3 styles are attached, and a corona made up of white and purple-striped filaments. Yellow-orange and maroon, ovoid fruit ca. 10 cm long, leathery-rigid rind with spongy white lining; many black seeds coated in sweet, slimy pulp (photo, p. 190).
Flowering/fruiting: Fruits most abundantly from October through May, while maracuyá (*Passiflora edulis* f. *flavicarpa*) peaks from May through October.
Distribution: Mexico to parts of South America and cultivated in Hawaii and

Africa. In Costa Rica, cultivated or escaped at 1,200–2,000 m.
Related species: Of ca. 450 species, 60 are considered edible. The cultivated yellow granadilla or maracuyá (*Passiflora edulis* f. *flavicarpa*) is the popular, tart passion fruit used in Costa Rican

glands

X ³/₄

Calyx lobe from flower of maracuyá (*Passiflora edulis* f. *flavicarpa*). Note the two small glands.

Passiflora ligularis fruits.

Passiflora edulis f. *edulis* flower.

refrescos. Another form of this genus found in Costa Rica is *P. edulis* f. *edulis* (flower photo above), a purple-fruited species; both forms have 3-lobed leaves with 2 small, kidney-shaped glands on petiole (near where it connects with blade), but *P. edulis* f. *flavicarpa* has 2 small glands on each calyx lobe (illus., p. 189). Another species found in Costa Rica is the giant granadilla (*P. quadrangularis*), with fruit that grows to 30 cm long.

Comments: This is one of about a dozen species of *Passiflora* that are cultivated for their fruits; others are grown as ornamentals. According to some religious folklore, the story of the crucifixion of Jesus Christ (the Passion) is depicted in this fascinating flower—the three styles being the nails of the cross, the five anthers Christ's wounds, the corona representing the crown of thorns, the tendrils the whips, and so on. For long-wing butterflies (*Heliconius* spp.), the plant has another significance—as a food plant for caterpillars. *Passiflora* species are known to contain alkaloids and cyanogenic glycosides. In Latin America, various plant parts are used medicinally; the most frequent use is as a sedative. The fruit is a source of vitamins A and C. Fruits are eaten fresh without preparation (in the case of granadilla), or made into drinks, sorbets, and mousses. Large bees pollinate some flowers, while hummingbirds visit red ones (*P. vitifolia*, p. 244). The plants are grown from either seeds or cuttings and do well in both hot lowlands and cooler tropical mountains, depending on the species or variety.

Piper nigrum
Family: Piperaceae

Black pepper, *Pimienta*

Description: Vine cultivated on posts, ascending to 4 m or higher. Leaves 9–16 cm long, often rounded. Opposite each leaf is a pendant flower spike of crowded, reduced, green flowers, lacking petals. Fruiting clusters ca. 15 cm long with 0.6 cm fruit, red when ripe; thin flesh, with one seed inside.

Distribution: Originally from India and Sri Lanka. In Costa Rica, cultivated in various parts of the country; usually in warm, wet areas below 500 m.

Related species: Possibly up to 2,000 species in the world; over 100 species in Costa Rica (see *Piper auritum*, p. 64). Kava (*P. methysticum*) is in this family.

Comments: On a global scale, this is the most popular spice. Although Costa Rica is not a major producer, at times black pepper has been a popular alternative crop. The fruit contains the alkaloids piperine and chavicine, plus ethereal oils. Black pepper comes from the dried not-yet-ripe

Piper nigrum

Rubus species.

Distribution: *Rubus* species are found all over the world, especially in temperate areas; also common and abundant in mountains of Central America, Colombia, and Ecuador.

Related species: Hundreds of *Rubus* species worldwide, with 13 in Costa Rica (3 of which are nonnative). The Rosaceae family includes a number of familiar temperate groups such as roses, cherries, strawberries, and apples.

Comments: The blackberry is not the first fruit that normally comes to mind when one thinks of tropical fruit, but in Costa Rica, *mora*, as it is known, is used to make one of the most frequently encountered natural fruit drinks. The berries are made into tasty jams and ice cream as well. One can find various species of blackberry in the mountains of Costa Rica, a hot spot being the Cerro de la Muerte region. Families there have been shredding their clothing and skin while picking berries for half a century. *R. miser*, and sometimes *R. costaricanus*, are sold commercially (F. Morales, pers. comm.). Agroforestry systems combining alder (*Alnus acuminata*) and blackberry are being tried out; if successful, this combination would allow farmers to begin selling blackberry crops many years before the slower growing alder trees are ready to harvest. The flowers are insect-pollinated, although hummingbirds sometimes visit; and the seeds are bird-dispersed.

Coffea arabica
Family: Rubiaceae

Coffee, *Café*

Description: Can grow to 8 m, but usually kept under 2 m. Opposite, entire leaves, dark green and shiny, to 20 cm long with tapering tip; pointed stipules; tiny domatia in leaf axils on underside appear as little bumps above. Fragrant, 5-lobed white flowers ca. 1 cm long, cauliflorous (coming directly off of stem), often in large bunches. Fleshy, red fruit to 2 cm long with two seeds.

Flowering/fruiting: Flowers March through May; fruits November through February.

Distribution: Originated in East Africa (Ethiopia and Sudan). In Costa Rica, mostly at 800–1,400 m. Extensive plantations in Central Valley, but also grown in mountains to the north and south. Occasionally found in forest, most likely from seeds defecated by birds.

Related species: Worldwide, dozens of species in genus *Coffea*; robusta coffee (*C. canephora*) and Liberian coffee (*C. liberica*) are two species regularly grown at lower elevations. Coffee family, Rubiaceae, is one of the major understory plant groups in Costa Rican rainforest, with ca. 430 species in the country (see sidebar, p. 197). The genera *Psychotria* and *Hoffmannia* are especially diverse.

Comments: While many people joke about their need for caffeine, this addictive beverage, coffee, is the "most widely used psychoactive drug in the world" (Simpson and Ogorzaly 1995). *Grano de oro* (grain of gold), as it is known in Costa Rica, is one of the top export products for the country; major producers worldwide are Brazil, Colombia, and Africa. People in Ethiopia, coffee's native land, supposedly chewed on the leaves and berries, but it was on the Arabian Peninsula, on the other side of the Red Sea, where cultivation began and the beverage was developed, possibly as long as a thousand years ago. In the latter half of the seventeenth century, coffee houses proliferated in Europe, and the Dutch succeeded in getting viable seed away from the Arabs to start plantations in Java. Shortly afterward, coffee was carried to the New World, eventually arriving in Costa Rica in 1808. The highest-quality coffee comes

Coffea arabica fruits.

from mountain-grown *C. arabica*, or Arabian coffee, which does best with some shade; various trees (often nitrogen-fixing legumes) are thus planted along with coffee. Besides Arabian, Costa Rican farmers cultivate two other coffee species (on a much smaller scale), robusta (*C. canephora*) and Liberian (*C. liberica*), both of which can grow at lower elevations. After the harvest, a machine removes the fruit pulp and the seeds are washed and left to ferment for up to a day

Coffea arabica in flower.

Coffea arabica plantation with *Dracaena fragrans* in foreground.

headache and colds, diuretics, and weight-control preparations. The flavor of coffee lends itself nicely to ice cream, mixtures with chocolate (i. e., mocha), and liqueur. Researchers studying coffee's make-up have found more than 700 compounds, caffeine being the most significant; a cup of coffee has over twice the amount of caffeine as a cup of tea. The stimulating effect of coffee is due mostly to the fact that caffeine blocks adenosine, which occurs naturally in our bodies and has a calming effect. When it is blocked, we become—and stay—wired and feel more energetic and less depressed. Caffeine overdose can lead to the jitters, gastrointestinal irritation, rapid or irregular heartbeat, and anxiety attacks.

for flavor and aroma enhancement. The seeds are then sun-dried, the remaining skin is removed, and the coffee "beans" are roasted. Caffeine that is extracted from coffee in the decaffeination process is put into soft drinks as well as medications for

Erythrina poeppigiana trees are often planted in coffee plantations. They provide shade for the coffee plants and enrich the soil with nitrogen.

COFFEE RELATIVES

If you walk through the understory of a Neotropical rainforest, many of the shrubs and treelets that you encounter are coffee relatives. The majority have rather inconspicuous small white flowers and basic, elliptical leaves that do not stand out. Some have showy flowers or fruits, however. Those with flowers that are tubular and white, and that emit sweet perfumes, attract hawkmoths or other insects; species with red blossoms are often hummingbird pollinated. Fruits range from pea to grapefruit size, usually contain pulp, and often have bright colors to attract birds or possibly emit an odor to draw in mammals. A minority of species have spindle- or cigar-shaped capsules with wind-dispersed seeds.

The coffee, or madder, family (Rubiaceae) is one of the largest in the world, encompassing well over 10,000 species. It is especially abundant in Central and South America. In addition to the shrubby life forms, there are species that grow as trees, epiphytes, vines and lianas, as well as creeping herbaceous plants. The key vegetative characteristics that distinguish the family are simple, opposite, usually entire leaves with stipules between the petioles.

Many Rubiaceae species contain alkaloids or other chemical compounds that are potential sources of medicines. Coffee, quinine, and ipecac are a few of the commercial products derived from the Rubiaceae. The Old World *Gardenia augusta* and various *Ixora* species (p. 150) are common house and garden plants.

Cinchona pubescens   **Quinine, *Quina***

Of the several alkaloids from species in the Rubiaceae family that have played important roles in medicine, the most famous is quinine because of its capacity to prevent and combat malaria. *Cinchona officinalis* and other species in this genus from South America were exploited heavily in the 1800s as a source of natural quinine. Plantations were established in Indonesia and later Africa; in 1944 quinine drugs were synthesized. Costa Rica appears to be the northern limit for *C. pubescens*, whose southern limit is Peru and Bolivia. There has been some speculation that the trees in Costa Rica are relicts from a time when the plant was cultivated by indigenous people. Scientists doing antimalarial research today are looking to wild cinchona species as sources of valuable

William A. Haber

Cinchona pubescens

genetic material because the microbes responsible for the disease are developing resistance to the synthetic medicines.

COFFEE RELATIVES

Genipa americana

Genipa, *Guaitil*

Genipa americana

The fruits of genipa are well-known as a source of dye and body paint among various indigenous people. The juice of immature fruit turns skin and cloth blue-black. Seeds are embedded in a jelly-like pulp that is attractive to mammals, including humans, who use it to make various medicinal or intoxicating concoctions. The larvae of a day-flying sphinx moth (*Aellopos titan*) feed on genipa leaves.

Posoqueria latifolia

Monkey apple, *Fruta* (or *Guayaba*) de mono

Posoqueria latifolia

Monkey apple is a small tree seen in many parts of Costa Rica, especially in wet lowlands. With its 15-cm-long, narrow white blossoms emitting an intense perfume at night, this is the quintessential hawkmoth flower. When touched, the anthers pop apart and the lowermost stamen dabs a visiting hawkmoth with a pollen mass. Monkeys and birds eat the yellow-to-orange, 5-cm fruit, which has angular seeds surrounded by pumpkinlike pulp. The bark has been used in infusions to treat diarrhea. A dusting of the dried flowers is supposed to repel fleas.

Psychotria elata

Hot Lips, *Labios de novia*

The "lips" of *Psychotria elata* are bracts that frame a head of buds and flowers. These bracts attract hummingbirds. *Psychotria* is a large genus with around 100 species in Costa Rica. *Psychotria pilosa* is a conspicuous shrub, seen on the Atlantic slope and in the Osa Peninsula region. *Notopleura uliginosa*, formerly classified as a *Psychotria*, is a common herbaceous species of the forest understory.

Psychotria elata

Psychotria pilosa

Notopleura uliginosa

Rondeletia amoena

Teresa

Rondeletia amoena

This attractive small tree is seen in all of the major mountain ranges of Costa Rica. The flowering branches can be used in cut-flower arrangements. This species is also called quina or palo cuadrado in Costa Rica. *Rogiera amoena* is another accepted scientific name for this species.

Citrus aurantium
Family: Rutaceae

Sour orange, *Naranja agria*

Other common name: Seville orange.
Description: Small tree to 10 m with long spines. Slight wing on petiole; simple leaf (though derived from compound leaf) to 10 cm long, shallow teeth or scalloping on margin. Crushed leaf is pungent; oil glands visible when leaf held up to light. Axillary, 5-parted, fragrant, perfect, whitish flowers. Juicy fruit ca. 8 cm across.
Flowering/fruiting: Flowers and fruits fairly continuously.
Distribution: Native to parts of southwestern Asia; various citrus have their origins there and toward the southeast to Malaysia. Now planted in many tropical and subtropical regions. In Costa Rica, planted and escaped in a variety of climates and elevations throughout the country.
Related species: Of the ca. 16 species in the genus *Citrus*, half a dozen are very popular, including *Citrus* x *paradisi* (grapefruit, photo below left). Other genera include *Fortunella* (kumquat), *Ruta* (rue, called ruda in Costa Rica), and *Zanthoxylum* (with many Costa Rican natives).
Comments: Sour orange is a common part of kitchen gardens in Costa Rica. To a foreigner, the fruit may look like a tangerine, but upon picking and tasting a sour orange, one learns the differences in shape and texture quickly. The plentiful juice is great as a fruit drink (add lots of sugar!) and in salad dressings and is, of course, a source of vitamin C. The fruit is technically a berry, and the many juice sacs are modified hairs—fleshy and glandular. This is a hardy species that tolerates wind and drought and thus makes a good rootstock for grafting other types of citrus. Sour orange is the main source of rind for making marmalade and liqueurs such as Cointreau.

Citrus aurantium

Citrus x *paradisi*

The flower petals produce neroli oil, an ingredient of perfumes, and the fruits and leaves contain saponins, which make them usable as a soap substitute. Sour orange is also popular in folk medicine; the most common uses involve teas and decoctions taken for cold and flu as well as for an upset stomach. A tea of the flowers supposedly has a calming effect; a decoction of the leaves of soursop (*Annona muricata*, p. 165) and sour orange is taken for asthma. Studies indicate that the oil has some antifungal and antibacterial properties. The oil in the peel can cause contact dermatitis and ingesting large amounts of the peel can cause intestinal pain, convulsions, and death in children. The origins of most cultivated citrus are not clear, but their development most likely involved selection and hybridization of mandarin (*C. reticulata*), pummelo (*C. maxima*), and

citron (*C. medica*). After traveling from Asia westward and eventually to Spain, citrus was brought to the New World by Columbus. Other citrus products include pectin extracted from the peel and used for making jelly and jams, potential sugar substitutes derived from flavonoids, and oil for flavoring toothpaste. Of the citrus species grown in Costa Rica, sweet orange (*C. sinensis*) is the only one that has been developed into a commercial export crop. Smaller orchards of trees with juicy sweet oranges occur in isolated mountain valleys, where people collect and sell them locally in the latter months of the year. Some of the interesting creatures that use citrus as a host in Costa Rica are thorn bugs (*Umbonia crassicornis*) and the larvae of both the giant swallowtail butterfly (*Papilio cresphontes*) and the citrus skipper (*Achelodes selva*).

Blighia sapida
Family: Sapindaceae

Akee, *Akí*

Other common name: *Seso vegetal.*
Description: Tree 10–20 m. Alternate, evenly compound leaves with ca. 4 pairs of glossy leaflets, 8–15 cm long. Inflorescences, to ca. 20 cm long, develop in leaf axils; small, fragrant, green-white, 5-parted flowers, staminate or bisexual. Wrinkly red, yellow-tinged, bulging, 3-lobed, dehiscent fruit, ca. 7 cm long; 2-cm, shiny, black seeds, each with an off-white aril.
Flowering/fruiting: Fruits various times of the year.
Distribution: Originally from West Africa and now planted in various tropical and subtropical areas, including Florida, Jamaica and other Caribbean islands, Central America, and parts of South America. In Costa Rica, most common in Caribbean lowlands, but

Francis X. Faigal

Open fruit of *Blighia sapida*.

Blighia sapida

also grown occasionally in Guanacaste (e.g., foothills of Rincón de la Vieja).

Related species: No other species in this genus found in Costa Rica, but there are many related, native lianas and trees of the family Sapindaceae. Also in this family are rambutan (*Nephelium lappaceum*, following species account) and mamón (*Melicoccus bijugatus*).

Comments: Since it is edible as well as colorful, the akee is a popular yard and street tree of the Caribbean region. The off-white aril, which looks like slightly spongy vinyl, is edible, however it is poisonous both before the capsule splits open and when it is over-ripe. The seeds and the pinkish material between the seed and the aril also contain the poison hypoglycin A. The consequences of eating akee at the wrong stage are brutal and often fatal—nausea and vomiting are followed by a period of sleepiness and what appears to be recovery, but 3–4 hours later there is violent vomiting, then convulsions, coma, and death. During this process, there is a change of blood pH and a drop in blood sugar, and internal hemorrhages in the intestines, brain, and lungs. The outside capsule, which has saponins, has been used for fish poison.

Nephelium lappaceum
Family: Sapindaceae

Rambutan, *Mamón chino*

Description: Tree to 20 m. Alternate, pinnately compound leaves; 2–3 pairs of leaflets, 5–20 cm long. Small white to greenish flowers. Fruit in pendant clusters—bright red to yellowish, ca. 5 cm in diameter with flexible, curved spines; one large seed with sweet-sour covering (aril).

Flowering/fruiting: Fruits during wet season.

Distribution: Origin in Malaysia; under cultivation in Southeast Asia, Hawaii, Florida, Mexico, some parts of Central America and South America. In Costa Rica, occurs in hot, humid areas.

Related species: Mamón (*Melicoccus bijugatus*), also cultivated in Costa Rica, is a smooth, green fruit that has similar type of edible flesh. Other edible relatives include lychee (*Litchi chinensis*), akee (*Blighia sapida*, p. 201), and pulasan (*Nephelium ramboutan-ake*).

Comments: This strange-looking fruit with a grapelike inner texture is a relatively recent introduction to Costa Rica. It is well known in its native lands, where it not only provides food, but also a red dye used in batik, hard reddish wood, and folk medicines. The rich oil in the seeds has potential for various uses, and the pulp is a good source of vitamin C. Rambutans are nice, portable snacks (just crack open the outer shell and eat the fruit), and a possible addition to sweet-and-sour sauces.

Nephelium lappaceum

Francis X. Faigal

Pouteria sapota
Family: Sapotaceae

Sapote, *Zapote*

Description: Tree to ca. 30 m with white latex; deciduous. Simple, alternate leaves that spiral the stem; blades ca. 25 cm long (but up to 40 cm long), tapering toward base and wider toward pointed tip; leaf stalk with swollen base. Green to white, 4–5 lobed flowers, less than 1 cm long, clustered along branches. Dull tan-brown fruit, 10–20 cm long, with reddish orange pulp; large, shiny, brown seed tapers at both ends and has light-colored scar.

Flowering/fruiting: Flowers beginning of wet season; fruits beginning in March, and on into wet season.

Distribution: Precise original range unclear—somewhere in southern

Pouteria sapota

Mexico and/or Central America. Cultivated in those regions and also in south Florida, West Indies, and South America; also grown in the Philippines. In Costa Rica, sea level to 1,000 m.

Related species: Two related edible fruits are star apple or caimito (*Chrysophyllum cainito*), and sapodilla, the fruit of the chicle tree (*Manilkara zapota*). Inhabitants of the Iquitos region of Peru harvest and eat approximately a dozen wild species of *Pouteria*.

Comments: On the outside, sapotes are dull and grayish tan and rather ugly, so vendors always display one cut open to reveal the eye-catching orange flesh and the dark lustrous seed. The pulp is either eaten as is or made into preserves or ice cream. One dessert recipe (Ross De Cerdas 1995) suggests combining the fruit with crude sugar, lemon and orange juice, and cinnamon. In Mexico, where the fruit is known as mamey, the ground-up kernel (boiled and roasted first) has been used for flavoring cocoa drinks. Various plant parts, especially the seed, play a role in folk medicine. The oil of the seed is reputed to prevent balding and is made into soap; it is also used for muscle pain and colds. The tree's latex is put on warts, although it may cause irritation. The seed releases cyanide when cut; it also contains casein and albumin. In addition to its toxicity, studies of the seed kernel show antibiotic properties and effects on the central nervous system. Sapote is a good-quality hardwood that can be used for furniture-making. A creative mind can turn the seeds into an assortment of toys for children; the seed is also useful when one is darning socks.

Solanum quitoense
Family: Solanaceae

Naranjilla, *Naranjilla*

Cross section of *Solanum quitoense* fruit.

Description: May grow to over 2 m, but usually ca. 1 m tall; purple tinge due to covering of purple hairs; in the form found in Costa Rica, various parts of the plant have thorns. Large leaves to 50 cm, with pointy-tipped lobes. Five-parted, starlike white flowers with yellow center, pubescent on outside, 4 cm in diameter. Orange fruit ca. 5–6 cm, covered with hairs that rub off easily; greenish pulp acidic, full of small seeds.

Flowering/fruiting: Flowers and fruits during most of the year.

Distribution: Probable origin in Andes of Colombia, Ecuador, Peru. Cultivated in Central America and northern South America. In Costa Rica, from lowlands to ca. 1,700 m in many areas, cultivated or escaped.

Related species: *S. sessiliflorum* (photo opposite page), on Costa Rica's Atlantic slope, is a lowland version of naranjilla with less of the purple tinge of *S. quitoense*; fruit is not green inside and has less-attractive flavor. Genus *Solanum* contains ca. 1,500 species, including well-known food plants: tomato (*S. esculentum*), potato (*S. tuberosum*), and eggplant (*S. melongena*), as well as a number of more obscure foods such as the pepino, or melon pear (*S. muricatum*).

Comments: Naranjilla, with its unique flavor, makes savory jam as well as hot or cold drinks and desserts. The hairs must be rubbed off the fruit before preparation. The contrasting colors of the purple in the hairs that cover the plant and the orange of the fruit make this an interesting ornamental. It does best in the high, humid elevations of the tropics; at lower elevations, it needs shade. Naranjilla is called lulo in parts of South America. Hybridization and grafting experiments are taking place to improve the crop and create more marketable fruits and juice. The plant was introduced to Costa Rica in the 1950s.

Solanum quitoense

Solanum sessiliflorum

Theobroma cacao

Cacao, *Cacao*

Family: Sterculiaceae (recently placed in Malvaceae)

Description: Tree to ca. 10 m, but maintained shorter. Alternate leaves to 40 cm long, reddish when young. Small, 5-petaled flowers, long-stemmed and cauliflorous (i.e., coming off of trunk); petals white with red streaks and wider yellow tips; each petal with a basal pouch where anthers develop; red staminodes protrude from flower center. Ribbed fruit ca. 20 cm long, with bulging midsection; color may be green, yellow, red-brown, or purplish; 2–5 dozen seeds (2.5 cm) inside, each covered with white-to-pink, sweet, edible flesh.

Flowering/fruiting: Flowers and fruits throughout the year, with peak flowering at start of rainy season.

Distribution: Found from Mexico to Amazon region; origin thought to be in Andean foothills of northern South America. In Costa Rica, hot, wet lowlands, in shade, on Caribbean coast and parts of southern Pacific coast, usually below 300 m.

Related species: Cupuaçu (*T. grandiflorum*) is popular in Brazil, where the pulp around the seeds is eaten. There are around 20 other species, four of which are found in Costa Rica. African cola nut (*Cola nitida*) is also in the family Sterculiaceae.

Comments: This "food of the gods," as the genus name denotes, which was frequently used in Aztec and Mayan ceremonies, was not at all like the delectable, sweet concoctions of today. A typical hot chocolate drink a thousand years ago was a mixture of cacao, hot chili pepper, and vanilla, with annatto (*Bixa orellana*, p. 172) for coloring and some cornmeal, added perhaps as a thickener. Explorers introduced the drink to Europe, where it became popular only after the chili was subtracted from, and sugar added to, the recipe. The Mayans were first to cultivate cacao, several thousand years ago, and they used it as currency in trade until the mid 1800s. Cacao's origin seems to be in South America, but it is not clear how it reached Central America and Mexico. Since the pulp around the seeds is sweet and edible, humans on the move may have carried it as snack food, or possibly for medicinal uses, and dispersed the seeds during their travels. In Costa Rica, most cacao is grown in the hot, wet lowlands, where either the remnant subcanopy forest or planted trees provide shade. In the late 1960s, Costa Rica was the top Central American cocoa exporter, but since then fungal attacks by *Moniliophthora roreri* (monilia, or watery pod rot) and *Phytophthora* spp. (black pod disease) have led to the demise of many plantations; the herbivorous cacao thrip (*Selenothrips rubrocinctus*) is another pest. Cacao grown in small groves near the forest may, in the

Theobroma cacao

Theobroma cacao

long run, do better than large monocultures that are more susceptible to rampant spread of disease. Small biting midges (Ceratopogonidae) are apparent pollinators. Pods chewed open by arboreal mammals begin to rot and form breeding places for the midges. Decaying matter under the trees may also contribute to sustaining midge populations, so adding cut-up, rotting banana stalks to plantations may increase pollination. Cacao's natural seed dispersers include bats and arboreal mammals such as monkeys. Although it is not a major source of livelihood, cacao is still grown on the Caribbean coast, and one can still find delicious, home-made, oily chocolate sticks for sale in that region. The first step in making commercial chocolate involves fermenting the pulp-covered seeds for up to seven days. Once they are dried and the pulp is removed, the seeds are shipped off to the importing country, where they are roasted and husked. The remaining parts, called nibs, make rich, chocolate liquor; cocoa powder also comes from nibs, but in a process that removes half the fat. Unsweetened chocolate is used in Mexican mole sauce, an exquisite blend of chilies, spices, and chocolate that is often served over chicken. The fatty oil in cacao is used in home remedies—topically on burns and as a rectal and vaginal salve—and commercially, in the form of cocoa butter, as an ingredient of suppositories. Cocoa butter is also used in a variety of cosmetics. The seed husks serve as both a mulch and as another source of cocoa butter and theobromine. Researchers have looked at various components of chocolate that have stimulating, euphoric, and even hallucinogenic effects, but these compounds occur in such small amounts that it appears any sort of "high" one gets from chocolate may be from the synergistic action of some of the hundreds of components; these include theobromine, caffeine, theophylline, and phenylethylamine (which may trigger migraines). Nutrients in cacao such as magnesium and the sensory aspects of aroma and texture may contribute to human cravings. Dogs should not be given chocolate, especially dark chocolate, because theobromine is poisonous to them, leading to diarrhea, rapid heart beat, and sometimes death.

Zingiber officinale
Family: Zingiberaceae

Ginger, *Jengibre*

Description: Leafy shoots 0.5–1.5 m tall. The aromatic ginger "root" is actually a rhizome, which is an underground stem. Alternate leaves, ca. 20 cm long by 2.5 cm wide, come off stem in a plane. The 5-cm cone-shaped flower head is on 20-cm stalk that arises from the rhizome; inflorescence made up of green and yellow (or orange) bracts and ca. 4-cm pale yellow flowers with maroon on the curved lip; labellum (lip) is made up of two sterile anthers forming a petal-like structure, the one fertile; protruding anther arches over and surrounds a hollow style. Fruit is a 3-parted capsule; good seed not always produced, so propagation is by cuttings from the rhizomes.

X ⅓

Typical ginger root of *Zingiber officinale*.

Mel Baker

Zingiber officinale in flower.

Zingiber officinale in post-flowering phase.

Flowering/fruiting: Flowers in rainy season.

Distribution: Unclear if precise origin is India, Pacific islands, south China, or elsewhere. Now found in cultivation all over the world in tropical regions. In Costa Rica, grows in most areas of the country, but does better in the hotter, more humid regions; sometimes escapes from cultivation.

Related species: Family includes many aromatic plants that are medicines, condiments, and dyes, including the Japanese myoga (*Zingiber mioga*), Indian arrowroot (*Curcuma angustifolia*), turmeric (*C. longa*), and cardamom (*Elettaria cardamomum*). Other *Zingiber* species, as well as *Alpinia* spp. (p. 156) and ginger lily (*Hedychium coronarium*, see p. 157), are ornamentals. *Renealmia* spp. (p. 158) are colorful tropical-forest understory cousins.

Comments: Well-known as a flavoring for cookies, Asian food, and ginger beer, ginger also has a reputation in folk medicine. A tea of grated ginger root, honey, and lemon is soothing to a sore throat. Country and indigenous people from Central and South America also use ginger for motion sickness, body aches, indigestion, and diarrhea. Some native people of northwestern Amazonia apply the root to aching teeth and boil it to make a contraceptive drink. Ginger has potential as a hypoglycemic agent (i.e., for treatment of diabetes) and is useful as a stimulant and in counteracting gas in the digestive tract. Recent scientific research shows that ginger contains compounds that are antioxidant and anti-inflammatory, and that, topically, it has potential as a skin tumor preventative (Katiyar et al. 1996). In Costa Rica, it is used in *chicheme*, a fermented corn drink. The botanist Pittier (1978) indicated that the use of ginger in Costa Rica appears to date back to precolonial times. He believed the country had the potential to produce ginger as a cash crop and to export it in the same fashion that Jamaica did. Today, Costa Rica is one of the major suppliers of ginger to both the European Union and the United States. The genus name *Zingiber*, which describes the rhizome, is derived from the Sansrkit *sringavera*, which means horn-shaped.

5. Living Fences and Reforestation

This chapter describes frequently seen trees used in fence rows as well as those planted in reforestation projects. In the tropics, many tree species grow readily from branch cuttings; and, over the years, farmers have stuck branches of various trees into the ground to see which species would take root. In Costa Rica, a handful of species is particularly dependable when grown this way, and their use to create living fences is widespread. Some of these, in addition to establishing a desired barrier, yield products useful in other ways. Those species most likely seen along roadways are described in this chapter.

Costa Rica is applauded as a country that protects much of its natural forests—some 25% is under some level of protection. Deforestation continues in some very sensitive regions of the country, however, especially the Osa Peninsula and the northern zone. In an attempt to offset some of this destruction, the government has established economic incentive programs to promote reforestation. Conservation organizations also work with landowners to encourage them to reforest their land.

The term *reforestation* is generally used in the country to mean planting trees, and reforestation may be done for various purposes—to restore forest cover to a cleared area, to provide a source of lumber, or to create a windbreak to protect pasture and crops, for example. Sometimes areas are reforested simply by fencing them off from livestock and letting the forest grow back. This approach can meet with varying success depending on the historical use of the land and its proximity to an established forest as a seed source.

Left, *Montanoa guatemalensis* windbreaks protect Monteverde pastures.

Another key to success, particularly in dry areas such as Guanacaste, is the prevention of grass fires that destroy pioneer tree saplings and shrubs.

Plantations of wood trees cannot, at this point in time, meet the country's demand for lumber, but it is still encouraging to see efforts made toward an alternative to cutting down primary forest. Species grown in these plantations range from the exotic teak and gmelina to pochote, laurel, and other natives that show promise as lumber. The advantage of reforesting with native species is that the trees fit into the habitat better and provide resources for native fauna.

In the late 1980s, OTS (Organization for Tropical Studies) researchers and DGF (Dirección General Forestal) forestry technicians at the La Selva Biological Station began to carry out a series of much-needed projects investigating the use of native (and some exotic) species in plantations and in reclaiming marginal lands. Before these experiments, exotic species were usually used because more was known about their growth patterns and funding was more available for studying their potential in Costa Rica. Some research is now demonstrating that certain native species, such as *Vochysia guatemalensis*, have much better survival rates in a variety of sites than well-known exotic plantation species such as *Eucalyptus deglupta*.

Branches of *Erythrina lanceolata* are planted to create living fences.

Spondias purpurea
Family: Anacardiaceae

Red mombin, *Jocote*

Other common name: Spanish plum.
Description: Tree less than 15 m, often shorter as living fence post, deciduous, with smooth gray trunk, sometimes with spines; aromatic sap. Alternate, compound, odd-pinnate leaves; 9–25 leaflets, entire or toothed, each to ca. 7 cm long. Small, 5-petaled, red-purple flowers in clusters along branches. Fruit oblong to 5 cm, red-purple or yellow-red-orange; flesh yellow; one relatively large seed.
Flowering/fruiting: Flowers from dry to early wet season; ripe fruit in late dry through wet season.
Distribution: West Indies and southern Mexico to South America. In Costa Rica, Pacific coast from Guanacaste to Osa Peninsula, 0 to ca. 1,200 m.
Related species: Two native forest species are jobo, or hog plum (*S. mombin*), with large leaves and yellow fruit and *S. radlkoferi*, with green fruits. Otaheite apple (*S. dulcis*), called *yuplón* in Costa Rica, is an exotic with large yellow to orange fruit. Mango (*Mangifera indica*, p. 164) and cashew (*Anacardium occidentale*, p. 162) are also in the Anacardiaceae family.
Comments: As a living fence post, jocote is seen most often at lower elevations. Branches are cut and planted at the beginning of the wet season, during a waning moon. This species is variable when it comes to the size, color, and flavor of the fruit, and if you bite into a jocote iguanero, you may not find it very appealing, even though it is eaten by black spiny-tailed iguanas. That variety, which is small (ca. 3 cm long), red, juicy and acidic, and ripe in the dry season, is probably closest to the wild type. The larger, more flavorful varieties, which are ripe late in the wet season, have rather scant flesh, but a wonderful, unique flavor. The most common way of eating the fruit is just as it is, off the tree, but it is also an ingredient of preserves (*jocotada*), vinegars, wine, and sauces. The young leaves have a pleasant acidic flavor and are eaten by people, livestock, and deer. The tree has various folk medicine uses, including a decoction of leaves and bark (or bark alone) taken for fever and diarrhea.

Spondias purpurea

Montanoa guatemalensis
Family: Asteraceae

Tree daisy, *Tubú*

Montanoa guatemalensis

Other common name: *Tobus.*
Description: To 15 m tall, but usually less than 10 m. Leaves ca. 14 cm long, soft pubescent and lighter-colored beneath, without lobes or with 3–5, the one at apex long and triangular with irregular teeth; leaf stalks to 9 cm long. Flowers toward branch tips, profuse at flowering peak; delicate fragrance; flower heads ca. 5 cm in diameter; 7–10 white ray flowers (look like petals); disc flowers (in center) yellow and brown. Many brown seeds (ca. 3 mm) in rounded heads; seeds lack plumelike pappus seen in many Asteraceae, including dandelion (*Taraxacum officinale*).
Flowering/fruiting: Typically flowers December through January.
Distribution: Mexico to Costa Rica. In Costa Rica, Pacific slope in moist to wet forest, 1,000–1,500 m from Talamanca to the Tilarán mountain range; in the Monteverde region, often on fairly dry ridges below 1,300 m.
Related species: Several other species occur in Costa Rica (*M. tomentosa*, *M. pauciflora*, and *M. hibiscifolia*); *M. tomentosa* has smaller flowers and more pubescent leaves, and it grows at lower elevations.
Comments: Tree daisy, or tubú, was promoted in the Monteverde area between 1989 and 1990 during a reforestation/windbreak project funded by the Canadian International Development Agency and carried out by the Monteverde Conservation League. Tens of thousands of the trees were planted. Today, around Christmas time, you can see rows of flowering tree daisies bordering many pastures. The seeds are probably eaten and dispersed by small rodents. The wood is good for fence posts.

Alnus acuminata
Family: Betulaceae

Alder, *Jaúl*

Description: Medium to tall tree, to 30 m; smooth grayish bark with horizontal lines; lenticels on twigs. Alternate leaves ca. 12 cm long; with toothed edge, prominent pinnate venation, gray-green beneath with some rusty pubescence. Monoecious, with tiny flowers in separate male and female inflorescences, the former pendant and 7–12 cm long, the latter much shorter. Woody, conelike

Alnus acuminata

fruit clusters to 2.5 cm long; seeds are tiny, winged nutlets.

Flowering/fruiting: Flowers in April.

Distribution: Mexico to South America. In Costa Rica, 1,100–3,000 m; in Talamanca, Central, and Tilarán mountains, both slopes, often at around 2,500 m on forest edge, in landslide areas and other clearings; sometimes in pure stands.

Related species: Related to temperate-zone alders (*Alnus* spp.) and birches (*Betula* spp.).

Comments: In the high mountains of Costa Rica, patches of alders on the slopes, and individual trees dotting pastures, bathed in cool, moist air, create a temperate-region atmosphere. Some farmers near the Los Santos Forest Reserve, in the Cerro de la Muerte region, are using alder in reforestation along with nonnative species such as cypress (*Cupressus lusitanica*) and eucalyptus (*Eucalyptus* spp.); it may become a more commonly planted species in this region with the decline of populations of the species of oaks traditionally used in charcoal production. Only fallen oak trees may be used legally now. Alder can be used for firewood, charcoal, and in construction and furniture. With its nitrogen-fixing qualities, it holds promise for agroforestry systems

x 1/3

Alnus acuminata branch with conelike fruiting spikes.

Alnus acuminata stand.

(e.g., planted with blackberries) and in the rehabilitation of old pastures. Alder is wind-pollinated and wind-dispersed. A similar and related tree species, *Ticodendron incognitum* (illus., p. 13), was confused with alder for many years; it was described only recently, in 1989, as a new species (and new family, Ticodendraceae).

Pachira quinata — Pochote, *Pochote*
Family: Bombacaceae (recently placed in Malvaceae)

Other common names: Spiny cedar, *cedro espinoso*.
Description: Huge trees to 30 m tall, as well as shorter living fence posts; the former with large buttresses and checkered, fissured, gray bark with some large flakes; distinctive horizontal and angular branching in large trees; trunk and branches have thorns, to 2.5 cm long, with expanded bases. Deciduous.

Alternate, palmately compound leaves with ca. 5 leaflets, longer ones to 18 cm long; petiole sometimes tinged maroon. Flowers ca. 11 cm across; 5 fleshy, arching petals, ca. 9 cm long, cream above with some scurfy red-brown beneath; many anthers, long and brushlike; delicate perfume at night. Brown 5-parted capsule, to 10 cm long, contains small seeds in cottony material.

Pachira quinata flowers.

Pachira quinata plantation.

Flowering/fruiting: Flowers December through April. Fruits in later dry season and into wet season.

Distribution: From Nicaragua to northern South America. In Costa Rica, dry to moist forest, generally from sea level to 500 m, but sometimes up to 900 m; from central Pacific area north to Guanacaste.

Related species: Another species, *P. sessilis*, occurs in the central and southern Pacific zones of Costa Rica. The kapok tree or ceiba (*Ceiba pentandra*, p. 41) and ceibo (*Pseudobombax septenatum*, photo, p. 43) are related. Durian (*Durio zibethinus*) of Malaysia and the red silk cotton tree (*Bombax ceiba*) of Asia are also in this family.

Comments: Pochote is often seen as a living fence post, along with gumbo limbo (*Bursera simaruba*, p. 219), in the dry areas of Costa Rica. It has also become popular as a tree-plantation species. The quality of the wood, which withstands rot and some insect attack, is often compared to tropical cedar (*Cedrela* spp., p. 58). Pochote is a source of lumber for construction, including doors and window frames; the tree, however, contains a hygroscopic substance that will rust nails where the wood is used in humid conditions. The flowers attract hawkmoths and bats. On a single tree, dozens of flowers open at night, then fall throughout the next morning. Orange-chinned parakeets, which are

X 1/6

Pachira quinata leaf.

Flowers on forest floor.

seed predators, dig out and eat seeds from immature capsules. Sometimes groups of silk moth (Saturniidae) larvae are seen on the tree trunks, and the large, colorful buprestid beetle, *Euchroma gigantea*, lays eggs in pochote and other Bombacaceae. *Pachira quinata* was formerly called *Bombacopsis quinata.*

Cordia alliodora
Family: Boraginaceae

Ecuador laurel, *Laurel*

Description: Tree to 25 m with straight trunk grayish-brown or lighter (due to lichen growth); tiers of branches; branches with hollow swellings at nodes. Alternate, simple, rough leaves to 20 cm long, including petiole; stellate pubescence on new growth and leaves, especially below. Flower clusters 15+ cm long; small, white flowers with 5-lobed corolla and 5 stamens; sweet fragrance. Fruit a 0.5-cm-long, one-seeded nutlet with corolla (which turns brown) still attached (illus. opposite page).

Flowering/fruiting: Flowers in the dry season; fruits mostly at end of dry season (March–April).

Distribution: Mexico to Argentina; also in West Indies. Planted as ornamental in Florida and in plantations in various parts of the world. Common pasture tree in Costa Rica, sea level–1,100 m, both slopes; also in secondary and older forests; wet and dry climates.

Related species: Ca. 300 species of *Cordia* worldwide, with 25 in Costa Rica. Borage (*Borago officinalis*), comfrey (*Symphytum* spp.), and heliotrope (*Heliotropium* spp.) are also in this family.

Comments: This widespread native is an established reforestation species and is a lumber industry favorite. The color of the wood varies, but it is usually an attractive mix of lighter colors with dark streaks. Because it is easy to work with, insect resistant, hard, and durable, it is used in furniture, flooring, tool handles, cabinets, and turned articles. It is also good firewood and has potential as pulpwood. Individual trees growing under various climatic conditions differ somewhat in appearance and wood characteristics. *Cordia* tolerates sun, and

Cordia alliodora

x 2

Cordia alliodora fruit with persistent corolla.

it will grow in old pastures; survival and growth rates are enhanced if the trees are planted in fertile soils; the typical growth rate is in the range of 1–2 meters per year. The seeds germinate relatively soon after dispersal, but they may be stored in refrigeration for up to three years. *Cordia* is sometimes grown in coffee plantations, and it is on the government-supported incentive program's approved-species list for reforestation in the northern zone of Costa Rica. The scientific species name, *alliodora*, refers to the garlic scent in the inner bark. Medicinal uses include drinking a tea of boiled leaves as a tonic and as a home remedy for lung-related ailments. The powdered seeds are made into a salve for skin diseases, and the leaves are used on wounds. The sweet-scented flowers, which open after dark, are pollinated by small moths; but a variety of insects, including bees, visit during the day to feed on residual nectar and pollen. The persistent corolla on the nutlets aids in wind dispersal of the seeds. The swollen hollow nodes in the branches house *Azteca* ants, along with mealybugs that secrete a sugary substance for the ants. Bruchid beetles attack the seeds.

Bursera simaruba
Family: Burseraceae

Gumbo limbo, *Indio desnudo*

Other common names: Naked Indian tree, *indio pelado, jiñote, caraña*.
Description: Tree to 25 m, deciduous, with resinous sap; photosynthetic, greenish trunks covered by copper-colored bark that peels off in thin, papery strips. Alternate, pinnately compound leaves with 5–13 entire leaflets to ca. 14 cm long, including the red-tinged stalks; lopsided bases; crushed leaves give off pleasant pungent scent. Plants with small, green-yellow flowers, either male or female; some perfect ones may be present. Fleshy, purplish-brown, 3-lobed capsular fruit, to 1.5 cm long; dehiscent, splitting from bottom up; contains a single 1-cm, 3-angled seed with thin scarlet flesh.
Flowering/fruiting: Flowers from March through May, and sometimes later; fruits during the dry season (but immature fruits present during much of the year).

Bursera simaruba

Bursera simaruba trunk.

Distribution: Mexico to Colombia, Caribbean, and coastal hammocks of Florida. Throughout Costa Rica, in dry to wet habitat, generally below 1,100 m; also planted as living fence post.

Related species: Eight other species of *Bursera* occur in Costa Rica; some have toothed leaflets and winged rachises. Frankincense (*Boswellia sacra*) and myrrh (*Commiphora abyssinica*), of wise-men fame, are in this family.

Comments: This is among the most popular species used for living fence posts. It is also a common large tree of the dry forest, where its distinctive coppery, peeling bark stands out in the deciduous landscape. Although it is associated with the dry forest, gumbo limbo actually grows in an array of habitats in Costa Rica, including La Selva and the Osa Peninsula. The larvae of the long, narrow-bodied brentid beetle (*Brentus anchorago*) develop in decaying trees. White-faced capuchins chew on new stem growth; this "pruning" may lead to a larger crop of fruit, which they, and peccaries, eat. Some birds consume the aril, and they may swallow the whole seed, perhaps to use as a grinding aid in the gizzard. A paper wasp, *Synoeca* sp., often makes its corrugated nest on gumbo limbo trees. The small flowers attract an array of insects, including flies, stingless bees, and small longhorn beetles. The trees are leafless during part of the dry season. The resin, bark, and leaves are used in home medicine. Teas of various plant parts are used for digestive tract problems; a bark decoction is made to treat syphilis, rheumatism, asthma, and urinary tract problems, and to relieve colds and fever; bark tea is taken by those wanting to lose weight; bark or leaf baths (or applications) are used for alleviating arthritis and back aches, insect stings, and sunburn. The resin, which is put on wounds, is also used as a glue substitute, varnish, and incense. Studies on gumbo limbo indicate that the dried bark has diuretic, cytotoxic, and antifungal properties. The light-colored, rather soft wood has been used as firewood and in the manufacture of matchsticks, packing crates, tool handles, and turned articles.

Casuarina equisetifolia   Australian pine
Casuarina cunninghamiana   Australian pine
Family: Casuarinaceae

Other common names: Whistling pine, *pino de Australia, casuarina*.

Description: These two similar species resemble true pines. Both grow to ca. 20 m and have rough bark. What appear to be needles are slender, grayish-green, grooved twigs. The ca. 6–10 leaves are whorled and joined to form sheaths that end in teethlike tips at each node. Male flowers are in spikes toward outer branch tips; female flowers, on side shoots, are in heads that develop into 1- to 2- cm-long conelike fruits made up of many paired bracts that look like baby bird beaks. Seeds are tiny and winged. *C. equisetifolia* is monoecious; *C. cunninghamiana* is dioecious. Old, recurved leaf teeth remain on older twig growth in *C. equisetifolia*, giving it a rough appearance, whereas on *C. cunninghamiana* the leaf teeth do not persist, so the twigs are smooth. In *C. equisetifolia* the scalelike leaves are 6–8 to a whorl, internodes ca. 6–8 mm long and

Casuarina cunninghamiana

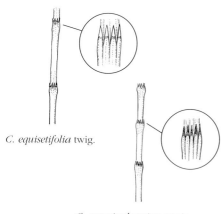

C. equisetifolia twig.

C. cunninghamiana twig.

In *C. equisetifolia* the internodes are ca. 6–8 mm long; in *C. cunninghamiana* the internodes are 3–6 mm long. In both species, leaf tips are visible at the nodes on the twigs. *C. equisetifolia* has 6–8 leaf tips encircling the twig; *C. cunninghamiana* 7–10 leaf tips.

0.6–0.8 mm wide. *C. cunninghamiana* has 7–10 leaves to a whorl (with darker leaf teeth), internodes 3–6 mm long and 0.4–0.7 mm wide.

Flowering/fruiting: Both species flower and fruit during various times of the year.

Distribution: *C. equisetifolia* is native to parts of Australia, Asia, and Polynesia; *C. cunninghamiana* is from eastern Australia. In Costa Rica, both are planted as windbreaks and as ornamentals in many parts of the country.

Related species: *Casuarina* species are not true pines; they are related to birches, bayberry, and oaks.

Comments: Forest scientists had high hopes of using *Casuarina* in numerous ways in reforestation projects throughout the world, but in most areas it has not been found to be very useful, and in some regions of the world it is a pest. In Costa Rica, it serves as a windbreak species and for firewood. Beneficial aspects of *Casuarina* include the abilities to fix atmospheric nitrogen, tolerate a range of conditions, and grow rapidly. *C. equisetifolia* is considered a pest in coastal southern Florida, where it is invasive and often displaces native vegetation. In most of the areas where it had been touted for holding coastal soils, it failed in hurricane conditions because of its shallow root system. Through a hand lens, the stems look very much like tiny replicas of horsetail plants (*Equisetum* spp., p. 444). Vegetatively, the two *Casuarina* species look similar, but if you look at the internodes—the space from one whorl of the toothlike leaves to the next—you'll see that they are thinner and shorter in *C. cunninghamiana* (illus. left). The name *Casuarina*, which comes from *kesuari*, the Malay word for cassowary, refers to the similarity of the branchlets to that bird's feathers. Casuarinas are wind-pollinated, and the seeds are wind-dispersed.

Cupressus lusitanica
Family: Cupressaceae

Mexican cypress, *Ciprés*

Other common name: Cedar of Goa.
Description: Resinous tree to 30+ m, branching from near base, often broader at base and narrowing toward top; bark somewhat lustrous reddish-brown to duller red- or gray-brown and flaking; twigs lustrous reddish-brown; pungent, conifer scent. Tiny, opposite, overlapping, pointed scalelike leaves form the green fingerlike foliage; sprays of foliage may or may not be flattened depending on the variety. Yellowish, pollen-bearing cones are tiny and at tips of branchlets; ovulate cones less conspicuous, tucked beneath scales near branches; 2-cm, mature, ovulate cones somewhat shiny maroon-brown, held close to the branches; each cone made up of 6–8 irregular interlocking plates (like puzzle pieces); plates with a spike-like projection in the center (illus. p. 224); many tiny brown seeds.

Flowering/fruiting: Recorded flowering in February.
Distribution: From central Mexico to Honduras, in mountainous regions. Long-cultivated in other parts of world. In Costa Rica, mostly planted at mid- to high elevations.
Related species: Various cultivars of *C. lusitanica* exist. Related species include Arizona cypress (*C. arizonica*), from southwestern United States and northern Mexico.
Comments: Most of the conifers you encounter in Costa Rica are not native but have been planted as ornamentals, as windbreaks, or for wood. Mexican cypress is often seen on farms in the Monteverde, Central Valley, and Cerro de la Muerte regions. It is a major windbreak tree species that protects dairy cows in pastures from harsh winds, thereby increasing milk production. The

Cupressus lusitanica

wood is made into furniture and posts and is useful in construction and as firewood. Toward the end of the year, trees—or their conical tops—are cut and transformed into Christmas trees. A drink made by steeping a branch in alcohol is taken to alleviate coughs and cold symptoms. The author has found morel mushrooms growing in association with these trees. *Cupressus lusitanica* was described in the 1700s in Portugal where it had been introduced; at that time, the origin was erroneously thought to be India.

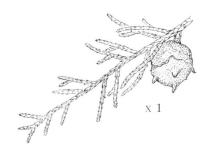

x 1

Cupressus lusitanica branchlet with developing cone.

Dracaena fragrans
Family: Dracaenaceae

Corn plant, *Caña india*

Other common name: Dracaena.
Description: Leafy stem to 6 m, usually not branching. Arching, sessile, straplike leaves usually less than 1 m long, with yellow stripe in the cultivar 'Massangeana'.

Dracaena fragrans

Inflorescences to 50+ cm long with dense groupings of short-tubed, fragrant, white to violet flowers with 6 tepals. Red-orange fruits, ca. 1.5 cm.
Flowering/fruiting: Recorded flowering in February.
Distribution: Native to Africa. In Costa Rica, planted in a range of climates; common as living fences along stretches of Inter-American Highway (often in coffee plantations) and road from San Ramón to La Tigra.
Related species: *D. americana*, one of two New World natives in this genus, is tall with narrow leaves and grows naturally in Costa Rica at around 100 m on southern Pacific slope. The colorful *D. cincta* 'Tricolor', which has narrower leaves than *D. fragrans*, is another of several common ornamental dracaenas found worldwide.
Comments: Tourists to Costa Rica will easily recognize the corn plant, which appears in office reception areas around the world. Its ease of care makes it an ideal interior foliage plant. It is also listed as one of the top ten plants capable of removing toxins

from indoor air, according to a NASA study that tested plants' ability to remove formaldehyde, benzene, and trichloroethylene. In Costa Rica, cuttings readily grow into living fences, and they help stabilize soil on steep slopes. *Dracaena* is sometimes harvested as a horticultural export crop since it can easily be shipped as stem sections that are propagated later in greenhouses. The common cultivar with yellow-striped leaves is 'Massangeana'. Molecular studies indicate that this and some related families are appropriately included under the family Ruscaceae (Grayum et al. 2001). The genus name, from the Greek word for female dragon (*drakaina*), refers to the red sap (dragon's blood) seen in some species (e.g., *D. draco*).

Erythrina lanceolata   Machete flower, *Poró*

Family: Fabaceae
Subfamily: Papilionoideae (= Faboideae)

Description: Small tree to 8 m; spines on branches and leaves. Alternate, compound leaves with 3 leaflets; leaflets to 18 cm long, smooth and light gray-green beneath, longer than wide with pointed tip and wedge-shaped base; pair of small yellowish stipels at base of each leaflet. Machete-shaped scarlet flowers, 5–7 cm long, along tips of branches; 1 large conspicuous petal, the rest hidden in the 1-cm-long calyx. Knobby pods, 15–30 cm long, split open to reveal hard, 1+ cm red seeds.
Flowering/fruiting: Flowers in dry season; fruits early rainy season.
Distribution: Honduras to Panama. In Costa Rica, midelevations (900–1,500 m), including forest patches along rivers near San José; botanists have also recorded specimens at lower elevations on the Osa Peninsula.
Related species: *E. costaricensis* grows at lower elevations, mostly in the southern half of Costa Rica and south to Colombia. Although very similar to *E. lanceolata*, it is often less prickly, its leaflets are broader and more rounded at the base, and its flower calyx longer. About a dozen species of *Erythrina* can be found in Costa Rica, most of them native; an introduced species, *Erythrina poeppigiana* (p. 33) is a large, showy orange-flowered tree seen in coffee plantations near San José.
Comments: Branches of most, if not all, species of *Erythrina* in the tropics will take root when planted in the ground. Some species of the

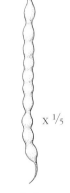

x ⅕

fruit

Erythrina lanceolata

Erythrina lanceolata flowers.

genus are known to contain alkaloids, especially in the raw seeds. The colorful seeds of erythrinas and some other members of the Fabaceae family are often used to make jewelry. The flowers attract hummingbirds, so the trees serve as hummingbird feeders.

Gliricidia sepium

Quick stick, *Madero negro*

Family: Fabaceae
Subfamily: Papilionoideae (= Faboideae)

Other common name: Mother of cacao.

Description: Usually a small to medium tree to 10+ m, deciduous; gray bark with lenticels. Alternate, compound leaves with odd number (5–19) of leaflets, each to 7 cm long, turned under at the base and light gray-green below with some pubescence. Fragrant 2-cm-long flowers, pink (sometimes more lilac or white) with a yellow center, in dense clusters along the branches. Pods 15 cm long, resembling large flat green beans, becoming dark brown; dehiscent; each pod contains up to 10 brown, 1-cm seeds.

Flowering/fruiting: Mostly flowers from January through March, occasionally in other months; leafless from January to about beginning of wet season.

Distribution: Mexico to northern South America. Introduced to Old World tropics, where naturalized in some areas (e.g., Philippines). In Costa Rica, on both Atlantic and Pacific slopes, including the Central Valley, usually below 1,400 m. Grows naturally in the dry forest of northwestern Costa Rica and sometimes forms clumps in pastures in that region. A population of stunted individuals forms a dense stand behind the turtle-nesting area at Playa Grande.

Related species: This is a member of the papilionoid or faboideae subfamily of legumes (Fabaceae), which includes *Erythrina* spp. (see previous species account), guachipelín (*Diphysa ameri-*

Gliricidia sepium

cana, p. 28), and black locust (*Robinia pseudoacacia*) of North America.

Comments: As one of the most widely planted living-fence-post species, madero negro is sure to be seen by anyone traveling around Costa Rica, although, depending on the time of year, it may or may not catch your eye. From a distance, in flowering season, it looks somewhat like a cherry or apple tree in bloom. Madero negro, which will grow in rocky, poor soil, is well-known as a multipurpose tree throughout its native range, as well as in faraway places such as Africa and India. It has a wide range of uses beyond serving as a fence post. It provides quality lumber for construction, furniture, and turned articles. The tree is also used for firewood and charcoal. The dried leaves maintain a pleasant odor similar to freshly-cut hay. The flowers are commonly eaten after being boiled and fried (with onions and eggs). The name *madre de cacao* (mother of cacao) is used in some cacao-producing regions where the trees are planted for their shade and nitrogen-fixing qualities. Its protein-rich leaves also provide forage for cattle and nourish the soil as a green manure. Some experiments that mixed madero negro into feed for livestock showed positive results in terms of weight gain. However, it contains one or more substances that are poisonous to horses and dogs, and various parts of the tree (roots, leaves, seeds) have been mixed with corn to poison mice. The presence of coumarin or other chemical compounds in the leaves makes the leaves unpalatable to some animals. Interestingly, from observations on some domestic animals, it appears that some individual trees are more palatable than others. In this species, the inconsistency of effects of the toxic, or at least unpalatable, substances can be seen in nature also—howler monkeys have been observed choosing certain trees to feed on while ignoring others, and saturniid moths (*Hylesia lineata*) were seen laying thousands of eggs on only one tree in a line of 24 (Janzen 1983). Chrysomelid beetles and the Guanacaste stick insect (*Calynda bicuspis*) feed on this plant. *Centris* and *Xylocopa* bees pollinate the flowers. In folk medicine, bark or leaf preparations are applied to wounds, bruises, and various skin ailments, and leaf decoctions are taken internally to treat cold symptoms. Some Costa Rican farmers recommend making cuttings in May at the time of the waning moon and planting around the time of the first rains.

Carapa nicaraguensis
Family: Meliaceae

Carapa, *Cedro macho*

Other common names: Crabwood, royal mahogany, *caobilla.*

Description: Tree to 40+ m, with buttressing in older trees; bark shed in circular plates. Very large (to more than 1 m), alternate, even-pinnate, compound leaves with 4–10 pairs of glabrous leaflets, to 40 cm long; new leaves reddish. Tiny, 4-parted, greenish-white to yellowish flowers in long inflorescences. Seed capsule, ca. 12 cm in diameter, splits into 4 sections, with ca. 8 large seeds; seed is rounded on outer side and has 3 flat inner surfaces.

Flowering/fruiting: Fruits in early to mid–wet season.

Distribution: Nicaragua to Ecuador. In Costa Rica, on both slopes to ca. 700 m, in intermittently flooded swamp forests and sometimes along streams.

Carapa nicaraguensis sapling.

Related species: Other members of this family native to Costa Rica include mahogany (*Swietenia macrophylla*, p. 259), Spanish cedar (*Cedrela odorata*, p. 58), and uruca (*Trichilia havanensis*, p. 59).

Comments: Carapa is a large tree that is dominant in some swampy regions of the Atlantic slope where older individuals are estimated to be more than 400 years old. Seed dispersal occurs via water channels and by rodents, although the latter are mostly seed predators. The seeds are also eaten by peccaries and attacked by moth larvae (*Hypsipyla*). The seeds occasionally wash up on beaches. Researchers predict that an already low genetic diversity in presently disconnected populations of carapa will continue to decrease as the lowland Atlantic forest becomes more fragmented. Portico S.A., a company that exports carapa in the form of royal mahogany doors, has been managing its forests near Tortuguero National Park in a way to sustain the harvest of trees—trees less than 70 cm in diameter are left in place, and, during extraction, measures are taken to lessen the impact on the surrounding forest. The growth rate and regeneration of carapa in logged areas is being compared to that in undisturbed forest. Studies indicate that carapa seedlings in new logging gaps grow more rapidly than those in undisturbed forest. Casting seeds into such gaps could be beneficial in sustainable harvest programs (Webb 1999). In addition to doors, the reddish-brown, durable wood of carapa, a mahogany relative, makes fine furniture and cabinets. Preparations of bark are taken for fevers and to get rid of worms; oil from the seeds is used for relieving insect bites and skin ailments, to eliminate head lice, and on arthritic joints. The oil is also used in making soaps, insect repellent, and candles. Alkaloids are present in both the bark and seeds. The name *C. guianensis* was generally used for this species in the past. That species, if it exists in Costa Rica, is very rare.

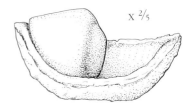

X ²/₅

Section of old open fruit of *Carapa nicaraguensis*, with one remaining seed.

Eucalyptus deglupta
Family: Myrtaceae

Rainbow bark, *Deglupta*

Other common names: Mindanao gum, bagras, *eucalipto, ocalipto*.

Description: To more than 30 m tall (to 70 m in countries where it is native); peeling bark gold, brown, army green, lime green, sometimes bluish-gray; somewhat pendant branches; menthol/eucalyptus scent in broken twigs. Simple, opposite or alternate leaves 15 cm long, lighter green below. Small heads of white flowers with many stamens. Dry, capsular 0.5-cm fruit.

Distribution: Native to Philippines and some other Pacific islands (Indonesia, New Guinea). Cultivated in plantations in many parts of the tropical world. In Costa Rica, humid regions, low to midelevations, in plantations and as an ornamental (e.g., Sabana Park west of San José).

Related species: More than 600 species, many cultivated. In addition to all of the plantation species, silver dollar gum (*E. cinerea*) is grown for its silvery leaves used in flower arrangements.

Comments: On an international scale, species of *Eucalyptus* are among the most popular trees grown in plantations. They grow rapidly and are a source of timber, fuel, tannin, and essential oils. In screening trials at La Selva, *E. deglupta* was one of the most successful trees at one of the sites, with an average height after two years of 7.2 m. Some coffee growers

Eucalyptus deglupta

Close up of bark.

in Costa Rica prefer *E. deglupta* over other species as a shade tree. Oils from various species, which are used in cold and cough medicines, insect repellent, soaps, shampoos, and other products, can be fatal if ingested in large amounts.

Some eucalypts are claimed to be the tallest trees in the world, reaching 150 m. Since eucalyptus leaf litter decomposes slowly and often inhibits plants beneath it from growing, it is not the best choice in forest restoration projects.

Pinus oocarpa
Family: Pinaceae

Pine, *Pino*

Description: A variable species, with a number of varieties; 15–30 m tall; branches sweeping upward; blocky bark, grayish with orange in fissures. Needles, ca. 20–25 cm long, stiff or somewhat flexible, in groups of 5. Monoecious with separate male and female (ovulate) cones; the latter develop into persistent, ovoid (to rounded) cones, ca. 6–8 cm long, on 2- to 4-cm-long stalks, often pointing backward. Seed and wing to 2 cm long.

Flowering/fruiting: Flowering recorded in February, March, August, and December.

Distribution: Mexico to northern Nicaragua. In Costa Rica, planted in mid- to high elevations.

Related species: Another reforestation species often used in Costa Rica is *P. caribaea*, known as Honduran pine; it has needles in threes or fours and is usually planted below 1,000 m elevation. There are ca. 110 species of *Pinus*, living in a wide variety of climates in the Northern Hemisphere.

Comments: The cones of this species stay on the tree for a long period and are stimulated to release seeds during prolonged dry weather or after a fire. Saplings can sprout from the root after a fire. *P. oocarpa* is found in a range of elevations, but does best at ca. 1,500 m. Some of the first plantings of the species in Costa Rica were made around 1950 in the Cartago region. The tree can be used in construction, as fire wood, and as an ornamental. It is tapped for resin in Mexico, Guatemala, and Honduras. *Pinus caribaea* is a popular plantation species at lower elevations. All pines in Costa Rica are introduced; in Mexico, however, there is a great diversity of native species—ca. 42.

Pinus oocarpa

Acnistus arborescens
Family: Solanaceae

Wild tree tobacco, *Güitite*

Description: Small tree, 2–10 m, with disorderly crown; corky, corrugated tan bark. Leaves 6–25 cm long, rather limp, often galled (warty appearance), lighter below. Small green-cream, bell-shaped flowers in clusters along the branches. Ripe yellow-orange berries, each under 1 cm, in groups of 6–15.

Flowering/fruiting: Fruits mainly in wet season

Distribution: Mexico to South America, and West Indies. In Costa Rica, generally from ca. 600 to 1,500 m, in pastures and along fence lines.

Related species: Botanists differentiated many species of *Acnistus* in the past, but most have been subsumed under this species.

Comments: This living fence post serves extremely well as a food for wildlife and as an orchid substrate. More than 40 bird species eat the fruits, which are edible for humans and can be made into a jelly. The flowers attract an array of insects as well as hummingbirds. Medicinal uses include using leaves externally in poultices or baths for hemorrhoids and bruises. Teas made from new growth or from flowers are taken for sore throat and coughs. The tree contains an antitumor substance, withacnistin.

Adrian Hepworth

Acnistus arborescens

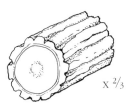

X $^2/_3$

Cross section of *Acnistus arborescens* branch.

Tectona grandis
Family: Verbenaceae

Teak, *Teca*

Description: To 30+ m (50 m in native habitat) with tan-gray furrowed bark; angled stems. Deciduous; opposite leaves to ca. 50 cm, even larger in young trees. Large, loose inflorescences; white corolla with 5 lobes. Wrinkly, old, inflated calyx encloses 2-cm, 1-seeded fruit (illus., p. 232).

Flowering/fruiting: Observed flowering June through August, and in November; fruiting in mid-wet to early dry season.

Distribution: Native to Southeast Asia. Cultivated in Latin America and Africa. In Costa Rica, most commonly seen in plantations and along fence rows in

Tectona grandis

have attracted a significant amount of foreign investment, although potential investors are wise to proceed judiciously since many teak projects have not yielded the promised rates of return, and others have turned out to be scams. United Fruit introduced the tree to Costa Rica in 1925; it is now grown in plantations in many regions of the country. Its origin is in the monsoon forests of Asia, which have a marked dry season. In its native habitat, both elephants and rivers have been used to transport giant teak logs. Girdling a tree and leaving it for a few years decreases its weight enough for the cut tree to float. Young leaves can make a brown, reddish, or lilac dye for coloring cotton. Some plant systematists think that *Tectona* (and *Gmelina*) should be placed in the mint family (Lamiaceae).

Pacific lowlands, but grows in a variety of other environments.

Related species: There are three other species in the genus, all native to Southeast Asia. Gray teak, or gmelina (*Gmelina arborea*), is also in this family, along with some ornamental shrubs such as lantana (*Lantana camara*, p. 103).

Comments: Teak produces an oily wood that is resistant to insect and fungus attack. It is prized for shipbuilding. The timber also makes fine paneling, flooring, and furniture, and, because it is weather resistant, it is appropriate for outdoor furniture and decks. The wood may be yellowish to dark brown with streaking. Teak plantations in Costa Rica

x 1

Tectona grandis fruit enclosed by inflated calyx.

Vochysia guatemalensis   *Mayo*
Family: Vochysiaceae

Other common name: *Chancho, barbachele.*

Description: To 30 m with straight, grayish-white trunk, bark with some papery flakes; dense crown; twigs furrowed. Leaves opposite or in whorls of 3, entire, wedge-shaped toward base, to 15 cm long. Fragrant yellow flowers in showy, terminal clusters; one of the 5 calyx lobes forming a spur; 3 small

petals and 1 fertile stamen. Dehiscent, 3-parted capsule ca. 4 cm long, with winged seeds.

Flowering/fruiting: Flowers March through June.

Distribution: Mexico to Costa Rica. In Costa Rica, lowlands to ca. 1,000 m; primary and secondary forest, both slopes.

Related species: There are 3 other species in Costa Rica, all with yellow

Vochysia guatemalensis

flowers; none have whorled leaves. *V. ferruginea* has noticeable rusty pubescence on the underside of leaves.

Comments: Planted individuals of this species are very common along the road to La Selva. Research on mayo in the La Selva/Sarapiquí area indicates that it has potential in reforestation and restoration. In an assessment of survival and growth rate done after three years of native-timber plantation experiments, *V. guatemalensis* was one of the top three species; the trees grew ca. 8 m in that time period. Another study carried out by M. R. Guariguata et al. (1995) showed that *V. guatemalensis* plantations have potential as a way of encouraging forest restoration. The shade of the young trees appears to suppress further growth of grass in old pastures; grass often inhibits seed germination and the survival of tree seedlings. These trees are also attractive perches for birds and bats that disperse seeds of forest tree species. While it has been assumed that medium to large bees are the pollinators of this tree, butterflies and hummingbirds also visit; as more observations are made, more will be revealed about the details of its pollination. This is not a prized wood, but it can be used in rough construction and for forming boards, plywood, firewood, furniture, and pulp. Mayo and palo de mayo are names given to the various species of *Vochysia*. Previously known as *V. hondurensis*.

6. Special Habitats

This chapter provides a sample of the many habitats that occur in Costa Rica and serves as a basic introduction to some of its many different plant communities. Each section focuses on a different botanically interesting region and describes a selection of the species that are typical of that habitat. The areas covered include the tropical rainforest surrounding La Selva Biological Station and the canals of Tortuguero National Park in the Atlantic lowlands in northeastern Costa Rica, the dry forests of the northwestern part of the country, the tropical montane forests in Monteverde, Poás Volcano National Park, and Cerro de la Muerte, and the Pacific and Caribbean beach and mangrove habitats.

Biologists often use habitat classifications that are more detailed and technical than what is presented here. The Holdridge life zone system—based on temperature, precipitation, and humidity—defines a world total of 116 life zones, 12 of which occur in Costa Rica. In the Monteverde region alone, for example, one can easily pass through four life zones on a hike from one side of the Continental Divide to the other.

The Osa Peninsula and environs have not been included in this chapter because it would be difficult to pick out a representative subset of the flora in such a habitat-rich area. Interestingly, its flora is actually more closely allied with the forests of South America than with other regions of Central America. Tree species occur there, for example, that are not seen elsewhere in Costa Rica. In this and other chapters, the author notes when a species occurs in the Osa. For a detailed look at these forests, see Quesada et al. 1997 and Weber 2001.

Left, the canals of Tortuguero (photo by Adrian Hepworth).

Wet Atlantic Lowlands

On the Atlantic coast of Costa Rica, warm air moves off of the ocean laden with moisture. The northeast tradewinds push it inland and upward, across the coastal plain and foothills; the rising, cooling air condenses and releases its load. This side of Costa Rica, which has a short and not very distinct dry season, experiences some of the highest rainfall in the country—up to 8 m during some years. The combination of hot temperatures and plenty of rainfall creates a typical rainforest, or tropical wet forest, environment. The forest canopy reaches 55 m in some areas and plant diversity is high. Featured in this section are La Selva Biological Station (p. 237) and Tortuguero (p. 247).

Buttressed tree and palms at La Selva.

La Selva Biological Station and Environs

The forest of La Selva Biological Station in the wet, northern Caribbean lowlands exemplifies what most people think of as tropical rainforest—a hot, steamy environment with enormous trees and an exuberant growth of typically tropical plants. This impressive habitat covered the entire Puerto Viejo de Sarapiquí region before it was deforested by loggers and farmers. La Selva's 1,600 hectares (3,900 acres) and the adjacent corridor of Braulio Carillo National Park serve as the primary research and education site for the Organization for Tropical Studies. Many of the species seen there also grow in Costa Rica's southern Pacific region.

This lowland forest has no distinct dry season; annual rainfall is 4 m, and the temperatures regularly reach into the high 80s Fahrenheit (ca. 30° C). Stepping into this forest, one is struck by the diversity and size of the trees and palms, and by the proliferation of epiphytes and climbers. Some of the hardwood tree species are threatened or endangered because their wood is highly prized and marketable. All of the trees and other plants play important roles in the complex interactions that take place in the forest; some are crucial to the survival of endangered animal species. The giant almendro tree, for example, provides food and nesting sites for the endangered great green macaw (see p. 241 for more information).

Welfia regia
Family: Arecaceae

Welfia palm, *Palma conga*

Other common name: *Palmito.*
Description: A subcanopy, nonspiny palm to 20 m; shiny trunk with conspicuous leaf scars; monoecious; new leaves tinted red. Pinnately compound leaves to 6 m, ascending and arching; leaflets, to 1+ m long, many and in a plane. Inflorescence (to 1 m long) located below the leaves; flowers develop in deep pits along the many thick, drooping fingerlike branches of the inflorescence. Red-brown to purple fruit ca. 4 cm long, almondlike with ridges on sides; fruits partially embedded in the pendulous fruiting structures.

Welfia regia

Distribution: Honduras to Ecuador. In Costa Rica, in wet, lowland forest below 800 m (including Osa Peninsula).

Comments: *Welfia regia* is second only to *Pentaclethra macroloba* (p. 239) in abundance in some of the primary forest study plots at La Selva. The crown of a single welfia can weigh 250 kg, and a leaf falling from adult plants can cause damage to its own and other seedlings. Bees and possibly beetles (weevils, Curculionidae) pollinate the flowers. The fruits ripen in wet season. Seed dispersers include monkeys, kinkajous, squirrels, and large birds, with rodents being the main seed predators, although they may bury and forget a few seeds and thus contribute to propagation. Large, young reddish welfia leaves in the understory are hard to miss when they are backlit by the sun. The heart of the palm is edible, and the leaves are used for thatch in some regions. The wood is hard and does not rot in salt water; thus the trunks make good supports for coastal homes. The current scientific name for this palm, *Welfia regia*, includes specimens that many botanists have referred to as *W. georgii* in the past.

Terminalia oblonga
Family: Combretaceae

Sura, *Surá*

Other common name: *Guayabón.*

Description: Large tree to 45 m with buttresses to several meters tall; smooth, tan- to flesh-colored trunk with peels of papery bark. Alternate, simple, entire leaves, 5–15 cm long and with translucent dots, tightly bunched toward branch tips; little tufts of rust-colored hairs in vein axils on underside. Spikes of small green and yellow flowers with honey scent. Winged fruit ca. 3 cm long.

Distribution: Mexico to South America.

In Costa Rica, to ca. 800 m, both slopes; it is found along rivers in dry areas.

Comments: The main distinguishing characteristics of this abundant tree are its light-colored, peeling bark and its high buttresses. Sura also grows in Guanacaste, usually near rivers. The fruits develop in the dry season. The seeds are wind-dispersed. Sura's heavy wood is appropriate for flooring, paneling, and furniture. There are many species in this genus, mostly in the Old World; only four are found growing naturally in Costa Rica, the others being the Indian almond (*T. catappa*, p. 331), found on beaches, and the less common roble coral (*T. amazonia*) and escobo (*T. bucidoides*), both occurring in forests. *T. ivorensis*, a West African species, surpassed native terminalias in reforestation tests at La Selva. It was one of the top three trees in terms of survival and growth in the first three years. The name *T. chiriquensis* was formerly used for *T. oblonga*.

Distinctive buttressed trunk of *Terminalia oblonga.*

Bauhinia guianensis
Family: Fabaceae
Subfamily: Caesalpinioideae

Monkey ladder, *Escalera de mono*

Description: A large liana, with distinctive, flattened, undulating stems, sometimes ladderlike. Has tendrils and alternate leaves to 10 cm broad, with heart-shaped base; leaf cleft part-way, or to base on younger plants. White, perfumed, 5-petaled flowers. Fruits flat and ca. 8 cm long.
Distribution: Southern Mexico to South America. In Costa Rica, lower elevations to ca.1,000 m; common at La Selva.
Comments: Monkey ladder pods dehisce explosively near the end of the dry season, making popping noises. The seedlings are abundant at the start of the rainy season. In northwest Amazonia, a preparation of the stem is used for kidney ailments, and the seeds are used as a diuretic. The plant is related to the orchid tree (*B. variegata*, p. 135) and *B. glabra*, a common liana of the dry, Pacific lowlands. *B. outimouta*, a liana with a more cylindrical stem, occurs at La Selva also.

Bauhinia guianensis

Pentaclethra macroloba
Family: Fabaceae
Subfamily: Mimosoideae

Pentaclethra, *Gavilán*

Description: Tree to 40 m with broad crown. Alternate, bipinnately compound leaves, ca. 30 cm long, with 15–30 pairs pinnae and many leaflets to 1 cm long. Erect, arching white spikes, 20 cm long, have a sweet, somewhat fermented scent; white styles, staminodia, and stamens are conspicuous, in contrast to the reduced, maroon calyx and corolla at base. Two-valved, curved, woody, brown pod, to 30 cm, held erect.
Distribution: Nicaragua to Panama and certain parts of South America. In Costa Rica, Atlantic slope to 500 m; a dominant at La Selva and common at Tortuguero.
Comments: Pentaclethra, the most common tree at La Selva, does well in swamps as well as in better-drained

Dipteryx panamensis fruit.

X ⅖

Fruit with fleshy
layer removed.

meters per year—when grown from seed and planted in open areas, but it will take hundreds of years to replace the giants that are being felled in the forest. The chances of survival for the many seedlings

that sprout naturally on the forest floor are low. In a study by Dave and Deborah Clark (1987), of 147 seven-month-old seedlings in their study area, only five were alive after five years, and all were under a meter tall. Human uses of almendro include making torches from the oily fruits (in the Darien region of Panama) and eating the seeds. Caribbean coastal dwellers of Nicaragua mix ground seeds with coconut water and, in special ceremonies, ingest this as an aphrodisiac. The English name tonka bean usually refers to a South American species, *Dipteryx odorata*, which is known as a source of tobacco flavoring, perfumes, and the anticoagulant coumarin. Some botanists used the name *D. oleifera* for *D. panamensis*.

Lecythis ampla
Family: Lecythidaceae

Monkey pot tree, *Olla de mono*

Other common name: *Jícaro.*
Description: Tree to 40+ m; impressive straight trunk with vertically fissured brown bark. Leaves simple, alternate, ca. 8 cm long, with pointed tip and fine-

ly toothed, wavy margin. Flowers ca. 3 cm in diameter, with 6 bluish or pinkish petals, hundreds of anthers. Large thick-walled, rounded fruit, taller than wide, to 20 cm in diameter, the operculum

Lecythis ampla fruit and seeds.

(lid) ca. 9 cm across; up to 50 4- to 5-cm, grooved, nutlike seeds attached to inner wall of "pot" by fleshy stalks.

Distribution: Nicaragua to Ecuador. In Costa Rica, from 50 to 500 m; Atlantic lowland wet forest in northern zone (e.g., Tortuguero and La Selva).

Comments: Since this is a rare and endangered species, to encounter one of these trees, with its large, impressive, furrowed trunk, surrounded by fallen "monkey pots," is a special event. Like its relatives, the Brazil nut (*Bertholletia excelsa*) and the sapucaia or paradise nut (South American species of *Lecythis*), it has edible seeds. Some animals prefer the fleshy stalks (funicles) that attach the seeds to the inner wall of the pot. Typically, the lid falls off while the giant downward-pointing fruit is still attached to a branch, making seeds accessible to bats that are attracted by those fleshy stalks. These dispersers carry the seeds to a roost, eat the funicles, and drop the seeds. Squirrels and monkeys also seek out the nuts. Some pots, with seeds intact, drop to the ground and attract deer, coati, agouti, and peccary. The flower has an interesting flap that curves over its center, where active pollen develops in 100–200 stamens. This flap is itself lined with many stamens, but these contain "food" pollen with which the plant attracts pollinators. As a large bee squeezes into this sandwich, its back becomes covered with the active pollen while it collects the food pollen. The hard, heavy wood of this tree is useful for fence posts, railroad ties, and farm tools. Some people of northern Costa Rica use the seeds as an ingredient in candies, while the indigenous Kuna people of Panama use them medicinally. The seeds may be toxic if they come from a tree growing in selenium-rich soil. Monkey pot is being tested as a possible reforestation species. *L. mesophylla*, the only other species of the genus in Costa Rica, occurs on the Osa Peninsula, along with nine other species of the family Lecythidaceae. The cannonball tree (*Couroupita nicaraguarensis*, p. 275) is also in this family.

Virola koschnyi

Family: Myristicaceae

Wild nutmeg, *Fruta dorada*

Description: Tree to 40 m, with medium buttresses; whorled, horizontal branching that looks like spokes of a wheel when you look up into the crown; bark pink-gray, sap red. Alternate, simple, entire leaves, 15–30 cm long with prominent venation (20–35 pairs of veins), tomentose beneath. Dioecious—clusters of tiny male and female flowers on separate trees. Two-valved, orange-brown fruit, ca. 3 cm, splits open to reveal dark brown seed covered with a netlike red aril.

Distribution: Central America and northern South America. In Costa Rica, to ca. 1,000 m at sites on both slopes,

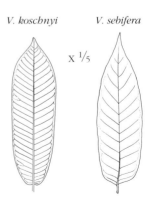

Leaves of two *Virola* species at La Selva.

Virola koschnyi fruit and leaves on forest floor.

including La Selva, Tortuguero, the Guanacaste foothills (on the Atlantic slope), Osa Peninsula, and Carara.

Comments: The striking fruits of this tree attract medium and large frugivorous birds, which swallow the aril-covered seeds and regurgitate them after removal of the aril. Monkeys may also eat the arils. Pollination is by small insects. This species, which has potential for reforestation, has been part of the TRIALS project at La Selva. Its lumber has a variety of uses, including furniture and general construction. Four other species occur in Costa Rica, one of which, *Virola sebifera*, is also at La Selva. It has fewer pairs of lateral veins in the leaf than *Virola koschnyi* (illus., p. 243). There are many species in the Amazon Basin. Resin from some species, especially *V. theiodora*, is famous for its use by Amazonian shamans in hallucinogenic snuff preparations and as an arrow poison. Various alkaloids in the resin cause a lack of muscular coordination, nasal discharge, visual distortion (seeing things larger than they are), nausea, and hallucinations. This family also includes true nutmeg (the seed) and mace (the aril), both from the same plant, *Myristica fragrans*, which originates in Indonesia.

Passiflora vitifolia Passion flower, *Granadilla del monte*
Family: Passifloraceae

Description: Liana that climbs to canopy, but flowers near ground level. Alternate, 3-lobed leaf, 7–15 cm long and broad, pubescent below, with toothed margin and 2 to several nectar glands at base of petiole. Axillary tendrils. Flower ca. 12 cm across, the 5 sepals and 5 petals intense red, with 3 green bracts below; corona made up of many thin red-to-white filaments; stalk in middle of flower is topped with 3 styles; 5 green anthers attached below. Fruit aromatic, maroon and green, mottled with white; to 10 cm long (illus. below), with many seeds in slimy pulp. **Distribution:** Nicaragua to northern South America. In Costa Rica, widespread, lowlands to ca. 1,000 m, wet or moist forest understory or edge; common on Osa Peninsula.

Comments: Of the 15 or so species of passion flowers at La Selva, this is the most conspicuous, due not only to its color but to the fact that it blooms in the forest understory. The young tendrils and leaves are food for the larvae of passion flower butterflies such as *Heliconius cydno* and *H. hecale*. Nectaries on the leaf stem attract ants and other insects. The fruit is probably eaten by mammals. Although hummingbirds have no problem probing through the many crowded, thin segments of the flower's corona, the corona does prevent nonpollinating visitors from accessing the nectar. There are more than 500 *Passiflora* species in the world, with nearly 50 in Costa Rica; species differ

X ¼

Passiflora vitifolia fruit.

Adrian Hepworth

Passiflora vitifolia

in color, size, and shape, and have correspondingly varying pollinators (bees, birds, bats, etc.). Some passifloras are cultivated for their fruit (p. 189). *P. coccinea*, a South American species that is sometimes used as an ornamental, also has red flowers.

Warszewiczia coccinea
Family: Rubiaceae

Chaconia, *Pastora de monte*

Other common name: Wild poinsettia. **Description:** Shrub or tree to ca. 10 m. Entire leaves opposite or 3 per node. Inflorescence ca. 50 cm long, with clusters of small orange corollas along its length; one flower of each cluster has a brilliant red 7-cm-long calyx lobe. Capsular fruit ca. 0.5 cm long with many tiny seeds. **Distribution:** Central America to Peru and Bolivia, and West Indies; also, as

ornamental in other tropical areas. In Costa Rica, a fairly common subcanopy tree in parts of La Selva, also on Osa Peninsula and other rainforest areas, in light gaps and forest edge to 300 m.

Comments: One of several members of the coffee family with red calyx lobes that make them attractive to hummingbirds and butterflies, as well

Warszewiczia coccinea

as to gardeners. Others are *Pogonopus exsertus*, which can occasionally be found in the wild in Costa Rica on the Pacific slope, below 800 m, and red flag bush (*Mussaenda erythrophylla*) of the Old World. Chaconia is not closely related to true poinsettia (Euphorbiaceae family). Some South American peoples use the root medicinally, externally for fungal skin infections and back pain, and as an aphrodisiacal perfume.

Tortuguero Canals

Tortuguero National Park, in northeastern Costa Rica, is well-known both for the great opportunities it provides for viewing wildlife along its canals and as a protected nesting beach for green turtles. It has a hot, humid climate with 5–6 m of rain per year. Plant-species diversity is high in the region, with vegetative communities that range from a variety of wetland areas to slope forest in the west, where the hills rise to 300 m. Over geologic time, river deltas have formed and merged throughout this region. Their deposits created the substrate of today's swamp forest, which in turn has built up more sediment.

This region shares some of the same vegetation seen in other lowland wet forests, such as those of La Selva, but the natural waterways, with their floodplain environment, predominate. Traveling through the canals, one can easily see floating and emergent aquatic plants as well as vegetation at the border of the swamp forest. The canal edges of Tortuguero are the perfect place to view, up close, large lianas in flower and fruit that normally are fertile only in the high canopy, where they are not easy to see. The following species accounts provide a small sampling of the fascinating botanical world of the canals.

Montrichardia arborescens
Family: Araceae

Description: Prickly stem 3–4 m long, thick (5 cm) at base; aquatic. Arrowhead-shaped leaf blade to 40+ cm long, with ca. 32-cm-long sheathing petiole. Hoodlike spathe to 18 cm enclosing typical Araceae spikelike inflorescence ca. 11 cm tall; lower, green, bulbous part of spathe surrounds pistillate (female) flowers; open white or green-cream section above exposes cream-colored staminate (male) flowers. In fruit, spathe falls off and cylindrical spadix enlarges to ca. 8 cm wide, with distinct, but densely packed, green sections ("berries"); seeds ca. 2.5 cm long.

Distribution: Guatemala south to northern South America, also West Indies. In Costa Rica, near Caribbean coast and San Carlos lowlands, to ca. 100 m, and southern Pacific mangroves; aquatic, in swamps and estuaries.

Montrichardia arborescens in flower.

Typical habit of *Montrichardia arborescens*.

Comments: *Montrichardia arborescens* forms dense colonies along sections of Tortuguero canals. The combination of its aquatic habit, spiny stem, and arrow-shaped leaf makes it easy to distinguish from other aroids. The buoyant seeds are dispersed via water. The roasted fruit and seeds are edible. Some of the reported medicinal uses in South America include ingesting the powdered root as a diuretic and applying the caustic sap to stop bleeding. One other species of the genus, *M. linifera*, occurs in South America. *Montrichardia* is related to philodendron (see chapter on aroids, p. 354) and taro (*Colocasia esculenta*, p. 167).

Philodendron radiatum   Dubia philodendron, *Cobija de pobre*
Family: Araceae

Description: Large climber to ca. 25 m, but sometimes at ground level. Large pinnately divided leaves (variable), to more than 1 m by 1 m, with overall heart shape; the linear (ca. 3 cm wide) leaf sections are forked. Green-whitish spathe, 20+ cm long, surrounds white spadix; deep rose-purple lines inside lower part of spathe. Fruits are berries.
Distribution: Mexico to Panama, sea level to ca. 700 m. In Costa Rica, usually below 300 m in wet forest, e.g., at Tortuguero and La Selva (latter site particularly rich in philodendrons, with ca. 33 species).
Comments: This is the common cut-leaf philodendron seen along Tortuguero canals. In the early evening—for one night only—a spadix heats up to 44° C (111° F) and volatilizes a scent compound that attracts scarab beetles (genus *Cyclocephala*). The beetles enter the lower portion of the spathe and feed on a sterile area of the

Philodendron radiatum

spadix, simultaneously depositing pollen on female flowers. While the beetles are feeding, the spathe constricts and traps them for 24 hours. On the following evening, male flowers on the upper part of the spadix release pollen and the spathe expands slightly. As the beetles climb up the spadix, pollen clings to their body and they fly off in search of a newly opened, heated spathe. The inflorescence of the Brazilian *P. bipinnatifidum* heats up to 46° C (115° F).

Desmoncus schippii
Family: Arecaceae

Matamba

Description: Scandent palm with prickly stems, climbing into canopy via hooklike spines (modified leaflets) on elongated rachis; short petiole sometimes with spines. Leaf rachis to more than 1 m long with extension (cirrus) to over 0.5 m with ca. 8 spines on each side; pinnately compound with usually 32 or more narrow leaflets (to ca. 4 cm wide). The many-branched inflorescence is made up of male and female flowers. Spiny bract persists in fruit. Yellow to red fruit, to 1.5 cm in diameter, develops in clusters, sometimes with dozens of fruits. Seed with 3 pores.

Distribution: Belize to Costa Rica. In Costa Rica, sea level–600 m, north Atlantic slope and perhaps Nicoya Peninsula; wet forest, sometimes in second growth. Occurs at Tortuguero as well as along rivers at La Selva.

Comments: This palm has a structure called a cirrus, an extension of the midvein of the leaf. The cirrus reaches out beyond the regular leaflets and is equipped with narrower, hooklike leaflets that aid in climbing. Some species of *Desmoncus* are beetle-polli-

Desmoncus schippii

nated, although it is not known if this is the case with *Desmoncus schippii*. The fruit is eaten by birds and monkeys. Matamba resembles Old World rattan palms, but it is not closely related to them. The flexible stems of some species of *Desmoncus* are used in basket or furniture making. *D. costaricensis*, a species endemic to Costa Rica, has wider leaflets and long, spineless leaf stalks. Another endemic, *D. stans* of the Osa Peninsula, has weaker, spineless cirri and thus does not climb.

The taxonomy for the genus *Desmoncus* is unclear, and even though lumping all Costa Rican climbers under *D. orthacanthos* (see Henderson et al. 1995) seems incorrect, more study is needed to precisely determine the number of species in Central America. The scientific name *Desmoncus schippii* is thus tentative.

Manicaria saccifera
Family: Arecaceae

Sea coconut, *Sílico*

Other common name: *Yolillo.*
Description: Palm, to 10 m tall, with several thick trunks. Large leaves, to 6+ m, erect or slightly arching, more or less entire—breaking up irregularly, especially with wind and age, with a jagged edge. Monoecious; inflorescence to 1 m surrounded by a fibrous, netlike bract. The 2- to 3-lobed fruit 6–7 cm in diameter; surface with pointy projections, each lobe with one seed.

Distribution: Along parts of Atlantic coast of Central America, northern Pacific coast of South America, and part of the Amazon region. In Costa Rica, at Tortuguero along canals and out toward ocean.

Comments: The giant leaves of this palm serve as long-lasting roof thatch in the Tortuguero region of Costa Rica. South American indigenous groups also use the palm leaves in sails for canoes, and they use the stem as a source of starch and the seeds as a source of oil; they produce medicine from various plant parts. The odd fibrous inflorescence bracts, which are ready-made hats, have led to the names monkey-cap or sleeve palm. The fruit/seed, which can float for two years, is sometimes carried to northern Europe by the Gulf Stream.

Francis X. Faigal

Manicaria saccifera

x ½

Manicaria saccifera fruit.

Raphia taedigera
Family: Arecaceae

Raffia, *Yolillo*

Description: Clumped stems, ca. 3 m or taller, covered with old leaf bases. Immense arching leaves, to ca. 15 m long, with hundreds of linear pinnae projected irregularly from the rachis; short spines on leaf margins and midribs. Monoecious; hanging flower clusters ca. 3 m long. Scaly 1-seeded fruit, 5–10 cm long, lustrous reddish-brown.

Distribution: Nicaragua to Colombia, and a patch in Brazil toward mouth of the Amazon. In Costa Rica, populations on both coasts in swampy or inundated areas (e.g., Tortuguero, Osa Peninsula, and near Gandoca-Manzanillo).

Comments: Raffia palms produce the conspicuous, large, arching leaves that form dense stands (called *yolillales*) along certain areas of the canals. The palms create deep shade and their falling leaves damage other plants growing beneath them. These factors, and the flooded conditions in which the palms grow, restrict the growth of other plant species. The flowers are wind- or insect-pollinated. Peccary and tapir eat the fruit. People use the leaf midvein of this palm to make poles to prop up banana plants, and they use the fiber of young leaves to make twine and baskets. The 25-m-long leaves of some African species are the largest leaves of any plant species. It is not clear whether *R. taedigera* is native or was introduced to Central America; all of the other 28 species in this genus grow in Africa and Madagascar.

X ¹⁄₂

Raphia taedigera fruit.

A stand of *Raphia taedigera*.

Pachira aquatica
Family: Bombacaceae (recently placed in Malvaceae)

Provision tree, *Jelinjoche*

Pachira aquatica fruit.

Pachira aquatica flower.

Other common name: *Poponjoche.*
Description: Tree to 20 m, sometimes with buttressed trunk. Alternate, long-stemmed, palmately compound leaves with up to 9 leaflets, pale beneath, each to 30 cm long. Phallic flower buds; 5 fleshy, straplike petals to 30 cm long, tan or greenish white; tube ca. 9 cm, ending in many reddish stamens 18 cm long; style 28 cm; flower is lilac-scented. Rounded, reddish-brown, 5-parted heavy fruit capsule, ca. 20+ cm, with angular seeds ca. 4 cm long.

Distribution: Southern Mexico to Peru and Brazil, cultivated in some other regions of the world. In Costa Rica, prefers wet places—in swamps and near rivers; sea level–1,100 m, both slopes.

Comments: Whether it is in flower or in fruit, this is one of the most conspicuous trees along the Tortuguero canals. The large blossoms, which produce a lilaclike perfume, are pollinated by bats. The fruit and seeds disperse by floating. Monkeys, and probably other mammals, eat the seeds. The soft wood is not particularly useful, but the bark can be used for cordage and yellow dye. The seeds, young leaves, and flowers become edible when they are cooked. Various parts of the tree are used medicinally as a tonic, for kidney pain, liver problems, and anemia. *Pachira aquatica* makes a good ornamental both because its showy flowers and fruits can be seen during much of the year and because it may flower when it is only a couple of meters tall.

Pterocarpus officinalis
Family: Fabaceae
Subfamily: Papilionoideae (= Faboideae)

Bloodwood, *Sangregado*

Other common name: *Sangrillo.*
Description: Tree to 30 m, with large, ropy buttresses that branch and extend well beyond the trunk; red sap. Alternate, compound leaves with up to 11 entire leaflets, each from 5 to ca. 20 cm long. Inflorescence to ca. 20 cm long; 5-petaled, yellow-orange flowers, with the standard, or banner petal, showy. One seed, ca. 3 cm long, enclosed in roundish 5-cm fruit that has netted venation and winged edge; ripe fruit green-yellow turning brown-black.

Pterocarpus officinalis

Distribution: Antilles and Mexico to northern South America. In Costa Rica, usually below 300 m, wet forest with periodic flooding; common in some low-lying areas of the Osa Peninsula, La Selva, Tortuguero, and southern Atlantic coast.

Comments: The impressive snaking buttresses of this tree stand out along the Tortuguero Canal, where the trees appear to have been left in areas where more valuable species were cut down. The rel-atively soft wood is vulnerable to damage by insects, but it is usable if treated. The leaves provide food for larvae of the butterfly *Morpho amathonte*. The flowers, which generally develop in the wet season, are bee-pollinated. The seeds may be wind- or water-dispersed. The tree's astringent resin, called *sangre de drago* (dragon's blood), is put on wounds; some Asian species of *Pterocarpus* are the source of kino, used in treating diarrhea.

Eichhornia crassipes
Family: Pontederiaceae

Water hyacinth, *Lirio de agua*

Description: Succulent, floating plant with leaves in rosette. Runners from leaf axils producing new plants. Feathery, comblike roots. Roundish leaf blades 7–12 cm across; leaf stems inflated. Erect inflorescence of lavender flowers 3–6 cm in diameter with 6 tepals, uppermost with nectar guide (darker blue-violet with yellow blotch). When it sets fruit, it produces a three-parted capsule.

Distribution: Probably originated in Brazil; has spread worldwide, especially in tropical and subtropical regions; in canals and other water bodies. In Costa Rica, Caño Negro, Palo Verde, Tortuguero, Osa Peninsula, and various rivers.

Comments: This is an attractive, but invasive, water plant. With its runners radiating out in all directions, water hyacinth reproduces rapidly and covers water surfaces, sometimes clogging waterways such as the canals of Tortuguero. In the past, there were 200 tons per acre on some water bodies in Florida, but the plant's growth is now controlled mechanically, by herbicide, or through biocontrol. The leaf stem bases are inflated and spongy, with air pockets in the tissue that aid in buoyancy. The leaf blades may catch the wind and help disperse the plants. Long-tongued bees pollinate the flowers in the Amazon.

Manatees feed on the plants. The root system provides habitat for aquatic creatures, including mosquito larvae, however eventually a hyacinth mat will create an oxygen-poor environment that is inhospitable to some forms of life. It is sometimes grown on sewage treatment ponds. Since it is so prolific, researchers are always seeking creative ways of disposing of it, such as using it as pig food or compost, or for methane production. Three other species occur in Costa Rica.

Eichhornia crassipes

Tropical Dry Forest

The term *tropical forest* may typically suggest a hot, rainy jungle scene, but in fact a number of different forest types occur in the tropics, including some that have little or no precipitation during half the year. In Costa Rica, this seasonal dry forest is found in the northwest section of the country, especially in the Guanacaste and Tempisque Conservation Areas. Patches of it also occur here and there to the south, on the Pacific slope. Prior to deforestation and subsequent farming and cattle ranching, this type of forest covered most of western Central America. Today, in addition to protecting what is left, conservation efforts led by biologists Daniel Janzen and Winnie Hallwachs have allowed vast tracts of pasture in Costa Rica to begin reverting back to dry forest.

The Costa Rican dry forest receives 1–2 m of rain a year. Many of the trees lose their leaves and produce flowers during the distinct six-month dry season, which lasts roughly from mid-November through mid-May. Much of the leaf flushing occurs just before or at the start of the rainy season in mid-May. A number of overexploited and now endangered species of tropical hardwoods grow in these forests, including ronrón (*Astronium graveolens*), mahogany (*Swietenia* spp.), lignum-vitae (*Guaiacum sanctum*), and rosewood (*Dalbergia retusa*). Cacti and other spiny plants are common. The diversity and abundance of epiphytes and ferns are much lower than in wetter, less seasonal tropical forests. In stream valleys and other moist pockets, trees can remain green year-round, creating shade and shelter for fauna, especially insects, during the dry season. Oaks (*Quercus oleoides*) dominate parts of the region, and their presence appears to be associated with particular types of volcanic rock and ash.

A mass blooming of *Tabebuia ochracea.*

Astronium graveolens
Family: Anacardiaceae

Ronrón

Other common name: Gonçalo alves, *jobillo*.

Description: Tree to 30+ m tall, 1 m in diameter; trunk light to dark gray and tan with bark flaking off, creating a patchy appearance; sap has pleasant turpentine odor. Compound leaves to ca. 30 cm long, alternate and odd-pinnate, with up to 15 leaflets, nearly entire to toothed; leaflets opposite or alternate, change to orange color before falling. Dioecious; tiny, yellow-green flowers in terminal clusters ca. 20 cm long; 5 segments of calyx expand and subtend the 1-seeded, 1.5-cm, oblong fruit.

Distribution: Mexico to Paraguay. In Costa Rica, sea level–1,000 m, in dry to moist forest; mostly in Guanacaste, but also reported from the Osa Peninsula and the forests around Orotina and Upala.

Comments: This precious hardwood, a threatened species, grows in secondary and primary forest and is protected in a number of Costa Rican lowland sites (Santa Rosa, Guanacaste, Palo Verde, Lomas Barbudal, and Carara). The marbled wood, which is extremely heavy and hard, is suitable for bowls and other turned objects, furniture and cabinets, beams, and flooring. It has also been used in making dampers for grand pianos. The wood is rot resistant, but not totally insect resistant. The starlike calyx on the fruit aids in wind dispersal. Parrots feed on the seeds. The leaves turn red-orange in the dry season and are reminiscent of the autumn colors of sumac, its northern cousin.

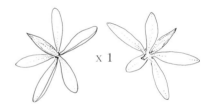

x 1

Astronium graveolens fruit.

Astronium graveolens with senescent leaves.

Astronium graveolens

MUSICAL INSTRUMENTS, FURNITURE, AND BOWLS

Dalbergia retusa in flower.
Fabaceae (subfamily: Papilionoideae)

Makers of musical instruments and fine furniture have long been attracted to tropical hardwoods. The wood is more difficult to work than that of other, softer-grained trees, but its inherent resistance to insect damage and the color and grain patterns of the wood itself lend added value to the objects these artisans make. In addition to ronrón (p. 256), rosewood and mahogany are two other hardwood species that occur in the dry forest and have a multitude of uses. Both are endangered, however, due to habitat destruction and their extraction from natural forests.

Rosewood (*Dalbergia retusa*) is a medium-sized tree to 20 m tall that has a twisted dark trunk with vertical fissures. It has alternate, compound leaves, clusters of white flowers, and flat, legume fruits to 15 cm long. When the wood is cut or burned, it gives off a sweet fragrance that may have originally suggested the name rosewood. Its very attractive

Bowl, made from ronrón, and rosewood box.

streaked wood, known in Costa Rica as *cocobola*, is used in turned articles and cabinetry.

Of all the precious Neotropical hardwoods, mahogany (*Swietenia macrophylla*) is perhaps the best-known and one of the most valuable. The beautiful high-quality wood, which is called *caoba* in Costa Rica, is made into furniture, cabinets, and musical instruments. The tree's bark is gray and black/brown and is vertically fissured into flat-topped strips; it has a checkered appearance in older trees, which can grow to 40 m. Mahogany has alternate, compound leaves and branching inflorescences made up of functionally unisexual, white to

Swietenia macrophylla fruit.
Meliaceae

William A. Haber

greenish, honey-scented flowers, 1 cm across. The distinctive capsular fruits, which take about a year to mature, are ca. 16 cm long. When the erect woody capsule dehisces, its 5 valves separate to release winged seeds. The 12-cm-long seeds have a high germination rate, and the seedlings grow well in light gaps in the forest and in old pastures.

A shootborer, the larva of the moth *Hypsipyla grandella*, often destroys the terminal growth of plantation trees, causing unwanted branching and thus lowering their commercial value. *S. humilis*, a rarer species of the Santa Rosa region, may or may not be a distinct species; it has sessile leaflets, while *S. macrophylla* has basal leaflets with distinct petiolules. *S. mahogoni*, which is native to Florida, Cuba, and Hispaniola, has smaller leaflets and fruits than those of the *Swietenia* species in Costa Rica.

X ⅓

Swietenia macrophylla seed.

Crescentia alata
Family: Bignoniaceae

Calabash, *Jicarillo*

Crescentia alata

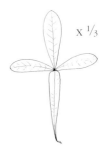

x ⅓

Crescentia alata leaf.

Description: Shrubby-looking tree ca. 8 m tall with crown of crooked branches. Leaves in small bunches alternating along branches; a mixture of compound leaves, made up of 3 leaflets with a winged petiole, and simple leaves. Flowers and fruits cauliflorous. Calyx 2-parted, soft tomentose, light green suffused with maroon; flowers bell-shaped, ca. 5 cm long, green marked with thick, dark purple lines; with skunklike scent. Spherical fruit 8–10 cm in diameter, yellow-green with acidic-sweet, fruity scent emanating from indented area in top of fallen fruit. Seeds small, heart-shaped.

Distribution: Mexico to the Guanacaste region of Costa Rica. In Costa Rica, it is found in pastures and in the savannas of Santa Rosa National Park; it sometimes appears in clumps called *jicarales*.

Comments: *Crescentia alata* (photo left) is very similar to the widely-cultivated *C. cujete* (jícaro, photo below), the source of calabash fruits used for making decorative and utilitarian items. *C. cujete* is readily distinguished from *C. alata* not only by its larger fruits (to 30 cm in diameter) but by its lack of compound leaves. Horses readily eat the fallen, ripe fruits of *C. alata*, and Janzen (1983) thinks that horses may have been dispersal agents during the Pleistocene era. The foul-smelling flowers attract bat pollinators. The bark is a good substrate for orchids and other epiphytes. Peeled and dried seeds are used in one version of the rice-based drink known as *horchata*. Hybrids of the two *Crescentia* species occasionally occur; there are some of these near the Hacienda La Pacífica outside of Cañas.

Crescentia cujete

Pithecoctenium crucigerum
Family: Bignoniaceae

Monkey comb, *Bateita*

Other common names: Monkey's hair-brush, *cucharilla*.

Description: Liana with fibrous bark and hexagonal, ribbed stem to 10 cm in diameter. Opposite, compound leaves with 2–3 leaflets, often heart-shaped at base, covered with tiny scales both sides; quite a range in leaflet size, from 3.5 to 17 cm; glands on underside, in vein axils near leaflet base; many-branching tendrils sometimes replace leaflets. Five-lobed flower, 4–6 cm long, with an elbowlike curve in the tube, white with yellow in throat, fairly fleshy, pubescent on outside; sometimes has musky odor; cuplike calyx ca. 1 cm long. Heavy, woody, deep brown, two-valved seed capsule, elongate and bulging (15–30 cm long, ca. 6 cm wide), with 6-mm "teat" at end; covered with stout blunt prickles; containing many seeds; the seed and the thin transparent wings surrounding it measure up to 10 cm wide.

Distribution: Mexico to Argentina. In Costa Rica, mostly on Pacific slope below 1,200 m, but may also be found on Atlantic slope.

Comments: In the dry season, the monkey comb pods are very noticeable as the liana loses its leaves, exposing the dark, oblong pods. The late Alwyn Gentry, who was a Bignoniaceae family specialist, commented on the beauty of the floating seeds in *Flora of Panama*, saying "with a breeze a seed remains airborne indefinitely." Monkey comb is one of the most widespread species of Bignoniaceae, partly due to the superb dispersibility of the seeds. The related *Amphilophium paniculatum* also has hexagonal stems, but a smooth fruit.

x ½

Pithecoctenium crucigerum seed.

Pithecoctenium crucigerum in fruit.

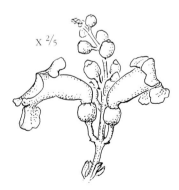
x ²⁄₅

Pithecoctenium crucigerum flowers.

Tabebuia ochracea
Family: Bignoniaceae

Tabebuia, *Cortez amarillo*

Other common names: Yellow cortez, *cortez.*

Description: Tree to ca. 20 m tall; grayish bark with shallow, vertical fissures. Opposite, long-stemmed, palmately compound leaves, the 5 leaflets with dense covering of star-shaped hairs on underside; leaflets varying in size, the largest to ca. 20 cm long. Dense, ball-like clusters of fragrant, yellow flowers, 4–8 cm long, with furry calyces at bases; corolla, flattened and a little curved, has 5 lobes, each ca. 2 cm long; pubescence and brown-red lines in throat. Long, brown, woolly, legumelike seed pods to 30 cm, dehiscent, with winged seeds.

Distribution: Central America and South America. In Costa Rica, dry to moist forest along Pacific slope to 1,000 m; more common below 500 m.

Comments: About four days after a rain shower (or a sudden drop in temperature) during the dry season, whole populations of this yellow *Tabebuia* burst into flower. During this dazzling display, which lasts about four days, the trees are abuzz with pollinating bees—mainly *Centris* and *Xylocopa* (carpenter) bees. *Pterocarpus* often flowers at the same time (photo below right). Tabebuia wood is heavy, hard, and long-lasting. Less common Costa Rican yellow-flowered tabebuias occurring in wetter forests are *T. chrysantha*, which has less pubescence, and *T. guayacan*, with a little pubescence on the underside of the leaflets where the

Tabebuia ochracea

Pterocarpus michelianus
Fabaceae (subfamily: Papilionoideae)

Tabebuia ochracea flowers.

side veins connect with the main veins. The latter has protuberances on the fruit, and it often has 7 leaflets instead of 5. There are three pink- to purple-flowered species in Costa Rica (see *T. rosea*, p. 22). There are about 100 species in the neotropics, many of them planted as ornamentals, including the yellow poui (*T. serratifolia*). *Tabebuia ochracea* is sometimes called corteza amarilla or corteza. *T. neochrysantha* is a synonym for this tree.

CACTI IN COSTA RICA

Acanthocereus tetragonus

Acanthocereus tetragonus flower.

About 1,500 species of cactus can be found in the Americas; only a few members of the Cactaceae family occur in the Old World. Cactus species range from British Colombia in Canada to Argentina, and can be seen from sea level to 5,000 m elevation. Typical arid-land cacti are easily recognized. The succulent, green, often-thick stems have spines, which are actually specialized leaves, and a waxy surface. These adaptations help the plant to store water and nutrients. The pleats seen on the stems of many species allow the stem to expand when water is abundant, and the spines and bristles condense water from the atmosphere. Cactus flowers are borne singly; they are often conspicuous due to their size, form, and/or color.

Stenocereus aragonii

Costa Rica is home to about 40 cactus species. Interestingly, the majority of these are epiphytic, and most of them grow in wet forests. Many Costa Rican cacti bloom nocturnally. Their scented flowers, which are principally white, are pollinated by hawkmoths or bats. Other species, with flowers of various colors, bloom during the day and attract hummingbirds or insects. The cactus fruits, which are fleshy with many seeds, are eaten by birds and mammals.

On the Pacific slope of Costa Rica, in the dry areas of Guanacaste, two terrestrial species stand out: *Acanthocereus tetragonus* (formerly *A. pentagonus*), which can be found from the southern United States to northern South America, and *Stenocereus aragonii*, an endemic (photos opposite page). They both have spines grouped in bunches, however the stem of the former commonly arches, while the latter is an erect, columnar species that grows to more than 6 m. Additional characteristics also distinguish the two: *A. tetragonus* has triangular to 5-angled stems to 15 cm in diameter; when a stem arches, it roots where its tip hits the ground. The flowers are as long as 20 cm, and it has edible red fruit. *S. aragonii* has 6–9 ribbed stems to 25 cm in diameter. Old plants are often candelabra-like, with a few stems arising from a tan-brown central trunk. It is more rounded on top than *Acanthocereus* and has distinctive whitish bands marking annual growth. *S. aragonii* has bat-pollinated flowers less than 10 cm long. It was formerly called *Lemaireocereus aragonii*.

Costa Rica's epiphytic cacti can be seen in both dry and wet forests. The epiphytic cactus with the most impressive flowers is a dry-forest, night-blooming cereus, called pitahaya (*Hylocereus costaricensis*), a species with branching triangular stems. In the wet season, the fragrant flowers open to a diameter of ca. 30 cm (photo right). The bright magenta fruits are edible, and pitahaya-flavored ice cream is sometimes sold at the Pops ice cream chain in Costa Rica. Species of *Epiphyllum*, which have flat outermost branches and flowers that can be more than 10 cm long, dangle from the canopy branches of large trees, including those of the cloud forest. A number of other genera resemble *Epiphyllum* but they bear flowers that are less than 10 cm long. Species of *Rhipsalis* have distinctive cylindrical or 4-angled, pendulous, stringy stems.

A number of other cactus species in Costa Rica, and elsewhere, produce edible fruit or pads, or provide other products. The red dye cochineal comes from a cactus (see *Nopalea cochenillifera*, p. 125), as does the narcotic peyote. Many cacti are grown as house plants—the Christmas cactus (*Schlumbergera* x *buckleyi*) for example—or as garden ornamentals.

M. & P. Fogden

Hylocereus costaricensis flower.

Hylocereus costaricensis

Curatella americana
Family: Dilleniaceae

Sandpaper tree, *Raspaguacal*

Curatella americana

Other common name: *Chumico de palo.*
Description: Small tree, 3–7 m, with twisted trunk; fissured, scaly, gray and brown bark; twigs flesh-orange beneath flaking gray bark, with conspicuous leaf scars. Pubescence on plant ranges from simple to stellate hairs. Simple, alternate leaves generally 10–20 cm long, with sand-papery texture, older ones yellowish-green with rusty spots; wavy edge sometimes with rounded teeth; venation prominent on underside. Clusters of cream-colored flowers, 1–1.5 cm in diameter, with 5 petals and many stamens. Branched clusters of dehiscent capsular fruit, 1–1.5 cm across, with two main compartments, each with 2 dark brown to black seeds with thin white aril.

Distribution: Mexico to South America, and Cuba. In Costa Rica, most common in northwest (e.g., Guanacaste), in dry, lowland savannas; also south to foothills near Carara (ca. 300 m) and in pockets even further south.

Comments: This small tree is common in the open areas of Guanacaste, along with nance (*Byrsonima crassifolia*) and, often, calabash (*Crescentia cujete* and *C. alata*). It survives fires, which actually stimulate its seeds to germinate. The seeds are bird-dispersed. Silica in the leaves make them useful for sanding wood or scrubbing pots and pans, and tannins in bark can be used for curing hides. The hard wood is good for posts, firewood, and charcoal. Some Mexicans add the toasted edible seeds to chocolate as a flavoring. Medicinal uses include a decoction of buds taken for asthma and smokers cough, and a bark or leaf decoction for bathing wounds. One study of the bark noted some antibiotic activity. This tree is called chaparro in parts of its range. Many of the species in this family (Dilleniaceae) are lianas (e.g., *Doliocarpus* species).

x 1½

Developing fruit of *Curatella americana.*

Hymenaea courbaril
Family: Fabaceae
Subfamily: Caesalpinioideae

Stinking toe, *Guapinol*

William A. Haber

Hymenaea courbaril fruits.

Other common name: West Indian locust.

Description: Tree to more than 35 m, with dense crown; trunk fairly smooth, grayish, without buttress; base of trunk may have gummy resin. Alternate, compound leaves with only 2 leaflets; leaflets asymmetrical, somewhat thick and lustrous above; leaf including petiole to 10 cm long; with translucent glandular dots. Clusters of fragrant, white (or purplish) flowers at ends of branches; flowers with 5 petals to 2 cm long, with long stamens. Fruits ca. 12 cm long and 4 cm thick; indehiscent, hard, woody, reddish-brown pods that stick out in various directions from branches and are borne on stout stems; seeds in powdery pulp that has a bad odor but is sweet and edible. Usually few seeds per fruit.

Distribution: Antilles and Mexico, to central South America. In Costa Rica, common along Pacific slope to ca. 800 m, in moist to wet forest, or along streams in dry forest; common at Manuel Antonio and Isla del Caño.

Comments: Stinking toe, or guapinol, lives in moist pockets of the dry lowlands, where it stays green for most of the year. The tree's flowers provide nectar for bats (*Glossophaga*). Some seeds do not mature because weevil larvae commonly feed on them. The seeds may originally have been dispersed by Pleistocene megafauna. Today, fruits on the ground are broken open by animals, or they rot in rainy season. Agoutis feed on the pulp and seeds, hording and dispersing some of them. Seeds that do germinate are vulnerable to rodents, which seek out the tender leaves. *Trigona fulviventris*, a species of stingless bee, collects resin from the trunk for nest-building material. A study in

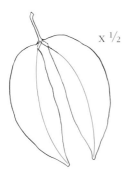

x ¹/₂

Hymenaea courbaril leaf.

South America showed that germination of seeds from broken, fallen pods increases when a type of fungus-growing ant (*Mycocepurus goeldii*) removes the pulp surrounding the seeds; seeds that are not cleaned usually become covered with fungus, and the seed-ger-

mination rate is low (Oliveira et al. 1995). Cleaned seeds last longer on the ground and are thus more available to animals that collect and disperse fallen seeds. The fruits are sometimes carried by ocean currents to faraway places; one that reached Martha's Vineyard had viable seed and was cultivated as a bonsai tree. The orange to yellowish resin that exudes from the tree is used for varnish and incense; amber that is found in the neotropics is usually fossilized resin of *Hymenaea*. The strong, hard wood has many uses, including firewood, posts, tool handles, furniture, and general carpentry. The bark, resin, and fruits have various uses in folk medicine—for gastrointestinal ailments, to get rid of worms, and for asthma, rheumatism, and hypertension. Researchers found that a leaf-bark extract lowers blood sugar in mice.

Parkinsonia aculeata
Family: Fabaceae
Subfamily: Caesalpinioideae

Jerusalem thorn, *Palo verde*

Other common names: Palo verde, *retamo*.
Description: Spiny shrub or tree to 10 m with open habit; trunks of young trees and branchlets green. Alternate, twice-compound leaves with the 1–3 pair of pinnae joined by a short spine-like rachis; each pinna has a flat, green rachis ca. 30 cm long with to 50 pairs of leaflets, less than 1 cm long, that often fall off. Clusters of fragrant, 5-petaled, yellow flowers ca. 2 cm in diameter, sometimes with red speckling. Legume pods, to 15 cm, with constrictions between the 1-cm-long seeds.
Distribution: Exact origin unknown, but thought to be in tropical America, where seemingly natural populations appear here and there. Now in cultiva-

tion from southern United States to Argentina, and Old World tropics, sometimes escaping. In Costa Rica, dry, lowland areas, or planted; appears native to Guanacaste (e.g., Palo Verde National

Compound leaf of *Parkinsonia aculeata*.

x ¹/₆

Park), where it withstands inundations as well as dry spells.

Comments: Early Costa Rican botanists did not believe this species was native, but naturalized. Pittier thought that it may have come from Spain, while Standley believed it to be Neotropical, but not native to Central America. Even when palo verde looses its leaflets, it still photosynthesizes by means of its green trunk and its rachises (the stems on which the leaflets are attached). Carpenter bee (*Xylocopa*) females visit flowers for pollen and possibly nectar. A paper wasp (*Polybia occidentalis*) regularly builds its nest in this tree, as well as in other thorny species of the lowlands of northwest Costa Rica. Since this is an open-growing tree with little foliage, it is planted more for ornament than for shade. It can survive drought and somewhat salty conditions. It is sometimes used for firewood, charcoal, posts, and basic construction. In northern Australia, palo verde is an invasive pest covering thousands of hectares. *Cercidium* is a closely related genus with which *Parkinsonia* is sometimes lumped (or visa versa). *P. florida* is a native of the southwestern United States.

Parkinsonia aculeata

Acacia collinsii
Family: Fabaceae
Subfamily: Mimosoideae

Ant acacia, *Cornizuelo*

Other common name: Bull-horn acacia.
Description: Small tree usually ca. 3 m tall, with tan-brown to copper thorns, ca. 4 cm long, in form of horns, straight or twisted. Leaves with 2–10 pairs of pinnae, with 13–20 pairs of narrow leaflets per pinna; each leaflet ca. 1 cm long; 2 or more glands on base of petiole. Stubby compact spike, 2–4 cm long, with minute bright yellow flowers. Two-parted, curved, 4- to 5-cm-long woody, brown pod; dehiscent, exposing 2.5-cm-long sticky yellow aril (looks like a caterpillar) containing dark brown seeds, 0.5 cm long.
Distribution: Mexico south to Colombia. In Costa Rica, on Pacific slope, in dry areas, sea level–1,000 m; common in Guanacaste region; formerly grew in Santa Ana area, Central Valley.

Adrian Forsyth

Acacia collinsii

Ants collecting Beltian bodies.

Comments: The ant acacia provides a classic example of ant-plant mutualism. Distinct bare areas, kept clean by the ants, surround the plants, protecting them from climbing vines and competitors. When a potential herbivore brushes up against a tree, it is greeted by a frenzy of stinging ants emanating from thorns at the leaf bases; it wisely chooses to eat a different plant. In exchange for these protection services, and in addition to their conical thorn shelters, the ants receive both sugary secretions from extrafloral nectaries on leaf stalks and protein and lipids via Beltian bodies at the tips of the leaflets. These food bodies are named after Thomas Belt, who, in *The Naturalist in Nicaragua* (1874), first described the relationship between ants and acacias. Almost a century later, in the 1960s, Daniel Janzen carried out in-depth studies of this ant-plant system. He determined that various species of

Pseudomyrmex ants may take up residence, but only one species of ant per tree. The queen makes an entry hole in a green thorn and creates a cavity where she lays eggs, starting a colony that will expand over the whole tree and beyond. Although most animals stay away from ant acacias, a few bird species are able to nest in the trees; and ant attacks do not phase the well-armored ant-acacia beetle (*Pelidnota punctulata*), which feeds exclusively on acacia. When the pod splits open, the aril around the seeds attracts birds, and possibly bats, both of which may disperse the seeds. Another Costa Rican bull-horn acacia is *A. cornigera*, which has a single oblong gland on the base of its petiole and an elongated tip on its fruit. Recent DNA studies define *Acacia* as an Old World genus; *Acacia collinsii* and some other New World species are being placed in the genus *Vachellia*.

Albizia saman

Family: Fabaceae
Subfamily: Mimosoideae

Rain tree, *Cenízaro*

Other common names: Saman, *genízaro.*

Description: Tree to 30 m, with nicely formed crown; deciduous in dry forest; dark, rough, fissured bark. Bipinnately compound leaves, 2–6 pairs pinnae, each with 2–8 pairs of leaflets ca. 3 cm long; leaflets asymmetric and sometimes squared-off at base, smooth above, lighter and pubescent beneath; nectaries between some of the pinnae. Brushlike flower heads ca. 5 cm in diameter; each flower has a small calyx (ca. 6 mm) and corolla (ca. 1 cm); showy stamens are longer, white at base and magenta-pink above. Indehiscent seed pods to 20 cm long by 2 cm, somewhat flattened, straight or curved, with sticky pulp around brown seeds. Seeds, ca. 1 cm, look somewhat like those of the Guanacaste tree (*Enterolobium cyclocarpum*, p. 50), but smaller and squared off on edges.

Distribution: Mexico to South America; grown as ornamental in other tropical regions, sometimes naturalizing. In Costa Rica, most common in the Pacific lowlands, from the middle of the country and north; dry to moist forest, to ca. 1,000 m; fairly abundant, but could become rarer soon due to exploitation of wood; occasional on Atlantic slope.

Comments: This large tree is most noticeable when it starts flowering, usually in the latter part of the dry season. One of the typical large, rounded-crown species seen in Guanacaste pastures, along with the Guanacaste tree and

Albizia saman

some figs, it has very long, horizontal branches and dark, fissured/checkered bark that cacti, orchids, and other epiphytes cling to. Hawkmoths take nectar from, and pollinate, the flowers at night. The fruits fall to the ground toward the end of the dry season. The original seed dispersers may have been Pleistocene mammals; today tapir may eat the fruit and pass some seeds, but many are eaten by rodents, and some never develop on the tree due to insect damage. A large percentage of mature seeds

WHERE DOES CHAN COME FROM?

Chan is a fruit drink found on many Costa Rican menus. Although this refreshing thirst-quencher has the consistency and appearance of toad eggs floating in dish water, contrary to common gringo belief its source is vegetable rather than animal. Small seeds of a type of mint plant, *Hyptis suaveolens*, are soaked in water; and, with time, the hard little black seeds soften and a slimy coating forms around each one. The drink is then ready to serve.

H. suaveolens plants, which may grow to more than 2 m tall, have square, hairy stems and short, tubular blue to lilac flowers. These aromatic plants often occur as weeds in various parts of Costa Rica below 1,200 m, but they are especially abundant in old fields in Guanacaste. While it is native to the neotropics, it has been introduced into other tropical regions, where it is sometimes a noxious weed, especially in Australia and Hawaii. The plant has various medicinal uses; dried and/or smoldering, it is purported to be a mosquito repellent.

Chan drink. Insert photo shows *Hyptis suaveolens*.

Couroupita nicaraguarensis
Family: Lecythidaceae

Cannonball tree

Other common names: *Zapote de concha, zapote de mico, bala de cañón.*
Description: Large tree to 50 m, without buttresses; trunk fissured with black and gray flat-topped ridges; large, conspicuous leaf scars. Leaves to 30 cm long, widest toward tip, wedge-shaped at base, grouped at ends of twigs; entire to slightly crenate, with some tiny glandular teeth where veins meet the leaf margin (especially noticeable in seedlings). Long inflorescences, sometimes branched, coming off of trunk and branches; flowers to 8 cm in diameter, yellow-cream with hundreds of yellow anthers in center ring and continuing along flap that curves over it; 6 petals, turning dark blue where bruised. Round brown fruit, 17 cm in diameter; many 1.5+-cm seeds in pulp that turns blue-green upon exposure to air.
Distribution: In certain parts of several Central American countries (El Salvador, Nicaragua, Costa Rica, and Panama). Rare in Costa Rica; at Lomas Barbudal and Palo Verde in the Pacific lowlands; some collections made in northern Caribbean slope region.
Comments: By no means a common tree but definitely a conspicuous one, this is a spectacular species that produces large, odoriferous flowers and cannonball-size fruits. The flower scent is reminiscent of freshly-baked, heavy chocolate cake. As with many Brazil nut family members, the center of the flower has an odd, bunlike structure where the stamens occur. Studies of a relative, *C. guianensis,* have shown that anthers in the hood provide a sterile fodder pollen as a reward to visiting bees, which get their backs dusted with fertile pollen from the anthers in the center ring of the flower. The seeds are probably dispersed by large mammals, or perhaps by floating in streams and rivers. *C. guianen-*

Couroupita nicaraguarensis

Fruits and flowers on forest floor.

sis, which has red and pink flowers and occurs in Panama and South America, is more commonly cultivated than *C. nicaraguarensis.* While *C. nicaraguarensis* has a continuous mass of stamens running from the ring in the flower center to the tip of the hood, in *C. guianensis* a section of the flap between the ring and the hood lacks stamens. Monkey pot tree (*Lecythis ampla,* p. 242) and Brazil nut (*Bertholletia excelsa*) are in this family.

Brosimum alicastrum
Family: Moraceae

Breadnut, *Ojoche*

Brosimum alicastrum

Other common name: *Ramón.*
Description: Buttressed tree to 40 m, gray to brown bark smooth or flaking in large square pieces; white latex; evergreen. Alternate, entire, smooth (very little if any pubescence) leaves to ca. 18 cm long, with pointed tip and 10–20 pairs of side veins connected by loop vein near leaf margin. Paired stipules at tips of branches that cover leaf buds eventually fall off, leaving scars that almost encircle the twig. Tiny unisexual flowers clustered in heads (less than 1 cm) that develop in leaf axils. The 2-cm, green to yellowish, round fruit has one large seed.
Distribution: Mexico and Central America, and some Caribbean islands. In Costa Rica, Pacific slope from Guanacaste to Osa Peninsula, to ca. 1,000 m, often near streams; some also

found on Atlantic slope. Grows in primary forest as canopy tree.
Comments: A common name often used for species of *Brosimum* is cow tree, since many produce a milky latex that is drinkable. *B. alicastrum* also produces edible seeds that are high in amino acids, so its common name breadnut is apt. The seeds may be boiled or roasted, eaten as is, or ground up to make tortillas. The Maya in Guatemala apparently cultivated *B. alicastrum*, and it was crucial to their survival, especially in years when the corn-crop yield was poor. Earlier in this century, Costa Rican country people also collected the fruits when there was a harsh dry season and the basic food crops of beans, corn, and rice failed. The flowers are pollinated by a variety of small insects. The fruit is eaten by pacas, agoutis, monkeys, and bats, with the latter probably being the main seed dispersers. The latex has been used medicinally for stomach disorders and in diluted form for asthma attacks. Livestock eat both foliage and fruit. The tree makes good firewood even when green; the wood is also used for furniture, tool handles, and floors. Six other species of *Brosimum* occur in Costa Rica; *B. utile,* on Caño Island, was probably planted.

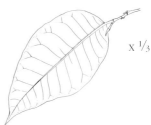

X ⅓

Brosimum alicastrum leaf.

Hyparrhenia rufa
Family: Poaceae

Jaragua

Description: To 2 m tall, in clumps. The long, sheathing green (sometimes reddish) leaf bases alternate with sections of smooth, yellow-tan stem, giving the plant a banded appearance; swellings above joints on stem. Leaf blades, green suffused with wine red, to 70 cm long. Inflorescence to 50 cm; long copper bristles on spikelets; color of fields appears greenish-gold to copper.
Distribution: Origin Africa, but naturalized in neotropics. In Costa Rica, pasture and roadsides to ca. 1,400 m; Pacific slope from Guanacaste south to Osa Peninsula.
Comments: This introduced forage species has taken over many open areas in Guanacaste and has been a bane to habitat restoration there. Jaragua is abundant on the upper plateau of Santa Rosa National Park in areas that were most likely evergreen oak (*Quercus oleoides*, p. 272) forest in the past. It does well where fields are burned annually, sprouting up after the fire; it covers the fields densely, provides forage for livestock, and prevents native shrub and tree species from moving in. The fires are very hot and kill most of the native species in the field. In habitat restoration projects, once annual burning ceases, some shrubs and trees begin to grow, and resulting shade helps diminish jaragua growth. An interesting result of a project in Guanacaste that involved composting of peels from an

Hyparrhenia rufa

orange juice factory was reduction of jaragua and subsequent growth of other plants (Jiménez 1998). There is hope of creating agroforestry systems in actively grazed fields of jaragua—experimental plantings of cenízaro (*Albizia saman*, p. 271) seedlings in cow pats and plain soil in such pastures showed that the seedlings fared better in the cow pats, which lent some protection to the seedlings while at the same time providing fertilizer (Barrios et al. 1999).

Calycophyllum candidissimum
Family: Rubiaceae

Lancewood, *Madroño*

Other common name: *Sálamo.*
Description: Tree to 25 m; musclelike trunk with peeling pink/gray bark flaking off in longitudinal pieces exposing red/orange beneath; branchlets reddish brown. Leaves simple, opposite, entire; blade ca. 10 cm long with wedge-shaped base and pointed tip. Clusters of

Calycophyllum candidissimum

sweet-perfumed white flowers ca. 1 cm across, with 4 reflexed petals; calyx lacks typical lobes, but may have single, expanded, 3- to 4-cm-long, green- cream, petal-like structure (showy part of the floral display) with 1.5-cm-long stalk. Capsular fruit to 1 cm, with winged seeds.

Distribution: West Indies and Mexico to northern South America. In Costa Rica, below 900 m elevation in Guanacaste, and pockets on southern Pacific slope (e.g., Buenos Aires).

Comments: The attractive trunks of these trees stand out in the lowland dry forest, as do their white-topped crowns during the flowering periods of October–January and May–June. Also known as lemonwood, the hard, durable wood is suitable for archery bows and tool handles, as well as firewood and charcoal. It is also used for turned articles and furniture. Medicinal uses include a tea of flowers taken for diarrhea and a bark decoction for stomach ulcers. Studies have found that compounds in the bark depress the central nervous system. The national tree of Nicaragua, it is occasionally planted as an ornamental. The flowers attract insects.

Manilkara chicle
Family: Sapotaceae

Chicle, *Chicle*

Other common name: *Níspero.*
Description: Tree to 30 m. Alternate, simple leaves to more than 20 cm long, gray-green on top and yellowish green beneath; fine veins perpendicular to the midvein. Small greenish white flower with 6 corolla lobes; flowers in bunches of 2–5. Light brown fruit to 4 cm long, usually with 2–5 seeds, each just under 2 cm long, with a scar that extends less than half the length of the seed.
Distribution: Southern Mexico to Colombia. In Costa Rica,

Manilkara chicle. Note slashes for extracting latex.

William A. Haber

moist to dry forest on the Pacific slope, 0–900 m.

Comments: Hundreds of years ago, Mayan and Aztec people had a chewng gum quite different from the vinyl resins that we chew today. Their gum came from a sticky latex extracted from the trunks of species of *Manilkara*, or chicle, trees. *Manilkara chicle* grows wild in Costa Rica and was tapped for its latex. Another species, *Manilkara zapota*, which has superior latex and is the better known source of natural chewing gum, is less common in Costa Rica. The latter has single vs. bunches of flowers in the leaf axils and seeds that have a scar that extends 3/4 or more along the length of the seed instead of ca. 1/2 the length as in *M. chicle*. The typical method of extracting the latex is by cutting diagonal slashes that intersect a vertical channel that carries the latex down the trunk to a container. Later the latex is boiled down and made into blocks. A variety of mammals seek out the fruit, which is also edible to humans.

Quassia amara
Family: Simaroubaceae

Bitterwood, *Hombre grande*

Description: Shrub or treelet to 6 m; bitter bark. Alternate, pinnately compound leaves with winged rachis, 5 leaflets, ea. 5–15+ cm long. Pendulous, pink-stemmed inflorescence, 25 cm long; a 3- to 4-cm-long tube formed by flower's petals, white inside, pink-red outside; protruding stamens. Up to 5, 1-seeded, blackish fruits, 1.5 cm oblong, on orange-red receptacle.

Distribution: Mexico to Brazil, and West Indies. Cultivated in many areas; precise origin unclear; sometimes naturalized in tropical areas of Old World. In Costa Rica, often in moister sites along the Pacific slope, and occasionally on the Caribbean slope; to ca. 700 m.

Comments: Over the centuries, indigenous people and settlers of the tropics have been familiar with bitterwood's

Quassia amara flowers.

Quassia amara fruits.

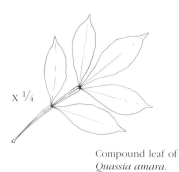

Compound leaf of
Quassia amara.

medicinal and insecticidal qualities. In Costa Rica, it is widely known for wood and bark preparations called bitters or, in Spanish, *gotas amargas.* Decoctions or alcohol infusions are popular tonics and are folk remedies for diabetes, diarrhea, internal parasites, various gastrointestinal complaints, and fever. In a controlled Costa Rican study, bark extract given to rodents before induction of gastric ulcers provided a protective effect on gastric tissues (Badilla et al. 1998). Other studies have discovered antimalarial and antileukemic properties. Pesticides have been made from bitterwood for use against insects and nematodes. The bitter compounds include alkaloids such as quassin and quassimarin. Due to its applicability as a natural pesticide and its use as a medicine, various studies (many through CATIE) are looking at the potential this species has as a nontimber forest product or crop. It can be grown from cuttings or seeds. The flowers are hummingbird-pollinated, and the seeds are bird-dispersed.

Bonellia nervosa *Barbasco*
Family: Theophrastaceae

Other common names: *Siempreviva, siempre verde, burriquita.*
Description: Shrub or treelet to 5 m tall. Alternate, simple leaves, to 8 cm long, narrowing toward base, spine at tip. Orange-red, 5-parted flower, 1 cm in diameter, has 5 staminodes that look like corolla lobes; pungent scent like perfumed soap. Leathery, orange 2.5-cm fruit with ca. 10 seeds.
Distribution: Mexico to Panama, dry hot lowlands. In Costa Rica, at various locations along the northwest Pacific slope, including Santa Rosa, Lomas Barbudal, and Palo Verde; below 200 m.
Comments: When the trees of the deciduous forest of northwestern Costa Rica lose their leaves in the dry season, *Bonellia* gathers moisture through its long taproot, leafs out, and takes advantage of the sunlight that reaches the understory; it then flowers. The leaves drop in the wet season. Even though it may be toxic, indigenous groups of western Mexico use the plant externally and internally for various ailments. Fishermen use the roots and fruits to make a fish poison; it is probably the presence of saponins and alkaloids that makes it effective. Pollinators are not known, but the pungent scent may attract some sort of insect. This species, the only member of its genus in Costa Rica, was formerly known as *Jacquinia nervosa.*

William A. Haber

Bonellia nervosa

Tropical Montane Cloud Forests

In general, the Atlantic slope of Costa Rica is hot and wet, with rain common on the plains and in the foothills. At higher elevations, along both sides of the Continental Divide, clouds, pushed by the northeast trade winds, envelop the forests. As these clouds move further west, they dissipate when they mix with the hot, dry air of the Pacific slope and a much drier rain shadow environment exists. Precipitation in the mountains, which generally ranges from 2.5–4 m annually, frequently occurs as wind-blown mist during the dry season. Temperatures are cool to cold. It is in this sort of setting that cloud forests flourish in Costa Rica. Featured in this section are Monteverde (p. 282), Poás (p. 306), and Cerro de la Muerte (p. 316).

Cloud forest, Cerro de la Muerte.

Monteverde

While there are pockets of cloud forest scattered throughout the mountains of Costa Rica, one of the best examples is in Monteverde, in the Tilarán mountain range. A profusion of epiphytes—plants that grow on tree trunks and branches—thrives here at elevations between 1,500 and 1,800 meters. In addition, the topography is such that there are a number of life zones compressed into a relatively small area, which adds to the diversity of flora and fauna. The elfin forest, on the windward (eastern) side of the divide, is made up of short, gnarly trees with dense wood. Several private forest preserves in the area, the principal ones being the Bosque Eterno de los Niños and the Monteverde Cloud Forest Preserve, make it possible to experience contrasting forest types, from the very wet to the highly seasonal moist, within a short distance.

Razisea spicata
Family: Acanthaceae

Pavón de montaña

Other common name: *Pavoncillo rojo.*
Description: Shrub to 2+ m; stem swelling above nodes. Opposite leaves with pointed tips; leaf blade and petiole together measure 25+ cm. Sometimes dull, dark purple on lower half of veins below; entire, or with irregular rounded teeth on leaf edge. Inflorescence to ca. 30 cm long; 5-parted calyx; tiny bracts associated with red, tubular flowers, ca. 5 cm long, that are 2-lipped and very narrow at base; 2 anthers and threadlike style pro-

Razisea spicata

truding. Two-parted, club-shaped, capsular fruit that ejects seeds when mature.

Distribution: Nicaragua to Peru. In Costa Rica, widespread, from lowlands to 2,000 m.

Comments: *Razisea spicata* is visited by long-billed hummingbirds as well as the nectar-robbing stripe-tailed hummingbird, which pierces the base of the flower. This is one of about a half dozen showy members of the Acanthaceae family that bloom in the understory of the cloud forest from September through the dry season. *R. spicata* also occurs at lower elevations in areas on both slopes (e.g., Tortuguero, Osa Peninsula). Other forest relatives in Monteverde include *Stenostephanus blepharorhachis*, *Justicia aurea* (often seen in light gaps), and *Poikilacanthus macranthus* (photos, clockwise from right). Also see *Aphelandra scabra* (p. 74), sky vine (*Thunbergia grandiflora*, p. 111), and polka-dot plant (*Hypoestes phyllostachya*, p. 108).

Stenostephanus blepharorhachis

Poikilacanthus macranthus

Justicia aurea

Schefflera rodriguesiana
Family: Araliaceae

Schefflera, *Papayillo*

Schefflera rodriguesiana

Other common name: Umbrella tree.

Description: Hemiepiphyte or tree to 12 m tall; taller in regions with less wind. Compound leaves; leaf stems long and variable in length, often red-tinged, angled upward, with 6–10 glossy leaflets spreading out from a central point; leaflets—to 20 cm, including the leaflet stalk—have wavy margins with tapering tips; 2-cm-long spur (ligule) on sheathing petiole. Thousands of tiny 5-petaled flowers, in clusters on long inflorescence branches. Small purplish fruits.

Distribution: Costa Rica and Panama. In Costa Rica, in wet forest at medium to high elevations; grows at Volcán Poás and Cerro de la Muerte, in human-made and natural disturbances.

Comments: This species is a major component of the wind-sculpted, elfin

Schefflera robusta

Oreopanax xalapensis

William A. Haber

forest on the continental divide. Schefflera is found growing on fallen logs in light gaps; it sometimes starts as an epiphyte and falls with a branch. The hemiepiphyte form perches in the tops of leeward cloud forest trees, which are taller than those of the elfin forest. The long-stalked leaflets shiver in the wind like the leaves of quaking aspen. This schefflera flowers mostly in rainy season. Small and large frugivorous birds eat the fruits. Twelve other schefflera species occur in Costa Rica. These include *S. robusta* (photo opposite page), which may also be a hemiepiphyte, and has longer ligules (ca. 9 cm) and leaflets that are more lanceolate. Another is *S. morototoni*, which has a rusty underside to the leaves and is a common tree along stretches of the Inter-American Highway south of San Isidro de El General. It also occurs in north-eastern Costa Rica. Ornamental scheffleras (including some Old World species), common ivies (*Hedera* spp.) that cover buildings, and ginseng (*Panax* spp.) are in the Araliaceae family, as are various other native hemiepiphytes and trees (*Oreopanax xalapensis*, photo opposite page). *S. rodriguesiana* was formerly known as *Didymopanax pittieri*.

Neomirandea angularis
Family: Asteraceae

Tora hueca

Description: Small tree, 4–8 m tall. Pubescent leaves, opposite with blades to 40 cm long, with many lobes and toothed margin; petiole to 25 cm. Many-branched inflorescence usually more than 30 cm across; lavender florets, in heads of up to 12. Black seed with tuft of white hairs.

Neomirandea angularis

Willow Zuchowski

Senecio sp.

Distribution: Only found in Costa Rica; in wet mountains along both slopes, from ca. 500 m to above 2,000 m.

Comments: This is a very showy tree in the rainy season that can be found on the forest edge in the Monteverde cloud forest and near the Santa Elena reserve. It generally flowers in the wet season. It is visited by bees and butterflies. It has quite a few epiphytic or hemiepiphytic cousins in the cloud forest canopy that appear as patches of pink-purple. The genus *Neomirandea* is related to *Ageratum* (p. 82), *Eupatorium*, and *Senecio* (photo left).

Langsdorffia hypogaea
Family: Balanophoraceae

Description: Parasite on roots of trees; lacks chlorophyll; has no leaves or roots— just an underground, branching structure referred to as a tuber, even though not technically one, covering more than 1 m square. Above-ground inflorescence is a brown conelike structure ca. 6 by 6 cm; monoecious or dioecious; brown scales triangular, ca. 3 cm long; tiny male flowers green-yellow with golden petals, cream anthers; female flowers more reduced; flowers from bottom up. Tiny, one-seeded fruits.

Distribution: Mexico to parts of South America. In Costa Rica, 1,200–1,600 m; Tilarán mountain range and south; in Monteverde, seen in Bajo del Tigre as well as swamp trail in Monteverde Cloud Forest Preserve.

Comments: Little is known about the natural history of this intriguing parasite. Studies have shown that its method of attachment to the host tree roots is rather unusual. The *Langsdorffia* "tuber" contains cells of both the host and the parasite. It is sometimes difficult to trace host roots to their source, but probable hosts of *Langsdorffia* include species of *Quercus*, *Ficus*, *Sapium*, and members of the palm family. The waxy substance secreted by the female flowers readily ignites. The inflorescence gives off a spicy to cabbagelike odor. The genera *Corynaea* and *Helosis* are also in this family.

Langsdorffia hypogaea

Begonia involucrata
Family: Begoniaceae

Angel-wing begonia, *Begonia*

Begonia involucrata

Description: Plant with jointed stem, often less than 1 m tall. Lopsided, pubescent leaves, dark green above, lighter and glistening below; to 20 cm long, with heart-shaped base and 2 to 5 lobes with pointed tips; stems and petioles pubescent and rust-pink; triangular stipules clasping stem. Monoecious; bracts enclose developing inflorescence; fragrant flowers in clusters; 1-cm-long white to pale pink tepals in female flowers; tepals in male flowers a bit longer; male flowers with many bright yellow anthers; females have 3 yellow, branched styles joined at base. Dehiscent, 3-parted fruit capsule with longest of 3 wings ca. 1.5 cm long; capsule brown and papery when mature, with many minute seeds.

Distribution: Panama and Costa Rica, possibly farther north. In Costa Rica, 1,000 to above 2,500 m, in all the major mountain ranges; stream- and trail-side, plus forest edge, primary and secondary forest.

Comments: Begonias are well-known as house and garden plants. In the Monteverde area, the main pollinator of this begonia is a stingless bee, *Partamona grandipennis*, which visits to collect the pollen that is produced in the anthers of male flowers. Female flowers, which have branching styles that mimic the male anthers, also attract bees, although the visits are fewer and shorter than at male flowers. This system is called pollination by deceit because the bee is tricked into visiting the female flowers, which offer no reward. *B. estrellensis*, a beautiful, epiphytic begonia that cascades down from tree limbs, and *B. cooperi* (illus. right), a shrub, are also found in the cloud forest. There are 900 begonia species in the world, and ca. 10,000 cultivars; 32 species occur in Costa Rica.

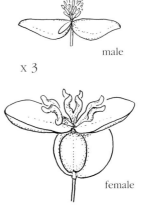

x 3

male

female

Typical *Begonia* male and female flowers; illustrated here in *B. cooperi*, a shrub.

EPIPHYTES

The tree trunks and branches of cloud-forest trees are festooned with carpets of mosses, ferns, orchids, aroids, bromeliads, and many other plants. The term *epiphytic* describes this growth form, in which plants grow on branches and tree trunks, and thrive without having roots firmly implanted in the ground. Contrary to a popular misconception, epiphytes are not parasitic, with the exception of some mistletoe species. In Monteverde's cloud forest, living and decaying epiphytic matter accounts for approximately 45% of the vegetative biomass. Plants that live in the canopy have evolved a variety of ways to get the moisture, nutrients, and sunlight they need for growth without having to develop a trunk. Many do not survive for more than a year if they drop to the forest floor.

Over time, live and decaying epiphytes and other organic debris accumulate to form a thick, spongy, water-absorbing carpet on tree branches. This organic carpet sometimes becomes so heavy it causes a branch to fall. Even so, some trees benefit from epiphytes by putting out nutrient-seeking roots from their branches directly into these carpets.

If you cannot climb into the canopy, the next best way to get a close look at epiphytes is to look for fallen branches. You can also view some of the larger species by using binoculars.

In Monteverde's cloud forest, the orchid family (Orchidaceae) comprises the largest number of epiphytic species (see especially miniature orchids, p. 425). Species in the blueberry family (Ericaceae) and the African violet family (Gesneriaceae) are also common. The blueberries sometimes occur as large shrubs; their reddish new leaves, along with colorful bracts of flowers, make them stand out in the canopy.

The photographs here depict examples of some of these common epiphyte groups. Also see the bromeliad, orchid, and fern sections of Chapter 7, as well as *Columnea lepidocaulis* (p. 294) and *Drymonia rubra* (p. 296).

Satyria meiantha
Ericaceae

Peperomia peltilimba
Piperaceae

Epiphyte-laden tree; cut log covered with epiphytic mat (photo insert).

William A. Haber

Guzmania nicaraguensis
Bromeliaceae

Acronia sanchoi (=*Pleurothallis sanchoi*), a miniature orchid.
Orchidaceae

Centropogon solanifolius
Family: Campanulaceae

Centropogon, *Gallito*

Other common name: *Pavoncillo.*

Description: Herb ca. 1 m tall, terrestrial, occasionally epiphytic; with latex. Toothed leaves to ca. 15 cm long, including petiole. Flowers clustered at top of stem; curved orange to reddish flower tube, ca. 4 cm long, bulging below yellow-green tip and at base, where it is enclosed by the pleated, green calyx with 5 toothlike lobes; a tube, formed by anther stalks that are joined, encloses the style. Fleshy, indehiscent fruit.

Centropogon solanifolius

Distribution: Costa Rica to Ecuador and Peru, but absent from Panama. In Costa Rica, ca. 800–2,800 m; in mountains throughout the country.

Comments: This plant belongs to a genus of colorful, mostly hummingbird-pollinated plants. There are 16 species of *Centopogon* in Costa Rica. In the Monteverde area, *C. costaricae*, with red flowers, occurs near the Continental Divide.

Centropogon granulosus

On the Atlantic slope one finds the larger *C. granulosus* (photo, below right), with scarlet and orange flowers that are more curved than those of *C. solanifolius.* Hummingbird flower mites (*Rhinoseius colwelli*), which ride from plant to plant in the nostrils of hummers, are often found in centropogons. Tiny fly larvae feed on pollen while *C. solanifolius* is in the bud stage; this destruction of a flower's pollen shortens the initial male, pollen-presentation phase that the flower goes through before the female, pollen-receptive phase. The family Campanulaceae includes the genus *Campanula*, which is popular in horticulture, as well as *Lobelia laxiflora* (p. 85).

x 1

Developing fruit of *Centropogon solanifolius.*

Clusia spp.
Family: Clusiaceae

Some common names: Clusia, autograph tree, *azahar de monte, copey.*

Description: Shrubs or trees, commonly hemiepiphytic, sending roots to ground; latex milky to bright yellow. Leaves simple, opposite, entire, may be thick and rigid; many parallel lateral veins; terminal bud protected by concave petiole bases pressed together. Dioecious; 4–9 medium-sized white to pinkish petals, 4–5 sepals; male flowers with many stamens. Fruit a fleshy capsule that splits open, sometimes tardily, along 5 or more sutures; many seeds with yellow to orange arils.

Distribution: Genus ranges from Mexico and the Florida Keys to South America. In Costa Rica, various species found from sea level to ca. 3,000 m—about 10 species in the Monteverde region.

Comments: The stiff rounded leaves and the star-shaped open capsules of

Hemiepiphytic habit of *Clusia.*

Clusia stenophylla has bright yellow latex, fragrant white flowers, and orange arils.

x 1

Male flower of *Clusia stenophylla.*

hemiepiphytic clusia species are common on the trails in the cloud forest. Some clusias will grow as terrestrial trees or as stranglers. Birds disperse the seeds of *C. stenophylla.* Moths pollinate its flowers at night, with hummingbirds and various insects visiting by day. *C. rosea,* a lower-elevation species, is seen both wild and in gardens in Costa Rica. The latex of some species is used both in folk medicine and as a caulking material. One of the common names for *Clusia,* autograph tree, derives from the fact that letters etched on the surface of the leaf remain there permanently. There are ca. 28 species of *Clusia* in Costa Rica. Mangosteen (*Garcinia mangostana*), gamboge (*Garcinia xanthochymus* and others), and St. John's wort (*Hypericum perforatum*) are also in this family, which is sometimes called Guttiferae.

Cojoba costaricensis
Family: Fabaceae
Subfamily: Mimosoideae

Angel's hair, *Lorito*

Other common name: *Cabello de angel.*
Description: Tree to 15 m. Fernlike, twice compound leaves with 4–8 pairs of pinnae, each with 7–14 pairs of leaflets to 2 cm; nectar glands between pinnae. Small, white, brushlike flowers in round heads. Fruit a red pod to 15 cm long, with constrictions between seeds. Mature, open pods spiral, revealing shiny, black seeds.
Distribution: Costa Rica and Panama. In Costa Rica, found in all major mountain ranges, 1,150–2,000 m; in the Monteverde region, common on windswept ridges.
Comments: The vivid red-and-black display of the twisted legumes on this tree appears to be a way of luring in seed dispersers, most likely birds, although it does not get many visitors. The flowers are pollinated by moths. *C. costaricensis* grows fairly well, although slowly, on former pastureland. The wood is hard. The foliage, flowers, and fruit make this a decorative tree, and both it and *C. arborea,* of lower elevations (sea level–1,500 m) make good ornamentals. *C. arborea* differs by having more pinnae, more leaflets per pinna (20–40 pairs instead of 7–14), and somewhat larger fruits; it can be seen planted along the Inter-American Highway near Palmares.

Fruit and flowers of *Cojoba costaricensis.*

Mucuna sp.

Family: Fabaceae
Subfamily: Papilionoideae (= Faboideae)

Mucuna, *Ojo de buey*

Other common names: Ox eye, *ojo de venado.*

Description: Large, extensive, draping liana that climbs by spiraling around other plants. Compound leaves ca. 18 cm long by ca. 20 cm wide with 3 leaflets, each ca. 10 cm long by 6 cm wide; side leaflets asymmetrical; bases of leaf and leaflet stems swollen cylindrically. Greenish, pealike flowers, arranged in chandelier-like cluster on long stem. Fruit a large golden-brown pod, ca. 15 cm, with long tip and wrinkled or irregularly ridged appearance, covered with stiff, irritating hairs that come off easily; pods contain 3 or more large black seeds, each 2.5–4 cm.

Distribution: Nicaragua to South America. In Costa Rica, widespread, ca. 500–2,000 m.

Comments: This is the most common liana in the lower part of the Monteverde

Mucuna sp.

Richard K. LaVal

Bat (*Glossophaga commissarisi*) visiting flowers of *Mucuna* sp.

Cloud Forest Preserve; it also occurs in the community of Monteverde, proliferating on the forest edge. The pods that dangle from the canopy become quite conspicuous when their bristly coats catch a ray of sunlight. Mucuna flowers dust pollen on nectar-feeding bats (e.g., *Glossophaga* and *Hylonycteris* species) as they probe for nectar. Morpho butterflies lay their eggs on the leaves, and agoutis eat and bury the seeds. Species of *Mucuna* that grow near streams have buoyant seeds that float downstream to the ocean (see more information about *Mucuna*

species, p. 336). *Mucuna* is a genus of perhaps 100 species, with seven in Costa Rica. For many years, this cloud forest species, which has no formal scientific name, has been incorrectly called *Mucuna urens*. Numerous studies have been conducted on the constituents of cow-itch (*M. pruriens*), which is originally from Asia and is grown as a soil enricher. Its seeds contain L-dopa, the compound used to treat Parkinson's disease. Another name given to various mucunas, because of the irritating hairs on the pods, is *picapica* (Spanish for *itchitch*).

Columnea lepidocaulis — Goldfish plant
Family: Gesneriaceae

Columnea lepidocaulis

William A. Haber

Other common name: Columnea.

Description: Epiphyte on tree trunks and branches; gray, speckled stem with scales. Opposite leaves to 9 cm long; shiny, leathery to succulent, sometimes with reddish tinge beneath. Five-parted calyx green with tinge of maroon; flowers 8 cm long, borne singly; 5 fused petals form tube, red-orange above, yellow along underside; tip of lower lip is orange; 2 side lobes folded back and up; long silky

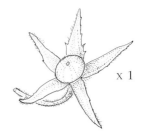
x 1

A *Columnea magnifica* fruit that is typical of the genus.

William A. Haber

Columnea glabra

hairs on upper side. Pubescent, white, spongy fruit, 1.5+ cm in diameter; many tiny seeds.

Distribution: Perhaps a Costa Rican endemic; mostly found in cloud forest, ca. 1,000–1,700 m, central mountains to Tilarán range (possibly farther north).

Comments: There are 17 species of *Columnea* in the Monteverde region, many with vivid flourishes of orange, red, or yellow when they bloom. Most species are hummingbird-pollinated. These plants are epiphytic, but they can survive as potted plants and thus are popular in greenhouses, where cultivars are given names such as 'Campus Sunset' and 'Bon Fire'. The flowers are relatively long-lived, going through staminate (male) and then pistillate (female) phases. *C. glabra* (photo above) and *C. magnifica* (photo right) also occur in the Monteverde region. Gloxinia (*Sinningia speciosa*) and African violet (*Saintpaulia ionantha*)

Columnea magnifica

are members of the Gesneriaceae. This family, which contains treelets, shrubs, and epiphytes, is one of the showiest in the cloud forest.

Drymonia rubra
Family: Gesneriaceae

Drymonia rubra

Drymonia conchocalyx habit.

Description: Branching, shrubby epiphyte. Opposite leaves to 15+ cm long, including petiole; one of the two at a node often larger than other. Flowers in leaf axils; light green, 5-parted calyx, to 5 cm long, with tapering tips; deep orange (to reddish) tubular flower, 5 cm long, with yellow and white throat and 5 fringed lobes; the 4 connected anthers have pores. Two-valved, light green to white, fleshy capsule to 2 cm long.

Distribution: Costa Rica and Panama. In Costa Rica, mountains on both slopes, ca. 700–1,700 m, occasionally lower.

Comments: *Drymonia rubra* is pollinated by long-billed hummingbirds such as the green hermit. The fruits are eaten by birds and bats. This epiphyte flowers during various months of the year and shares the cloud forest habitat with *Drymonia conchocalyx* (photos above), which has pink-purple and white flowers enclosed by a pinkish calyx. There are ca. 20 *Drymonia* species in Costa Rica, several of which

Drymonia conchocalyx flower.

are used in horticulture. The genus includes species with a variety of flower forms and colors corresponding to different groups of pollinators (bees, moths, birds, or bats).

THE "LITTLE AVOCADOS"

The common name aguacatillo, meaning little avocado, applies to many species of trees in the Lauraceae family. There are about 130 Lauraceae species in Costa Rica, and more than half of them, ca. 75, occur in the Monteverde region. The majority of trees in this family produce small to medium fruits that are critical to the survival of the resplendent quetzal, whose altitudinal movement is determined largely by which species of Lauraceae are fruiting where. Quetzals, toucans, three-wattled bellbirds, and black guans disperse the seeds.

x ¼

Ocotea tonduzii
Lauraceae

One of these avocados, *Ocotea tonduzii*, is a common tree in the lower cloud forest. It grows to 30 m and produces erect, 1-seeded, 1.5-cm-long black fruits set in red cupules. At least 18 species of birds seek out its fruits, and it serves as an important food for the resplendent quetzal during that bird's breeding season. There is some confusion regarding the taxonomy of *O. tonduzii*; it is sometimes lumped under the name *O. insularis*, which grows on Cocos Island.

Ocotea tenera, a small tree that grows to 8 m tall, most commonly occurs at elevations below the Monteverde Cloud Forest Preserve, and sometimes occurs in Atlantic lowland forest. The leaves have a strong tarlike odor, and domatia are present in or near vein axils on the leaf underside. This is one of the showiest avocados when it is in full fruit. Most of its relatives are larger trees, many of them forming part of the cloud forest canopy. An *O. tenera* tree usually has all male- or all female-functioning flowers, but it is able to switch sex over time (a form of temporal monoecy). The mechanism for this is not known. Are individuals somehow communicating (e.g., via pheromones) with one another?

Adrian Hepworth

Resplendent quetzal eating a little avocado fruit.

Ocotea tenera
Lauraceae

Chusquea patens
Family: Poaceae

Cañuela

Chusquea patens

Description: Arching, branching bamboo ca. 5–10 m tall; solid, green stems 2–4 cm in diameter. Tan, pubescent sheaths, 30 cm long, on new shoots; tufts of leafy stems at the nodes, with narrow leaves ca. 0.5 cm wide by 15–20 cm long, gray-green below, yellow-green above; adventitious roots from lower nodes. When it flowers, inflorescences develop at the ends of leafy branchlets.

Distribution: Costa Rica and western Panama; cloud forest, 1,200–2,400 m, in openings and along forest edge.

Comments: Approximately 20 species of wild bamboo in Costa Rica belong to the genus *Chusquea*. Several chusqueas grow in the Monteverde region; *C. longifolia* is similar to *C. patens* but has wider leaves (0.6–1.4 cm). As with many other bamboos, members of the genus *Chusquea* flower after long intervals, with nearly all the plants in a population doing so at the same time and then dying back. The seed crop produced starts another set of plants. Occasionally, a small subset of individuals will flower out of sync. Two other bamboo genera occur in Montevede: *Arthrostylidium merostachyoides* is a shorter bamboo of the Monteverde cloud forest. The elegant, densely growing *Rhipidocladum pittieri* grows at lower elevations, such as on the Bajo Tigre trail. Also see common bamboo (*Bambusa vulgaris*, p. 147) and batamba (*C. subtessellata*, p. 325).

Panopsis costaricensis
Family: Proteaceae

Palo de papa

Other common name: *Papa.*
Description: Large tree, to 40 m, with dense foliage. Alternate, tonguelike, wavy, leathery leaves, spiraled along, and bunched toward tips of, branches; leaf to ca. 20 cm long, with short blunt tip. Flower spikes to 18 cm long; flower 1 cm long, including exerted style; tepals narrow, white, and curled back. Clusters of green to brown lemon-shaped fruit, to 6.5 cm long, woody with thick rind.

Distribution: Costa Rica and western Panama; mostly in mountains from 1,000–2,000 m; in Monteverde area, from 1,300 to 1,700 m.

Panopsis costaricensis in flower.

Comments: Like macadamia (*Macadamia integrifolia*), this species is a member of the mostly Australian Proteaceae family. Thus, with its edible nut, it could be considered a "Neotropical macadamia." The flavor cannot compete, however, with true macadamia. In the forest, rodents gnaw through the hard outer rind to get to the softer seed. The flowers, which are fragrant at night, are most likely moth-pollinated. The wood is suitable for construction and furniture-making. *P. costaricensis* was formerly lumped under the name *P. suaveolens*, a South American species.

Panopsis costaricensis fruits.

Hillia triflora
Family: Rubiaceae

Tres flores

Description: Epiphyte, sometimes shrub-like, located low or high on trunks or branches. The 4-cm stipules often not seen since they readily drop off. Opposite leaves ca. 11 cm long, somewhat fleshy, entire, with midvein thickened at base on leaf underside. Tubular red flowers, 7 cm long, often in threes; 6 short, curled-back

Hillia triflora

corolla lobes; 6 anthers; tube bulging two-thirds of the way up from narrow base, then constricted near mouth. Narrow capsular fruit to 10 cm long; many small seeds, each with a tuft of hair.

Distribution: Southern Mexico to Colombia. In Costa Rica, found mostly in cloud forest, 500–2,000+ m.

Comments: Members of the genus *Hillia*, which is in the coffee family, have medium to long, tubular flowers. Some species have fragrant, white flowers that attract hawkmoths, while others have reddish flowers that lure hummingbirds. *Hillia triflora* flowers are most conspicuous from November through the dry season in the Monteverde region, but the plant may also be found in flower in rainy season. All *Hillia* species have tiny seeds that are carried on the wind via a tuft of hair. The lower-elevation *H. allenii* has a shorter, single corolla that is paler (sometimes salmon or yellow) with longer corolla lobes, to 1.2 cm long. The uncommon *H. longifilamentosa*, which is not found in the Monteverde area, has anthers that stick out beyond the flower opening. *Hillia triflora* was formerly called *Ravnia triflora*. There has been some confusion in the taxonomy of similar, related species.

Pouteria fossicola
Family: Sapotaceae

Sapote, *Zapote*

Description: Canopy tree to 40 m tall, with sticky white latex. Alternate leaves rather stiff, to 14+ cm long, narrowing toward base, bunched at twig tips. Cream-colored, 1 cm flowers in clusters along twigs. Fruit 7–12 cm long, green to brown, with warty texture; 1–2 seeds, lustrous brown with tan scar (illus., opposite page).

Distribution: Nicaragua to Panama. In Costa Rica, in cloud forest near

Monteverde at 1,400–1,600 m, and on the Osa Peninsula.

Comments: These wild, edible sapotes produce the largest seeds in the Monteverde region forests. Both arboreal and terrestrial mammals relish the fruits. The flowers, which appear from the late dry into the wet season, produce a scent at night and are probably moth-pollinated. The unripe fruit gives off hydrogen cyanide when damaged. A large canopy tree of the cloud forest, *Pouteria fossicola* also occurs at lower elevations on the Osa Peninsula. There are more than 30 species of *Pouteria* in Costa Rica— *Pouteria sapota* (p. 203), is the commercial zapote or mamey; star apple, or caimito (*Chrysophyllum cainito*) and chicle (*Manilkara* spp., p. 278) are also in the Sapotaceae family.

Pouteria fossicola

Pouteria fossicola seed.

x ¹/₃

Red-tailed squirrel eating *Pouteria fossicola* fruit.

GAP SPECIALISTS

As one walks through cloud forests, it is not unusual to come upon open areas, both large and small, that receive a lot of light and warmth on sunny days. These gaps are created when strong trade winds snap off and uproot trees, or when landslides occur on steep, saturated slopes. The newly opened areas allow sun-loving plant species to grow. The increased light gives the saplings of some trees that had been growing in forest shade a chance to accelerate their growth. Other plants, whose seeds require certain environmental conditions before they will germinate and may have been sitting in the soil for years—even decades, perhaps—now have the opportunity to sprout. And other gap specialists like *Cecropia*, whose seeds are freshly deposited by birds and bats, also take advantage of the warm, sunny conditions. In addition to *Cecropia* (p. 45), the following five species are some of the most conspicuous colonists found in gaps within the Monteverde cloud forest.

Bocconia frutescens
Papaveraceae

Tree celandine (*Bocconia frutescens*) is a fast-growing weedy species that is present in most light gaps in the cloud forest. Its small black seeds can lay dormant in the soil for many years. The foliage and the display of small seeds, each hanging out of a little oval frame, are attractive; the contrasting colors (black and red) of the seed display lure small fruit-eating birds. This plant is rich in alkaloids. Research shows the presence of the antimicrobial and anti-inflammatory compounds chelerythrine and sanguinarine.

Phytolacca rivinoides
Phytolaccaceae

Various species of pokeweed may be seen around the world. They normally grow as large shrubs, with the exception of the South American species *Phytolacca dioica*, which grows to be a 20 m tree. *Phytolacca rivinoides* has tiny black seeds that sit in the soil for several decades, if necessary, waiting for a gap to occur; only then, as the light, temperature, and other conditions suddenly become favorable, will it germinate. Various pokeweeds are used as food (young plants) and medicinally, but caution must be taken since toxic substances are present in various parts of the plant.

Piper umbellatum
Piperaceae

Piper umbellatum and its relatives are common components of tropical forests. Their fruits play an important role as food for bats and, to a lesser extent, birds. The finger-like spikes, which are characteristic of the family Piperaceae, are made up of hundreds of minute, tightly packed flowers or fruits. The flowers are probably insect-pollinated. *P. umbellatum* was formerly classified as *Pothomorphe umbellata* or *Lepianthes umbellata*.

William A. Haber

Witheringia meiantha
Solanaceae

The nightshade and tomato family, Solanaceae, encompasses a variety of tropical sun-loving herbs, shrubs, and treelets, as well as a few lianas. Many of these grow well in forest gaps and abandoned pastures. *Witheringia meiantha* is one of several witheringias in the Monteverde area that have orange-red fruits that are bird-dispersed. The fruits have laxative qualities in birds, speeding up the passage of seeds (Murray et al. 1994).

Urera sp. in flower.
Urticaceae

Ortiga is a Spanish name given to a variety of plants with irritating hairs, most of them in the stinging nettle family, Urticaceae. This *Urera* species, a small tree, is mildly urticating compared to others such as *U. baccifera*. Tiny, bright orange, bird-dispersed fruits form in clusters along the branches. Although called *U. elata* for many years, the correct species name is not clear.

Poás Volcano National Park

Forests in Costa Rica's higher mountains, at elevations of 2,500 to 3,500 m, share some plant species with lower cloud forest areas such as Monteverde. Many of the plants that occur on these peaks and volcanoes, however, have adaptations that enable them to live in the cold temperatures or, in the case of the active volcanoes, to grow in freshly compacted ash.

In Costa Rica's central volcanic range, clouds shroud Poás Volcano on many days, but during the early morning hours and on clear dry-season days, the chances of seeing into the crater, which is 300 m deep by 1.5 km wide, are good. At an elevation of just more than 2,700 m, Poás has a chilly climate, with temperatures ranging from around freezing to 18° C (65° F). Volcanic activity and noxious gasses have limited the growth of vegetation near the crater, but hardy shrubs and stunted trees, such as *Clusia*, survive. As one moves away from the crater, species typical of other high mountains in Costa Rica appear, and plant diversity increases.

Bomarea hirsuta
Family: Alstroemeriaceae

Bomarea

Description: Spiraling vine without tendrils. Leaves to 13 cm long, alternate, elliptic-lanceolate, with twisted base; variably pubescent on leaf underside. Red-orange flowers, in large umbels, at end of stems; tube ca. 2–2.5 cm long, with the 3 petals a bit longer than the 3 sepals; tiny spots sometimes present on

Bomarea hirsuta

inside of petals. Three-valved capsules, splitting open to reveal seeds with red-orange coating.

Distribution: Costa Rica to Ecuador, 1,800–3,500 m. In Costa Rica, Tilarán Mountains and ranges to the south.

Comments: Flowering bomareas appear as bursts of color along the edge of mountain forests. These climbing vines twist around, and drape down from, surrounding vegetation. Since all the highland species have red-orange flowers, they can be difficult to distinguish; some have dark blotches inside the flower. Bomareas are hummingbird-pollinated and the seeds are probably bird-dispersed. Two of the ten Costa Rican

x ²/₃

Fruit capsules of *Bomarea* sp.

species, *B. obovata* and *B. edulis*, have pink and green-yellow flowers and occur mostly below 1,000 m. The Alstroemeriaceae family is closely related to Amaryllidaceae.

Myrrhidendron donnellsmithii
Family: Apiaceae

Arracachillo

Other common name: *Cachillo.*

Description: Giant herb to shrub, 2–5 m tall. Alternate, compound leaves with toothed, celery-like leaflets; leaf stalk often maroon and sheathing, expanded toward base. Large clusters of tiny white or greenish flowers in heads similar to those of Queen Anne's lace, with honeylike scent reminiscent of elderberry flowers. Spicy, fragrant, cylindrical, ribbed seed capsules to 2 cm, held erect in clusters; remnants of flower parts at tip of capsule.

Distribution: From ca. 2,000–3,800 m in central volcanic range and Talamancas in Costa Rica and across the border into Panama; also north to Chiapas, Mexico; often in clearings in humid mountain areas and along streams.

Comments: It's not clear whether this giant celery relative should be classified as a herb, a shrub, or a tree, although it does appear in the book *Arboles de*

Myrrhidendron donnellsmithii

Costa Rica (*Trees of Costa Rica*). Bees, including the high-elevation bumble-bee, *Bombus ephippiatus*, visit the flowers, which can be seen during most of the year. A smaller species, *M. chirripoense*, which is less then 50 cm tall, occurs in the Chirripó region of Costa Rica.

Weraubia ororiensis
Family: Bromeliaceae

Weraubia ororiensis

Description: Terrestrial plant or epiphyte to 75+ cm tall. Leaves to ca. 50 cm long, 5 cm wide, green mottled with maroon. Erect inflorescence with green flowers ca. 3 cm long; long red bracts below groups of flowers. Capsules ca. 3 cm long.

Distribution: Costa Rica and Panama. In Costa Rica, most commonly seen at 2,000–3,300 m, in central and Talamanca Mountains.

Comments: This is a bromeliad frequently seen at Poás and in the Talamanca Mountains. It is most likely hummingbird-pollinated. *W. viridis* is similar, but the sepals are deep red. These two species were formerly considered to be in the genus *Vriesea*.

Pernettya prostrata
Family: Ericaceae

Pernettya, *Arrayán*

Description: Shrub 0.5–2 m, sometimes epiphytic. Alternate, shiny leathery leaves, 1–2 cm long, with teeth on margin. White to rose flowers, which look like miniature globose Chinese lamps, have 5- to 7-mm-long tube with 5 tiny lobes. Calyx remains are at the base of the purple-black berries, which are less than 1 cm.

Distribution: Mexico to Andes of South America. In Costa Rica, generally 2,000–3,800 m in open areas near the central volcanoes (Poás, Irazú),

Cerro de la Muerte (and rest of Talamancas), and below 2,000 m at Rincón de la Vieja.

Comments: *Pernettya prostrata* is variable and sometimes split into various species or varieties. It is somewhat similar to another high-elevation blueberry shrub of Costa Rica, *Vaccinium consanguineum*, which is larger and has a more cylindrically-shaped flower; it also has four corolla lobes instead of five, and the remains of the calyx are at the tip of the fruit instead of at the base.

Pernettya prostrata

Although they are eaten by bush tanagers and sooty robins, pernettya fruits are poisonous to humans; some South American species cause either hallucinations or (possibly permanent) insanity. *P. prostrata* is closely related to wintergreen (*Gaultheria procumbens*) of temperate North America.

Ulex europaeus
Family: Fabaceae
Subfamily: Papilionoideae (= Faboideae)

Gorse, *Chucero*

Other common name: *Corona de Cristo.*
Description: Prickly shrub, 1–2 m tall but grows taller in other parts of the world; shoots upon which leaves and flowers are attached end in spines. Simple, 1-cm, rigid leaves are sharp and spinelike; seedlings have leaves with 3 leaflets. Two-cm-long yellow, scented flowers with fuzzy calyx. Brown, hairy pods, 1–2 cm long with a few seeds.
Distribution: Originally from western Europe, naturalized in Australia, New

x 1¼

Ulex europaeus branch.

Ulex europaeus

Zealand, Hawaii, certain coastal areas on both sides of the United States, some high mountains of South America. In Costa Rica, at Poás Volcano and nearby fields.

Comments: Gorse is an invasive weed that takes over in natural or man-made disturbances. It was possibly introduced to Costa Rica around 1920 along with grass seed from Europe, where it serves as a fodder and hedge plant. The seeds, which are heat tolerant and can remain dormant for many years, bear oil bodies (elaiosomes) that are attractive to ants, which sometimes further disperse the seeds after they have been ejected from the pod.

Escallonia myrtilloides
Family: Grossulariaceae

Escallonia, *Cipresillo*

Description: Small tree 2–6 m, with dense foliage and pagoda-like branching; shaggy bark. Alternate leaves rather resinous, 1–2 cm long, with fine teeth, underside whitish, wedge-shaped at base. Small, whitish-green flowers. Capsular fruit.

Distribution: At 2,000–4,000 m elevation, Costa Rica to South America. In Costa Rica, occurs on Poás, Irazú, and Barva volcanoes, as well as in the Talamancas (e.g., Cerro de la Muerte).

Comments: A key identifying feature of this tree is the pagoda-like branching—the main branches, which are densely covered with branchlets, sweep down and then upward. Another less-common species, *E. paniculata*, with larger leaves, is also found in Costa Rica. Many escallonias occur in the high Andes of South America, dominating some forests. Some have been taken

into cultivation in temperate areas of the United States and Great Britain for use as hedge plants, and today there are many garden cultivars. Escallonias are related to hydrangea and saxifrage; they are sometimes put in their own family, Escalloniaceae.

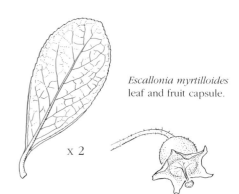

Escallonia myrtilloides leaf and fruit capsule.

x 2

Escallonia myrtilloides

Gunnera insignis   Poor man's umbrella, *Sombrilla de pobre*
Family: Gunneraceae

Other common name: *Higuera.*

Description: Robust herbaceous plant, at times with stout stem to 30 cm tall. Rosette of long-stemmed leaves with very rough texture; blades 1–2 m, rounded, lobed, and toothed; large, red scales at base of leaves. Reddish inflorescence, 1 m, triangular in outline, with many branches bearing minute flowers. Tiny one-seeded fruits.

Distribution: Nicaragua to Colombia. In Costa Rica, ca.1,000–2,600 m; wet forest in all of the major mountain ranges; often on steep landslides and exposed ridges.

Comments: These makeshift umbrellas, which occur in poor, wet, acidic soils, have a fascinating symbiotic association with a cyanobacterium (genus *Nostoc*) that lives in the leaf stems. Cyanobacteria, colloquially referred to as blue-green algae, fix nitrogen in

Gunnera insignis

Giant *Gunnera insignis* dwarf the author.

Gunnera insignis cover this steep embankment.

exchange for nutrients from the plant. It appears that the cyanobacteria enter in the plant's seedling stage via special mucus-secreting glands at nodes on the stem; scales at the base of the petioles may also transfer *Nostoc*. A smaller species, *G. talamancana*, which has more deeply incised leaves, appears at higher elevations in Costa Rica. The genus *Gunnera* is found in far-flung parts of the world. Some species are used as ornamentals. Although *Gunnera insignis* is quite large, some species in this genus are very small (10 cm). At times this plant is classified under the family Haloragaceae.

Elaphoglossum lingua
Family: Lomariopsidaceae (*Elaphoglossum* is now in Dryopteridaceae)

Paddle fern, *Helecho lengua*

Other common name: Tongue fern.
Description: Terrestrial or epiphytic fern; stems blackish and long-creeping. Leaves long-elliptical, smooth, simple and entire, very thick and leathery; deep green to yellowish green, raised midrib a lighter color; blades to 30 cm long with long stalks; fertile fronds narrower, with sori (clusters of the spore-producing bodies) covering lower surface.
Distribution: West Indies, Costa Rica to Peru. Widespread in Costa Rica; occurs in Central, Talamanca, and Tilarán ranges, 1,300–2,800 m, growing in open and forested areas. Common on banks along forest trails at Poás Volcano.
Comments: This genus, which is mostly epiphytic, is the largest fern genus in Costa Rica; it includes more than 120 species. Most have tongue- or paddle-shaped leaves and parallel veins. See parsley fern (*Elaphoglossum peltatum*, p. 443). The genus *Bolbitis* is closely related to the paddle ferns.

Elaphoglossum lingua

Psittacanthus schiedeanus
Family: Loranthaceae

Mistletoe, *Matapalo*

Description: Large, epiphytic shrubs with angled, ridged stems. Opposite or subopposite, flexible, leathery leaves of various shapes, asymmetrical, ca. 10 cm to much longer. Large clusters of flowers; orange base and orange-yellow tip, with pink-red pedicels; thin buds ca. 8 cm; when flower opens, narrow petals curl back and anthers protrude. Rubbery 2-cm, purple-black fruit on orangish stalk.
Distribution: Mexico to Panama. In Costa Rica, usually at high elevations, 1,500–3,000 m, occasionally lower; common near major volcanoes in Central Valley (on way to Poás, Irazú, etc.), in Braulio Carillo, and in the Talamancas in the southern part of country.
Comments: As one travels toward the Poás or Irazú volcanoes, masses of red, orange, and yellow light up pasture trees. These are showy mistletoes that cling to, and tap into, their hosts. *P. schiedeanus* is one of nine Costa Rican species of *Psittacanthus* in the Loranthaceae family; mistletoes with reduced flowers either belong to the Viscaceae or one of several minor mistletoe families. Mistletoes attach to trees by haustoria, special structures that

Psittacanthus schiedeanus

penetrate the host tissue to obtain water and nutrients. Since these plants have chlorophyll and are able to make some of their own food through photosynthesis, however, they are technically described as hemiparasitic. In some Loranthaceae species, an interesting pattern forms at the

point where the mistletoe attaches to the host. When the mistletoe dies, this woodrose, or *rosa de palo*, remains. These often inspire artisans to create carvings out of the floral patterns in the wood. Mistletoes flourish in some pastures around Costa Rica. *Psittacanthus* species are among the showiest with their hummingbird-pollinated flowers; *Phoradendron* species, in the Viscaceae (or Santalaceae) family, have reduced flowers, but may have orange-tinted leaves (e.g., *P. robustissimum* in Monteverde, photos below). Different species of mistletoes seem to favor particular families of trees as hosts. Among the hosts that *P. schiedeanus* grows on are the oak (Fagaceae), winter's bark (Winteraceae), myrtle (Myrtaceae), and birch (Betulaceae) families. In Costa Rica, you may hear the word *parásitas* (parasites) applied to all epiphytes, but it is only the mistletoes that parasitize the trees they grow on. The genus *Psittacanthus* has not been well-studied regarding chemical activity, but other mistletoes are known to have poisonous and/or potential medicinal properties. Mistletoes that are displayed at Christmas time in North America and Europe include *Phoradendron leucarpum* and *Viscum album*.

Phoradendron robustissimum growing on tree.

Phoradendron robustissimum in fruit.

Monochaetum vulcanicum
Family: Melastomataceae

Melastoma of the volcanoes

Other common names: *Escoba real, milflores.*

Description: Shrub 1–2 m, with peeling, rust-colored bark; young stems red and quadrangular. Opposite, 1.5-cm leaves with 3 prominent veins; sparse hairs on leaves. Rosy pink-purple, 4-petaled flowers to 3 cm across, with red and yellow spurred anthers. Fruit a reddish capsule, 0.5–1.3 cm, that dries and dehisces; seeds tiny.

Distribution: Endemic to Costa Rica, 1,800–3,000+ m, near central volcanoes and Rincón de la Vieja.

Comments: This shrub is abundant along the main path to the Poás Volcano crater. Pollen-gathering bees appear to be the main pollinators. The similar *M. amabile* of the Talamancas does not have 4-angled branchlets.

Monochaetum vulcanicum

Monochaetum vulcanicum growing on rim of Poás Volcano.

Cerro de la Muerte

A mountain range called the Talamancas rises high into the clouds in the southern half of Costa Rica. It includes the tallest peak in the country, Chirripó, which rises to 3,819 m and where a low temperature of –9° C (16° F) has been recorded. Its slopes encompass a number of indigenous reserves, as well as habitat for the resplendent quetzal. Much of the area is within the Amistad Biosphere Reserve, an international park shared with Panama. The Inter-American Highway, which passes through the northwest corner of the mountain range when heading south from San José, gives travelers a glimpse of the extraordinary vegetation of this region. Although there is a peak called Cerro de la Muerte along this route, the term is used here to refer more loosely to this section in the Talamanca Mountains, from about 2,500 to 3,500 m. Paramo vegetation, which occurs on the highest points of the Cerro, is a mix of compact herbs and shrubs well-suited for life in cold-to-freezing, windy areas that receive a high level of ultraviolet radiation. Oaks dominate the forest at elevations below 3,300 m in the region. For more extensive coverage of plant species in this area, see Alfaro and Gamboa, 1999.

Even though many plant species are found at both Poás and Cerro de la Muerte, the separate collages of photographs presented here are intended to indicate species that stand out in the respective sites. A visit to one of these areas on a clear day brings the rewards not only of a great botanical experience, but also of superb panoramic views. (Don't forget your sunscreen!)

Cirsium subcoriaceum
Family: Asteraceae

Plume thistle, *Cardo*

Description: Giant prickly herb to 3 m; hoary/woolly stem ca. 6 cm in diameter. Leaves, to 75 cm, with spine-tipped lobes and deep indentations. Many spines on overlapping bracts that surround yellow (or pinkish) flower heads at top of stem, each head ca. 6 cm long, dense and packed with many thin-tubed, yellow-beige to pink flowers; bracts hoary and clothed in a spider-web-like pubescence; bronze-silver headdress of plumes (2.5 cm) radiating out from top of each 0.6-cm-long seed.

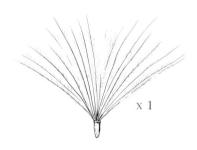

x 1

Plumed seed of *Cirsium subcoriaceum.*

Adrian Hepworth

Cirsium subcoriaceum

Distribution: Mexico to Panama, 2,000–3,400 m, sometimes lower. In Costa Rica, high elevations of Central and Talamanca mountain ranges; common on Cerro de la Muerte.

Comments: These oversized thistles are hard to miss as you travel the road over Cerro de la Muerte. Hummingbirds and bumblebees visit the flowers; the volcano hummingbird (*Selasphorus flammula*) uses the silky hairs of the seeds to line its nest. *C. mexicanum*, a shorter, purple-flowered species, occurs in Costa Rican pastures below 2,000 m. Thistles (*Cirsium* spp.), of which there are more than 200 species in the world, are closely related to artichoke (*Cynara scolymus*); both belong to one of the largest flowering plant families, the Asteraceae.

Blechnum auratum
Family: Blechnaceae

Tree blechnum

Description: Terrestrial fern with "trunk" to 2 m; resembles a cycad; old dead fronds often form skirt around trunk. Separate sterile and fertile (spore-bearing) fronds, 1-pinnate, to more than 1 m (stalk plus blade); ca. 50 pairs of pinnae (entire), those on sterile fronds to 13 cm long (by 0.8 cm wide); fertile pinnae a little shorter and narrower; 3-cm-long scales at base of each frond stalk, clumped and looking like coarse hair; the scales are bicolorous, with a dark central stripe and lighter margins.

Distribution: Costa Rica to Bolivia, in paramo and high-elevation (2,400–3,200 m) cloud forest.

Comments: Along the highway over Cerro de la Muerte, you pass bowl-like marshy areas that are dotted with the giant *Blechnum auratum* and the tall bromeliad *Puya dasylirioides* (p. 318). The rhizome of *B. auratum* (also known as *B. buchtienii*) is sold as a substrate for orchids and other epiphytes. A blechnum species with a shorter trunk, to just 1 m, is *B. loxense*, found at similar elevations but with smaller leaf blades, to 30 cm long, with fewer and shorter pinnae, up to 25 pairs to 6 cm long.

Adrian Hepworth

Blechnum auratum

Puya dasylirioides
Family: Bromeliaceae

Puya dasylirioides

Description: Towering terrestrial bromeliad to 2+ m when in flower/fruit; flowering spike bulges toward top, becoming brown and more candlelike in fruit. Leaves ca. 50 cm long, erect and stiff, glaucous beneath, with short, curved spines along edge. Upper bracts golden, overlaid with tan-brown hairs. Flowers ca. 3 cm long, light blue with bright orange-yellow anthers. Small, 3-parted, capsular fruit with tiny seeds.

Distribution: Endemic to Costa Rica, in open bogs and meadows, 2,500–3,400 m.

Comments: Puyas and the giant blechnum ferns they usually grow with create picturesque bogs high in the Talamanca Mountains. Puya flowers

Puya dasylirioides flowers.

between March and August. The genus *Puya*, which mainly occurs in the Andes, includes species that grow taller than 10 m and may live for more than 100 years.

Brunellia costaricensis
Family: Brunelliaceae

Brunellia, *Cedrillo*

Other common name: *Cedrillo macho.*
Description: Tree ca. 8 m tall, to more than 25 m; stout branches with slight wintergreen odor when broken. Compound leaves 30–40 cm long with 7–13 leaflets, each to 16 cm long with rounded teeth on margin; leaves dark shiny green above, light gray-green below; older leaves blood red to orange, stiff and leathery; 9–15 pair secondary veins. Small flowers without petals, followed by small, dehiscent, star-shaped, 5-parted fruits, 1.3 cm across, in clusters ca. 15 cm across; contain dull orange seeds.

Distribution: Endemic to Costa Rica, 1,500–2,900 m, Tilarán, Central, and Talamanca ranges.

Comments: The colors in the old leaves of brunellia are reminiscent of New England maples in autumn. The seeds are probably bird-dispersed. The soft wood is not very durable. *B. costaricensis*

is one of five species found in Costa Rica. *B. standleyana*, which is found from 850–1,900 m, differs in having 16–21 pair secondary veins and larger inflorescences.

Adrian Hepworth

x 2

Fruit

Brunellia costaricensis

Centropogon talamancensis
Family: Campanulaceae

Adrian Hepworth

Centropogon talamancensis

Description: Terrestrial herb, but somewhat scandent, to ca. 2 m long; white latex. Alternate, short-stalked leaves to 15 cm long, finely toothed. Five-lobed, pink-fuchsia flower tube 4–5 cm; pink stamen tube with a navy blue tip; white tuft of hair in male phase, protruding stigma in female stage. Fruit a pear-shaped, 2.5+-cm capsule, bulging, ribbed and bumpy, bent downward.

Distribution: Endemic to Costa Rica; mostly Talamanca Mountains, 2,800–3,000+ m, but recorded in central range as well.

Comments: Magnificent hummingbirds (*Eugenes fulgens*) pollinate the flowers, but the shorter-billed fiery-throated hummingbirds (*Panterpe insignis*) may be seen robbing nectar from holes made by slaty flowerpiercers (*Diglossa plumbea*) or bees. Flowers harbor hummingbird mites (see *C. solanifolius*, p. 290). The flower has separate male and female phases, and experiments have shown that pollen removal shortens the male phase. Fifteen additional species of *Centropogon* occur in Costa Rica. *C. gutierrezii* (illus. right), which occurs in the central volcanic range (including Poás) and could be confused with *C. talamancensis*, has fringes and odd, wart-like projections on the ends of the corolla lobes.

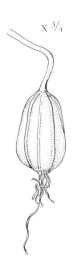

x ³/₄

Centropogon gutierrezii seed capsule.

Cavendishia bracteata
Family: Ericaceae

Colmillos

Description: Epiphytic or terrestrial shrub, to 3 m, sometimes growing as a liana. Larger leaf blades ca. 12 cm long, narrowing at tip; usually 5–7 major veins ascending from base; leaf is coppery purple when young, later light to dark green and leathery; petioles suffused with red-purple. Broad, notched, deep pink bracts, to 3.5 cm long, associated with tight cluster of up to 20 tubular flowers at branch tips; glossy, waxy corolla to 2 cm long, constricted toward tip; color variable; basal part light pink or orange, upper part translucent white, looking somewhat purplish. Berries 1+ cm.

Distribution: Mexico to Bolivia. In Costa Rica, ca. 800–3,000+ m; thrives along continental divide from north to south, and seen in adjacent forest down slope on both sides.

Comments: This blueberry relative, which may flower at any time of the year, is the most likely of the 20 species in the genus *Cavendishia* to be noticed since it turns up in most mountainous areas in Costa Rica. Permutations of the flower color, growth form, and leaf size, however, make it look somewhat different from one site to another. The fruit is edible. The conspicuous floral bracts are characteristic of the genus. Plants in the Ericaceae family are abundant in Costa Rican mountains, where they often grow as epiphyte shrubs. One tree-size species, *Comarostaphylis arbutoides*, is seen in groves in some parts of the Talamancas.

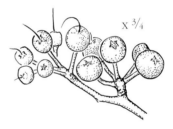

X ³⁄₄

Cavendishia bracteata fruits.

Cavendishia bracteata

THE IMPRESSIVE OAKS

Oak forests are a principal feature of the landscape between 2,000 and 3,300 m in the Talamanca Mountains. At these elevations, behind a veil of mist, one sees giant, gnarled trees laden with moss, bromeliads, and other epiphytes. Oaks dominate in this zone and may make up 80% of the trees in a given area. *Q. costaricensis* is the principal species in the highest forest, mixing with *Q. bumelioides* (formerly known as *Q. copeyensis*) and/or several other species that become more dominant in other parts of the mountain range. Bamboos such as *Chusquea talamancensis* are common in the oak-forest understory.

All oaks belong to the genus *Quercus* in the Fagaceae family. Although they are more typical of northern temperate forests, some oak species occur as far south as Colombia. Costa Rica has 15 species.

When they are in fruit, oaks are easy to recognize because they all have acorns, which are nuts with cupules. A flowering oak has separate male and female flowers, the latter being very small and inconspicuous. The male flowers are grouped in pendant spikes called catkins or aments. The simple leaves spiral the stem, are tough and/or leathery, and may have a few irregular rounded or bristle-tipped teeth. Twig tips have clusters of scaly buds. The large oaks of the Cerro de la Muerte region range from 25 to 40+ m tall.

A number of mushrooms, especially in the genus *Boletus*, commonly grow near oaks. Some of these fungus species have a mycorrhizal relationship with the tree roots, in which both the fungi and the tree benefit through an exchange of nutrients. A collaborative effort between researchers from the United States and Costa Rica who are surveying macrofungi of the Talamancan oak forests has led to the discovery of some mushrooms that are unique to that region (Mueller 2002).

The oak forests in the Talamanca range are patchy; they are sometimes ravaged by fires, and for several decades beginning in the 1950s, they were decimated by settlers and lumber companies. Some Costa Rican oak was transformed into European wine barrels and furniture during that period, and much of it was processed into charcoal. Today it is illegal to cut standing oaks for charcoal production, but fallen trees can be scavenged and processed. The Los Santos Forest Reserve and La Amistad Biosphere Reserve offer some protection to these threatened forests. In addition, researchers from national and foreign universities, with support from institutions such as CATIE, INBio, and the Costa Rican Ministry of the Environment and Energy, are seeking sustainable forestry methods to lessen future human impact on this fragile environment.

Oak forest at Villa Mills.

La Georgina restaurant in Villa Mills, on the Inter-American Highway, is a popular rest stop where one can see—and experience—cold, high-elevation oak forests. Several lodges, as well as the Cuericí Biological Station, provide facilities for more in-depth studies.

See Kappelle (1995, 1996) for in-depth information on Costa Rican *Quercus* forests.

Fuchsia paniculata
Family: Onagraceae

Fuchsia, *Achiotillo*

Fuchsia paniculata

Description: Shrub or treelet, 2–5+ m. Leaves opposite or in whorls of 3, entire or with red-purple teeth, ca. 7.5–18 cm long. Large terminal flower clusters, with pink-purple flowers, to 2 cm long, that differ in size from plant to plant, depending on whether female (producing fruit) or male; female flowers smaller than male; each has 4 showy sepals and 4 petals. Fruit a dark purple berry ca. 0.5–1 cm.

Distribution: Mexico to Panama. In Costa Rica, Central and Talamanca ranges, 1,700–3,000+ m, common in forest and pasture areas in Cerro de la Muerte, Irazú, and Turrialba.

Comments: Fuchsias are in the evening primrose family. Both wild and cultivated fuchsias have a distinctive flower form—often with a set of showy reflexed sepals that look like petals—that makes them recognizable. *Fuchsia microphylla*, a shrub with single flowers in the leaf axils, and *F. splendens*, which has pendant 5-cm pink-red and green flowers, are two other native high-elevation species. *F. jimenezii* is common on the Continental Divide at Monteverde and has also been collected at Tapantí and Braulio Carillo. Fuchsia fruits are edible. There are a large number of showy fuchsia cultivars, many with double flowers.

Chusquea subtessellata
Family: Poaceae

Batamba

Other common name: *Cañuela.*
Description: Bamboo to 3 m, erect, in clumps; much branching in upper half of stem; stem 1–2 cm in diameter, orange-yellow with tan sheaths. Yellow-green leaves 2–12 cm long, ca. 1 cm wide. Dense, nonspreading inflorescences, turning black when in seed.
Distribution: Costa Rica and Panama. In Costa Rica, usually above 3,000 m (some down to 2,200 m) in Talamanca Mountains from Cerro de la Muerte region to border with Panama.
Comments: This is a dwarf bamboo that grows well in high, open, burned-over areas. It will resprout after being burned. It is wind-pollinated. Although mass flowering may occur, sporadic flowering of individuals is seen in most years. Batamba was formerly called *Swallenochloa subtessellata.*

Adrian Hepworth

Chusquea subtessellata

Podocarpus oleifolius
Family: Podocarpaceae

Podocarp, *Cipresillo*

Other common name: *Cobola.*
Description: Large trees, to more than 25 m. Glabrous, leathery, entire, alternate leaves, ca. 6 cm long and ca. 1 cm wide, with pointed tip; upper side of leaf with groove and one central vein; the leaves spiral the stem. Separate male and female cones; a fertile scale of the latter becomes fleshy and surrounds seed, developing into a purplish fruit.
Distribution: *Podocarpus* species occur in the Old and New Worlds; from Mexico to the Andes; also Africa, Australasia, and Japan. *P. oleifolius* occurs from Mexico to South America. In Costa Rica, 1,100–3,100 m, from Guanacaste Mountains south to Talamancas.

Comments: The pine and cypress trees growing in fence rows or plantations in Costa Rica are exotics, but podocarps seen in the Talamancas

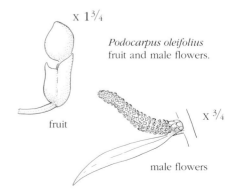
x 1³⁄₄
Podocarpus oleifolius fruit and male flowers.

fruit

x ³⁄₄

male flowers

Podocarpus oleifolius

Podocarpus oleifolius foliage.

are native gymnosperms. Sometimes used in charcoal production, the wood is also suitable for light con-struction. Podocarp has potential as a reforestation species in high mountains. *Podocarpus* forest is a feature of the high Andes, where much has been cut for use as timber. Of the three species in Costa Rica, one, *P. guatemalensis*, occurs in lowlands and hills, up to 1,000 m, on the Osa Peninsula and in the northern part of the country. The Asian *P. macrophyllus*, used as an ornamental in the United States, looks somewhat like yew (*Taxus* spp.). *P. oleifolius* has also been referred to as *P. macrostachyus*.

Jamesonia scammaniae
Family: Pteridaceae

Description: Terrestrial fern growing in clumps. Erect, rigid, 1-pinnate leaves generally 15–35 cm tall and less than 0.5 cm wide; many tiny, rounded, inrolled pinnae all about the same size, appearing like a stack of coins on either side of the leaf midrib; the tight little fiddlehead at top is characteristic for the genus (the leaves never completely unfurl); upper side of pinna has a glandular varnished look; underside is cupped and woolly (as is rachis). Sori form along veins.
Distribution: Costa Rica to Bolivia, in high mountains. In Costa Rica, 2,800–3,800 m, in paramo on Cerro de la Muerte and Chirripó National Park.
Comments: This is an odd-looking fern that grows in the higher, colder parts of the Cerro de la Muerte. The fiddlehead at its tip is always present and is not covered by long hairs as it is in *J. rotundifolia* and *J. alstonii*, the other two species that occur in Costa Rica. *Jamesonia* commonly hybridizes with *Eriosorus*, which has fronds with a more typical fern shape; the appearance of the offspring is a blend of the two morphologies.

Adrian Hepworth

Jamesonia scammaniae

Beach and Mangrove

Much of the natural vegetation along Costa Rica's coasts is distinctive in comparison with that of other habitats within the country, but its coastline shares many species with tropical seashores of other countries. Plants growing near the ocean often have seeds that float to other beaches before sprouting. These plants are salt tolerant, able to grow in the sun and sand, and can endure high winds. Some, such as the mangroves, can cope with both unstable soils and having their roots inundated by brackish water. Overall, beach plants are hardy, and they have evolved interesting ways of dealing with an environment that, at times, can be quite inhospitable.

Most of the examples included in this section are beach and mangrove species, along with one common tree, frangipani (*Plumeria rubra*), that grows in rocky coastal areas. Along the northern Pacific shore of Costa Rica, many dry forest plant species, including cacti, creep down to the coast, where they grow on rocky sections as well as in areas just behind the upper beach.

Adrian Hepworth

A beach on the Pacific coast of Costa Rica.

Plumeria rubra
Family: Apocynaceae

Frangipani, *Juche*

Plumeria rubra

Other common names: *Cacalojoche, flor blanca.*

Description: Tree usually less than 15 m tall, but may be larger; with white latex; stout, tan-gray branches with conspicuous leaf scars; young plants look like fat, flexible sticks emerging from ground. Deciduous; long, alternate leaves to 40 cm. Terminal flower clusters with a few to dozens of sweetly scented flowers, white with yellow in throat; other color forms—purple, pink, yellow—exist; flowers 5-lobed, lobes ca. 4 cm long; tube ca. 2 cm long. Seed capsules ca. 25 cm long; new fruit yellow-green, warty, forming upside-down V; old, dehisced fruit golden-tan inside, weathered gray outside; 3-cm seeds with 3- to 5-cm wings, resembling those of a maple.

Distribution: Mexico to Colombia, and West Indies; ornamental in New and Old World tropics. In Costa Rica, sea level–1,200 m, Pacific slope.

Comments: This tree of the dry to moist, often rocky, areas of Costa Rica, is common along the Pacific coast, as well as in gardens around the world. In Hawaii, the flowers are favorites of lei makers. The scent is used in making perfumes. Latex and plant infusions are used medicinally; they are applied externally for syphilitic sores and skin afflictions, and taken internally as a laxative. The latex can cause skin blistering and irritation. The bark contains plumierid, a glycoside. Research on *P. rubra* indicates

x ½

Plumeria rubra seed.

that it contains compounds that lower blood sugar, reduce pain, and are cytotoxic (i.e., kill cells). The flowers which are seen year-round, are most profuse in April and May. They are pollinated by deceit—the white color and the perfume draw in nectar seeking hawkmoths, which carry pollen from one flower to another, but upon probing the flower, the insects are not rewarded with any nectar. The colorful black and yellow striped larvae of the frangipani sphinx moth (*Pseudosphinx tetrio*) may be found feeding on the leaves at night; and, upon close inspection of the flowers, you may find the white larvae of another sphinx, *Isognathus rimosus*. *P. rubra* is the national flower of Nicaragua, where it is called sacuanjoche.

Plumeria rubra flowers and fruits.

Sphagneticola trilobata
Family: Asteraceae

West Indian creeper, *Botón de oro*

Adrian Hepworth

Sphagneticola trilobata

Other common names: Creeping daisy, trailing wedelia, beach marigold, *clavelillo de playa*.

Description: Low-growing plant, with stems to 2 m long running along ground, rooting at nodes. Thick, glossy opposite leaves, upper surface sometimes rough, 5–12 cm long, on short petioles; irregular teeth on leaf margin with 3 lobes above middle of leaf (lateral ones like large teeth), and lower part of leaf narrowing. Flower heads 2.5 cm across, yellow to golden, with ca. 7 ray flowers that look deceptively like petals. Golden brown seed heads, 1.3 cm across, with individual seed ca. 0.5 cm long.

Distribution: Coasts and wet, disturbed areas in Florida and the New World tropics; often escapes in areas where cultivated. Widespread on both coasts of Costa Rica, along beach and roadsides.

Comments: This attractive beach "weed" has been adopted as an ornamental by some people; with its creeping habit, it creates a ground cover that can be kept short by mowing. Its various medicinal uses included chewing flowers for toothache; using a decoction of plant parts to relieve cold symptoms, amenorrhea, and urinary and stomach ailments; and preparing external baths and poultices for sore feet, backs, and arthritic joints. *S. trilobata* purportedly causes abortions in sheep. Studies on a closely related species in the genus *Wedelia* show compounds with activity against certain tumors and the AIDS and herpes viruses. Synonyms are *Complaya trilobata*, *Wedelia trilobata*, and *Thelechitonia trilobata.*

Terminalia catappa
Family: Combretaceae

Tropical almond, *Almendro*

Other common name: Indian almond, *almendro de playa.*
Description: May grow to more than 25 m, but often smaller; whorls of branches in distinct tiers, separated by branchless segments of the trunk—especially noticeable in young trees. Alternate, clustered colorful leaves to 33 cm long, obovate and often wedge-shaped at base; yellow-green to red/orange/brown. Has axillary spikes of small flowers that are starlike, with yellow centers; odd scent—mixture of musty honey and old smelly socks (or, to some, rotting starfish). Fruit 6 cm long, green to reddish yellow, bulging and edged all around with a keel; in the very center of the fruit is a seed 3-4 cm long surrounded by a hard shell which is inside a fibrous husk surrounded by acid to sweet, white to reddish flesh (all within the green to reddish skin).
Distribution: Originally from Malaysia and the Andaman Islands, off the coast of India; cultivated throughout Old and New World tropics, including south Florida and the Keys. In Costa Rica, naturalized on beaches of both coasts; also planted in parks and yards for ornament, shade, and nuts.
Comments: This is a hardy seaside tree that can tolerate high winds, salt spray, and sun. It provides a welcome bit of shade at the beach. The outer flesh of the fruit is edible—it's normally acidic, but sweet in some individuals. The kernels in the very center have an almond or filbert taste; best way of getting at them is to sun- or oven-dry the defleshed fruits. This species could be developed as a signifi-

Terminalia catappa

x ¹/₂

Terminalia catappa fruit.

cant food source by cultivating greater sweetness and larger kernels. The seeds are relished by scarlet macaws—with their strong bills they have no problem opening the hard shell that encases the seed. Monkeys, agoutis, and bats eat the fruit. The seeds are dispersed by animals, including people, and by ocean currents.

Various other products from the tree include an inky, black dye from the fruit; tannin from the bark, roots, and green fruit; firewood and lumber; and medicine. The bark and leaves are used internally for fevers and diarrhea, and externally for skin afflictions. Related species in Costa Rica include sura (*Terminalia oblonga*, p. 238), a lowland rainforest giant used as lumber, and the mangroves *Conocarpus* (p. 342) and *Laguncularia*. The common names in English and Spanish are confusing since this is not the typical almond that we eat (*Prunus dulcis*); there is also an almendro tree (*Dipteryx panamensis*, p. 240) in Costa Rican rainforests.

Ipomoea pes-caprae
Family: Convolvulaceae

Beach morning glory

Other common name: *Pudreoreja de playa*.
Description: Prostrate stems, 10+ m long, creeping along and rooting in sand; has white latex. Somewhat succulent leaves to 10 cm, often slight indent at tip. Five-lobed, funnel-shaped flowers, ca. 5 cm long by 5 cm across, deep rose to purple. Brown capsule, 2 cm in diameter, with 4 pubescent, 0.8-cm seeds flat on 2 sides, rounded on third.
Distribution: Coasts of Old and New World tropics. In Costa Rica, just above high-tide line, both coasts, and Cocos Island.
Comments: *Ipomoea pes-caprae* superficially looks like sea bean (*Canavalia rosea*), with which it frequently mingles, but the latter has leaves with 3 leaflets (illus. right) and smaller, pink-purple flowers typical of the pea-family, along with 10-cm-long pods bulging with "beans." Beach morning glory flowers, which may be seen at various times of the year, are visited by bees and other insects. The seeds float in ocean cur-

rents and are common in the tide-line litter. The cooked tubers may be used as emergency food, although they are not safe to eat in large quantities. A decoction of the plant is ingested for rheumatism and kidney ailments; skin ulcers and swellings are treated with poultices or decoctions. A leaf-extract ointment is effective in treating jellyfish stings. There are a variety of wild and cultivated vines and lianas among the ca. 500 species of *Ipomoea* (p. 86), including sweet potato (*Ipomoea batatas*).

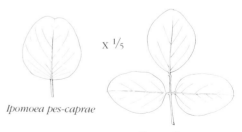

x ¹/₅

Ipomoea pes-caprae

Canavalia rosea

These two similar beach creepers can be distinguished by their leaves and their flowers.

Ipomoea pes-caprae

DRIFT SEEDS

Beachgoers often gather shells they find scattered in the sand along the water's edge, but another world of collectibles exists in the drift line on the upper beach. Among the driftwood, lost sandals, and pieces of plastic dolls that waves deposit, there are seeds of various shapes and sizes that may have drifted in ocean currents for days—or even years— before washing ashore.

Only a very small percentage of plant species has seeds or fruits that can survive months or years at sea. Some of these float, or drift, seeds, as they are known, are buoyant because of air spaces in the seed. Others float simply because they do not weigh much, or because they have a corklike or fibrous covering that keeps them from sinking. It is easy to find twenty or more different species of drift seeds along Costa Rica's beaches. Among them are members of the bean family that have hard, attractive seeds that will last for many years in collections or as jewelry. Some of the more distinctive drift seeds that might be found are presented here.

Entada gigas (**Sea heart**)   Fabaceae

These heart-shaped seeds, to 6 cm in diameter; are a lustrous deep red-brown. They come from an impressively long segmented, woody pod, to 2 m, of a canopy liana that occurs in low-elevation wet areas. The plant is found from Mexico to South America, in the West Indies, and in West Africa. Seeds have drifted as far north as Massachusetts, Norway, and Greenland; some were uncovered in postglacial peat bogs in Sweden. The seed can be split and fashioned into a small box for storing jewelry.

Merremia discoidesperma (**Mary's bean**)   Convolvulaceae

This seed, also called crucifixion bean, comes from a forest liana with yellow funnel-shaped flowers that is found from 200 to 1,000 m in the Osa Peninsula, as well as on the Atlantic slope around Monteverde and in other parts of Central America and Mexico. It travels to the ocean via streams. A cavity in the seed allows it to float for long periods of time; it has been found as far away as the Pacific islands and Norway. The 3-cm seed, borne in a translucent papery capsule, becomes smooth and lustrous with weathering. It is black with tinges of maroon or brown, has a scar (hilum) where it is held in the capsule by a black strap that runs around the seed, and a cross imprinted on the other side. The latter feature has endowed the seed with many supposed powers—pregnant women of the Hebrides would hold on to a seed during labor to ensure a smooth birth; people in other countries carry the seed for protection and good luck.

text continued on page 336

Sea heart

Mary's bean

Gray nickernut

True sea-bean

Sea purse

Tea mangrove

Calatola

Ticodendron

DRIFT SEEDS

Mucuna species (**True sea-bean**)   Fabaceae

One can find various species of *Mucuna* on beaches in Costa Rica. The seeds come from lianas that are often seen growing along rivers and forest edges. They have compound leaves with 3 leaflets and clusters of pealike, greenish or purplish flowers (usually bat-pollinated) on long stems. The legume fruits and the calyx surrounding the petals often have urticating hairs. Sea beans, which are ca. 3 cm in diameter, frequently reach the beaches of northern Europe.

Dioclea reflexa (**Sea purse**)   Fabaceae

This liana has clusters of beautiful violet flowers. It is pantropical, occurring from lowlands to ca. 1,000 m in forests and along rivers. The seeds can drift for at least two years and have traveled as far as Europe. The rich chestnut or chocolate-brown 3-cm seed is round and somewhat compressed, with one squared-off end. It has a narrow scar (hilum) going about three-quarters of the way around. Seeds of *D. wilsonii* are so similar to those of *D. reflexa* that definitive identification based on the seeds alone is difficult.

Calatola costaricensis (**Calatola**)   Icacinaceae

Found in Central America and northern South America, this forest tree, to 15 m, grows from lowlands to 2,000 m on both slopes. Its inner bark turns black-blue when cut. The flesh of the fruit, which is eaten by mammals, surrounds a sculptured 7-cm-long endocarp (bony layer). The seed is edible, although it may cause nausea, an effect that is less likely if the seed is roasted. People of several South American groups chew the leaves to blacken their teeth and prevent cavities. The endocarps probably fall into rivers and then make their way to the ocean. They can float for at least one and a half years, although those found after that length of time usually no longer have viable seed.

Ticodendron incognitum (**Ticodendron**)   Ticodendraceae

This 3-cm seed is not common, but a rare special find. It comes from a tree that belongs to a newly described tree family, Ticodendraceae (see illus., p. 13). The trees, which grow in Costa Rican forest from ca. 700 to 2,000 m, also occur from Mexico to Panama. The toothed leaves are reminiscent of alder or birch. The fruit is oval and has a green rind. At times, the seed may be misidentified as a small, worn-down *Pelliciera* fruit.

Caesalpinia bonduc (**Gray nickernut**), p. 338
Pelliciera rhizophorae (**Tea mangrove**), p. 343

Hippomane mancinella
Family: Euphorbiaceae

Manchineel, *Manzanillo*

Hippomane mancinella

Description: Tree to ca. 15 m with spreading crown; grayish trunk contains copious (caustic) latex. Alternate, yellow-green to dark green, smooth, shiny leaf blade ca. 8 cm long, with reddish gland at base and small, gland-tipped teeth along edge; petiole ca. 3.5 cm long. Inflorescence ca. 5 cm long with separate, tiny male and female flowers. Round, fleshy, green (to yellow with reddish tinge) fruit ca. 3 cm in diameter; an odd-looking spiky pit (illus., p. 338) in the center contains the seeds.

Distribution: From southern Florida, West Indies, and Mexico to northern South America. In Costa Rica, only on Pacific side, on beach or close to it.

Comments: Manchineel is known for its toxic properties and accidental poisonings. The fruits, which resemble apples, are plucked and eaten by many a naive beach-goer, who may also make the mistake of collecting the tree's wood to burn in a campfire. Eating the fruit results in swelling and pain from the mouth

Hippomane mancinella fruit and flowers.

through the digestive tract; vomiting and diarrhea may be followed by shock, and even death if treatment, usually stomach-pumping, is not received. Latex on the skin is irritating and blistering; in the eyes, it may cause temporary blindness. Smoke from burning wood causes respi-

Pit from fruit.

x 1

ratory irritation in addition to burning of skin and eyes.* If you come into contact with any part of the plant, wash with sea water, or preferably soap and water. Supposed antidotes include arrowroot (*Maranta arundinacea*) taken internally or applied externally, and the fruit pulp of calabash (*Crescentia cujete*) and the juice of colpachi leaves (*Croton niveus*), both used externally. On certain beaches, manchineel is very common, forming a line of trees above the high-tide mark. In parts of Florida, it has been eradicat-ed. The fruits float in the ocean, which wears them down to a tan, ribbed, corky globe, and eventually to the spiky stone that encloses the seeds. The attractive wood, once dried and cured, makes fine furniture. Indigenous people of the West Indies applied the latex to their arrows. Related species include cassava (*Manihot esculenta*, p. 175), sandbox tree (*Hura crepitans*), castor bean (*Ricinus communis*, p. 177), croton (*Codiaeum variegatum*, p. 133), and poinsettia (*Euphorbia pulcherrima*, p. 134).

*For details about specific poisoning incidents, see the interesting article in *Biotropica* (Howard 1981).

Caesalpinia bonduc
Family: Fabaceae
Subfamily: Caesalpinioideae

Gray nickernut, *Hembra y macho*

Other common name: Gray nickerbean.
Description: Sprawling, viny shrub, 1–2 m; whole plant, including alternate, compound, light green leaves, armed with hooked prickles. Leaf 40+ cm long, twice compound, 5–9 pairs of divisions in leaf, each with ca. 6 pairs of opposite leaflets to 6 cm long. Flowers yellowish, petals ca. 1 cm long. Prickly, bulging pod, ca. 7 cm long, becoming stiff and brown; splits open above, revealing 1 or 2 smooth, more-or-less oval seeds, ca. 2 cm in diameter, gray, sometimes with pink or olive tinge and fine, concentric lines.

Distribution: Thought to be from Southeast Asia originally, it is now found in both Old and New World tropics. In Costa Rica, above high-tide mark, Pacific coast.

Comments: *Caesalpinia bonduc* is an unpleasant, prickly plant to bump into at the beach, but it has attractive seeds that are often found in the drift-line litter (photo, p. 335). They have been used in jewelry and as substitutes for marbles. There are various beliefs linked to the seeds, including their value as a charm against the evil eye. The seeds float for long distances and have been known to wash up on European shores and Pacific

Caesalpinia bonduc

islands; they will still germinate after two years at sea. The seeds are ground up after roasting—they are poisonous unroasted—and made into decoctions used for treating diabetes, dysentery (boiled in milk), and fevers; the seed preparation was once considered a quinine substitute. Topical ointments for hemorrhoids have also been made from powdered seeds. The genus *Caesalpinia* is being split up, and *C. bonduc* will become *Guilandina bonduc*.

Hibiscus pernambucensis
Family: Malvaceae

Mahoe, *Majagua*

Other common name: Beach hibiscus.
Description: Tree with full crown and low branches, usually small, but up to 10 m. Leaves alternate, usually heart-shaped but sometimes rounded, ca. 13 cm across, with long petioles; yellow-green above, whitish below; small, hairy, pinkish mounds where blade meets petiole; old leaves yellow. Flower yellow with 5 overlapping petals 7 cm long, turning reddish. Dehiscent, 5-parted seed capsule, to 3.5 cm across, with small brown seeds.
Distribution: Mexico to Ecuador. In Costa Rica, both coasts, on beach or sometimes in colonies along edge of mangrove forests and Tortuguero waterways.

Comments: The beach hibiscus of Costa Rica and other New World tropical countries is very similar to the Old World mahoe (*H. tiliaceus*), which has a flower with a reddish center. Some sources consider there to be a single pantropical species, *H. tiliaceus*. The bark of this group of trees is known for its useful fiber—it is stronger when wet, which makes it appropriate for nets and ropes. Many parts of the plant are mucilaginous. The leaves are made into a poultice or decoction for relief of hemorrhoids and skin diseases.

Hibiscus pernambucensis

Mangroves

Mangrove forests, or *manglares* as they are known in Costa Rica, are zones where tides create changing water levels and river water mixes with the ocean's saltwater. Although mangrove tree species belong to several different plant families, they are usually studied as a group because of their shared habitat and way of life. Adaptations for living in this brackish environment include specialized plant tissues that store salt in old leaves, prevent salt from entering, or simply expel it. The trees also have distinctive roots that provide both physical support and a mechanism for gas exchange. The trees' seeds, many of which drop during the rainy season when water levels are higher, do not go through a dormant stage and may actually begin sprouting before dropping from the tree. An individual species' salt tolerance and capability of establishing itself in loose, saturated soils determines where it will be found in the various zones of a mangrove forest. Once they establish themselves, mangrove roots play a role in land-building because they collect sediment and debris.

Different combinations of mangrove species prevail in various regions of Costa Rica. In certain Pacific Coast areas, tea or red mangrove dominates. At Curú Wildlife Refuge on the Nicoya Peninsula, an area with tea, red, and black mangroves is easily accessible via a foot bridge. Bloodwood (*Pterocarpus officinalis*, p. 252), a tree that can tolerate inundations and some salinity, may be seen growing with mangroves in the Atlantic coastal region as well as on the Osa Peninsula and around Golfito.

Both the lichens and algae that grow on the trees and the leaf litter that accumulates below them support an array of small organisms that in turn support fish, which not only feed among the trees, but use the habitat as a nursery. Egrets, herons, and other wading birds that forage in the surrounding mud flats by day roost in the trees at night.

Human attitudes about mangrove forests run the gamut from seeing them as natural treasures to considering them waste areas that should be filled in. Direct products extracted from mangroves include bark, for tanning and medicine, and wood. Indirect products include fish and shellfish, and honey from bees taking nectar. The trees also provide buffering effects during tropical storms.

Four of the common mangrove species—black, gray, tea, and red—are described here. White mangrove, *Laguncularia racemosa* (Combretaceae), also grows in Costa Rica.

Avicennia germinans
Family: Acanthaceae

Black mangrove, *Palo de sal*

Description: Shrub, or tree to 20+ m; gray trunk; bark sometimes checkered or fissured; pencil-like pneumatophores, to 30 cm, emerge from root system. Simple, opposite, narrow (to 3 cm) leaves, to 10 cm long; lighter beneath; salt crystals noticeable on lower surface. Small, 4-lobed, tubular flowers, ca. 1 cm in diameter, with fragrant honey-mushroom odor. Flat, green, 1-seeded asymmetrical fruit, ca. 2 cm, with pointed ends.

Distribution: South Florida and Gulf states, West Indies, Mexico to South America; also in Africa. In Costa Rica, on both coasts.

Comments: The form of a mangrove tree's roots often gives away its identity. Pneumatophores—pencil-like breathing roots that obtain oxygen from the air— are characteristic of black mangrove. The pneumatophores stick up from long roots that travel horizontally close to the mud surface; lateral roots extend

Avicennia germinans

out from the horizontal roots. The leaves of this species secrete salt. The young, sprouted fruit is supposedly

Pneumataphores of *Avicennia germinans* with *Rhizophora mangle* roots in background.

edible after it is cooked well, but it is toxic when raw. Unlike red mangrove propagules, which can root in an inundated area, the floating seedlings of black mangrove need to be stranded on substrate in order to root. Black mangrove can tolerate a very high salt content in its soil. Nectar from the flowers yields good honey. The bark produces dyes, and it is the basis for home remedies for dysentery (internally) and sores and hemorrhoids (externally). The hard wood makes decent lumber, as well as charcoal. *A. bicolor*, another species in this genus that occurs in Costa Rica, has a wider leaf (3.5–8 cm) and smooth, elliptical fruit with rounded ends. Until recently, *Avicennia* was classified in Avicenniaceae or Verbenaceae.

Conocarpus erectus
Family: Combretaceae

Buttonwood, *Mangle negro*

Other common names: Gray mangrove, *mangle botoncillo, mangle torcido*.
Description: Tree often under 10 m, with furrowed bark, gnarly trunk; when growing on beaches, it reclines toward the ocean. Alternate, somewhat thickened leaves ca. 8 cm long, with minute pits (domatia) near midvein; glands at base of blade. Dioecious; tiny, honey-scented, greenish-white flowers in round heads (ca. 0.7 cm). Numerous winged nutlets packed into reddish-brown, conelike head, ca. 1 cm in diameter.
Distribution: South Florida and Mexico to South America, and West Africa. In Costa Rica, on beach and landward edge of mangrove swamps, both coasts and Cocos Island.
Comments: The small pockets, or domatia, in the vein axils on the leaf may house mites. The dense, hard wood is good for charcoal and firewood, as well as for turned articles

Conocarpus erectus

Conocarpus erectus infructescence.

x 2

and in maritime construction. The tree's twisted branches and trunk yield attractive driftwood. The bark is used for tanning skins. Decoctions of various plant parts are used for treating fever, venereal diseases, and diabetes, or as a tonic. White mangrove (*Laguncularia racemosa*), sura (*Terminalia oblonga*, p. 238) and tropical almond (*T. catappa*, p. 331) are in this family. Buttonwood is often referred to as a mangrove associate because it grows in less-inundated areas than typical mangroves.

Pelliciera rhizophorae
Family: Pellicieraceae

Tea mangrove, *Mangle piñuela.*

Description: Tree to 20 m; trunk fluted, ropy, and cone-shaped at base. Leaves to ca. 15 cm long, alternate, simple (glandular teeth may be present on margin), fairly thick, with fine fibers evident when leaf is torn. Flowers with 5 cream to reddish, petal-like sepals that are 2 cm long; 5 white petals, 6 cm long, and 2 large, 7-cm, light green to pink bracts below flower (photo, p. 344); stamens and style pro-truding ca. 6 cm from center of flower; scent somewhat sour. Red-brown, bulb-shaped, furrowed fruit to 10 cm long, including beak; 1 seed. (See photo, p. 335).

Distribution: Nicaragua to Colombia, most common on Pacific coast. In Costa Rica, in Pacific coastal mangrove forest.

Comments: The buoyant part of the fruit is the rounder part, so when it drifts, the beak, where roots emerge,

Pelliciera rhizophorae

Pelliciera rhizophorae flower.

points downward. The flowers are seen in the dry season, but also in June and July. They are visited by hummingbirds and insects, and probably have nocturnal pollinators. Tea mangrove's cone-like base is made up of layers of adventitious roots whose outer, spongy covering is important in gas exchange. Recent reclassification places this family within the Tetrameristaceae.

Rhizophora mangle
Family: Rhizophoraceae

Red mangrove, *Mangle*

Other common name: *Mangle caballero.*
Description: Tree 3–25 m, but usually less than 12 m; this species varies in size probably due to a combination of factors such as nutrients, salinity, and flooding. Has arching stilt roots that extend from trunk and branches; conspicuous lenticels on roots play role in gas exchange. Thick, opposite leaves, ca. 15 cm long, peppered with dark dots below; new growth covered by long pointy hood formed by stipules. Inflorescence with up to 6 flowers, each ca. 2 cm across with 4 yellow-green sepals and 4 cream, woolly petals. Fruit 2–3 cm long, with one seed. Seed germinates while still on branch, and an embryonic root that looks like a green bean with a brown tip extends to ca. 30 cm.
Distribution: Southwest Pacific islands, Florida and Antilles, and Mexico to Peru; also in West Africa. In Costa Rica, forms dense mangrove forests called *manglares* on both coasts in salty or brackish water; also along waterway to Tortuguero, and often in river estuaries.

Comments: Red mangrove is found where rivers and streams meet the sea, in the estuaries where fresh and saltwater mix and the tides cause fluctuations in the water level. The lower part of the root system is submerged or in saturated soil, so gas exchange occurs via lenticels on the aerial prop roots. Seeds germinate while on the tree, and the resulting seedlings drop off after 3–6 months of growth; they can float and remain alive for at least 12 months. These sea pencils, as they are sometimes called, can take root either vertically or horizontally by contacting a substrate. At first, roots grow from the brownish tip, but stilt roots begin to appear while the plant is still under a meter tall. Many insects and crustaceans depend on mangroves for food and shelter. Various kinds of caterpillars feed on the leaves, shoots, and roots, beetle larvae feed on twigs, and an arboreal crab (*Aratus pisonii*) eats the leaves. An isopod, *Sphaeroma terebrans*, often inhabits the prop roots; its boring behavior causes some of the roots to die, but those that survive branch as a result of the isopod damage. Sediments settle out around the roots of red mangrove, which results in land-building. Although some bees visit the flowers, they are thought to be wind-pollinated. The reddish-brown wood, which is hard and heavy, is useful in making boat frames, fence posts, fuel, and charcoal. The tannin content of bark is 20–30%, which makes it good for tanning hides. Preparations of red mangrove have been used in infusions to treat diarrhea and leprosy, and for lowering fevers; they are also used externally in baths for skin conditions. Another *Rhizophora* species that one may encounter in Costa Rica is *R. racemosa*, which grows only on the Pacific coast. *R. racemosa* has larger inflorescences with up to 70 flowers, lacks the uniform covering of black dots on the underside of the leaf, and often has a better-defined trunk. It grows in more protected areas of mangrove forest, such as inside meanders and in the inner forest zone.

Rhizophora mangle

Caribbean Beaches

The flora of the Costa Rica's Caribbean beaches has some elements that make it distinct from the typical Pacific beach vegetation. Some of these plants are only seen on the Caribbean, while others are just more abundant and noticeable there.

Hymenocallis littoralis
Family: Amaryllidaceae

Spider lily, *Lirio de playa*

Description: Plant with 60-cm-tall flower stalk and pear-shaped bulb at base. Tonguelike leaves, 0.5–1 m long. Aromatic flower with tube to 20+ cm long, 6 green stamens with orange anthers; white, 11-cm-long tepals; funnel-like, membranaceous cup formed by fusion of bases of anther filaments. Three-lobed capsular fruit.

Distribution: Mexico to South America. In Costa Rica, near Caribbean coast, above high-tide line or behind beach; also on the Nicoya Peninsula.

Comments: Spider lily bulbs contain some powerful alkaloids, including tazettine, which lowers blood pressure, and lycorine, which is emetic. Hawkmoths pollinate the flowers. The seeds, which are 75–90% water and sometimes begin developing while still in the fruit capsule, are probably water-dispersed. There are ca. 50 species of spider lilies in the New World, some cultivated in tropical gardens. *H. littoralis* is related to swamp lilies (*Crinum* spp.).

Adrian Hepworth

Hymenocallis littoralis

Chrysobalanus icaco
Family: Chrysobalanaceae

Coco plum, *Icaco*

Description: May grow to 5 m tall, but grows as a sprawling 1.5 m shrub on beach dunes. Alternate, simple, smooth leaves are thick and leathery, ca. 5 cm long. Small, greenish white flowers. Fruit to 4 cm long, whitish purple-pink to black, with soft white flesh and one pit containing an oily seed.
Distribution: From southern Florida to northern South America, including West Indies; along coasts but sometimes in freshwater swamps. In Costa Rica, both coasts; may be found in Tortuguero and points south on the Caribbean coast, and on the Osa Peninsula.
Comments: This is an attractive beach shrub that is grown as a small ornamental tree in some yards. The edible fruit is a bit unpleasant since it has an astringent quality that sucks moisture from one's mouth. Its taste improves if one soaks and peels the fruit before stewing it or making jam. The inner kernel of the seed contains about 20% oil that can be ingested or used in candle making. The fruit is high in tan-

Francis X. Faigal

Chrysobalanus icaco

nins, and it and various other plant parts are made into teas for dysentery and chronic diarrhea.

Costus woodsonii
Family: Costaceae

Dwarf French kiss, *Caña agria*

Other common names: Bitter cane, red cane.
Description: Herbaceous plant, 1–2 m, spreading by rhizomes. Spiraling stem; smooth, shiny leaves, to 15 cm long, with somewhat heart-shaped base. Topped by an inflorescence shaped like a pine cone. Thick overlapping, glossy, red bracts, each associated with one flower (or fruit); small calyx with tubular, 3-lobed corolla; 1 fertile petal-like stamen; corolla red,

with tubular, yellow to orange labellum or lip, to 3 cm long. Spongy white fruit topped by calyx; black seeds with white aril.
Distribution: Nicaragua to Colombia. In Costa Rica, along Atlantic beaches, behind strand vegetation.
Comments: Costa Rica, Panama, and Colombia are particularly rich in species of *Costus*. *C. woodsonii* grows adjacent to Caribbean beaches, but many other of the 24 Costa Rican

Adrian Hepworth

Costus woodsonii

species grow inland, in lowland or mountain forests. *C. woodsonii* has leathery bracts that not only physically protect the flower buds and developing fruit, but have extrafloral nectaries. The nectaries attract ants that deter seed predators such as the picture-winged fly (*Euxesta* sp., family Otitidae). The flowers are said to be edible, and the plant's acidic juice is used medicinally as a diuretic and for other ailments like fever and cough. The common name for any costus in Costa Rica is caña agria. In English, spiral ginger is a general name given to costus species, although the species used in horticulture have fanciful names, as does *Costus woodsonii*: dwarf French kiss. Also see crepe ginger (*Cheilocostus speciosus*, p. 128), a nonnative ornamental.

Coccoloba uvifera
Family: Polygonaceae

Sea grape, *Papaturro*

Other common name: *Uva de playa.*
Description: Low-branching tree, generally shorter than 10 m, with gnarly, smooth trunk. Alternate, simple, rounded, stiff leaves 10–25+ cm across, with a reddish midvein; reddish-brown sheaths where leaves join stem. Dioecious; inflorescences to 30 cm long, with small white, honey-scented flowers. Purple fruit 1–2 cm long with one seed.
Distribution: Florida, West Indies, Mexico, Central America, and various parts of South America. In Costa Rica, on both sea coasts, but more common on the Caribbean side.
Comments: Punta Uva (Grape Point) on the Caribbean coast is named after the sea grape. The cuplike structure of the flower swells after flowering and becomes part of the fruit. The clusters of red-purple fruits resemble grapes,

but they are not closely related. These tasty berries can be eaten raw or made into preserves or jelly. Planted as an ornamental, sea grape will tolerate salty, dry conditions. Its astringent properties make a decoction of the bark and/or roots an effective home remedy for diarrhea. The hard, heavy wood is suitable for furniture-making, firewood, and posts. The red sap from the bark has been used for tanning and dyeing hides. More than a dozen species of *Coccoloba* are native to Costa Rica, generally occurring in forests, below 1,000 m. The Polygonaceae family includes common North American weeds such as *Polygonum* and *Rumex*, along with buckwheat (*Fagopyrum* sp.), rhubarb (*Rheum*), and the ornamental coral vine (*Antigonon* spp., p. 149).

Francis X. Faigal

Coccoloba uvifera

Morinda citrifolia
Family: Rubiaceae

Noni, *Yema de huevo*

Other common name: Indian mulberry.
Description: Small, shrubby tree, often 2–3 m but sometimes to 8 m tall. Very large, simple, opposite leaves, to 35 cm long, with prominent, ladderlike lateral veins and light midvein. Hard, green, conelike flower heads, ca. 4 cm long, in leaf axils; 5-lobed, 1-cm long, white flowers. Ripe fruit with foul cheese odor; the compound fruit is heavy, green-white, somewhat translucent, potato-shaped, 7 cm wide by 13 cm long; pentagonal sections, each with a green to brown "eye"; small seeds, ca. 3.5 mm.
Distribution: Natural range appears to be Polynesia, Australia, India, and Malaysia; introduced to Hawaii and cultivated in parts of tropical Asia. Naturalized on the Caribbean shores of Costa Rica.

Comments: The most striking aspect of this tree—the odd-looking, lumpy, and malodorous fruit—is sure to be noticed by the Caribbean beach-goer. The ripe fruits, while not very appealing, are edible raw or cooked, as are the young leaves. The fruit is sometimes fed to pigs. There are many medicinal uses, especially in Southeast Asia and the Pacific Islands. These include using the juice as a gargle for sore throat; taking a decoction of leaves and bark for tuberculosis; applying young fruit, with salt, on deep wounds; and using leaf poultices to relieve pain and reduce inflammation. Noni fruit shampoo is said to be insecticidal. In Samoa, the root is used to alleviate toothache. A number of alternative medicine sites on the Internet tout the "Tahitian noni" fruit as a virtual panacea, and there are companies that sell the juice. While there may be

Francis X. Faigal

Morinda citrifolia

some benefits to ingesting noni, there is a lot of hype in the marketing. The wood is attractive and has been used in some regions for turnery. In India, the plant is seen mainly as a source of dye—the root yields reds, purples, and yellows. Bees and ants visit around flowers; perhaps they are scavenging nectar from old flowers. The fruit mass, which may be eaten by bats, is formed by succulent calyces that expand around the individual, one-seeded fruits. Most of the ca. 80 species in the genus are Old World. A couple of horticultural cultivars with variegated leaves exist. *M. panamensis*, also of Costa Rica, has a smaller fruit cluster, ca. 3 cm in diameter, leaves less than 20 cm long, and grows in the rainforest. There may be confusion between the two species because both are reported from similar sites on Costa Rica's Caribbean coast.

Stachytarpheta jamaicensis
Family: Verbenaceae

Blue snakeweed, *Rabo de gato*

Other common names: Jamaica vervain, *verveine*.

Description: Semiprostrate shrub, 0.5–1.5 m tall; mostly glabrous, squarish stems. Leaves usually opposite, ca. 8 cm, with large teeth, lower third of leaf tapered into winged petiole. Flower spike, ca. 25 cm, with small purple or blue flowers, 1–1.5 cm long; flowering beginning at base of spike, proceeding upward. Small, dry, 2-seeded fruit, to 0.5 cm.

Distribution: Southern coasts of United States; Mexico to South America, various Caribbean islands. Introduced and spreading in Old World tropics, it

has become a pest in Hawaii. In Costa Rica, found on the Atlantic coast, often on the leeward side of coconut palms; on Pacific coast, along roadside and beach areas of Osa Peninsula; also in various lowland or foothill areas around country.

Comments: Blue snakeweed is used on various skin conditions, and decoctions are taken for fever and coughs and to expel worms; for menstrual disorders; and as a diuretic or tonic. A root decoction is used to induce abortion. "Brazilian tea" is made from dried leaves. The related porterweed (*S. frantzii*, p. 107) is often seen as an ornamental.

Stachytarpheta jamaicensis

7. Typical Tropical Groups

There are some plants that one instantly associates with the tropics—the philodendron, with its large heart-shaped leaves, for example, or the pineapple-like bromeliads—but plants such as these are representative of families of plants that have a tremendous amount of variation within them. This chapter focuses on six groups of typical tropical plants and describes not only the similarities within each group but also their range of diversity. Not all of the plants included are common, and some of those included may be difficult to find in the field but nevertheless illustrate some interesting aspect of the family or group. The plant groups included are:

Arum and Philodendron Family
(Araceae)

Bromeliad and Pineapple Family
(Bromeliaceae)

Palm Family
(Arecaceae)

Heliconia and Bird of Paradise Families
(Heliconiaceae and Strelitziaceae)

Orchid Family
(Orchidaceae)

Ferns and Lycophytes
(various families)

Left, *Heliconia wagneriana*

Arum and Philodendron Family (Araceae)

Philodendrons and their relatives have luxuriant foliage that, with its strong identification with rainforests, often makes it a core element in designs for tropical gardens and interior plantings. Similarly, the bright red, plasticlike spathes of the flamingo flower (*Anthurium andreanum*) and the white funnels of the calla lily (*Zantedeschia aethiopica*) add a certain flair to flower arrangements. Despite the many horticultural uses this plant family has come to represent, its more utilitarian members are intrinsic to the daily lives of millions of people around the world. Certain Araceae such as taro (*Colocasia esculenta*), and some *Xanthosoma* species, provide edible underground stems that have been cultivated as dietary staples for thousands of years.

Aroids, as members of this family are known, are easy to recognize when they are flowering. The flowers themselves are small and not particularly showy, but they occur in a distinctive inflorescence called the spadix, which is associated with a bract called the spathe. Variations of this spadix/spathe characteristic help define the different genera. Growth forms of these plants range from terrestrial to epiphytic and climbing; some species, such as water lettuce (*Pistia stratiotes*), are aquatic. The fruits of most aroids are berries that are eaten by birds or mammals, including bats.

The entire leaves, which have sheathing stalks, are typically alternate along a stem, but may be clustered at the base of the plant. The leaves vary from simple and elliptical to many-lobed or compound. The family contains more than 100 genera, with up to 3,000 species worldwide; ca. 249 species occur in Costa Rica. While almost all aroids are tropical, skunk cabbage and Jack-in-the-pulpit are two well-known temperate species.

Aroids attract pollinators through fascinating mechanisms; they may produce heat and odors, offer a source of chemical compounds that pollinators use as building blocks for pheromones, or have sterile stamens called staminodia that pollinators use as food. Certain philodendrons, dieffenbachias, and xanthosomas heat up and release pungent scents that attract nocturnal scarab beetles. The large *Dracontium* and Old-World *Amorphophallus*, which are measured in meters rather than centimeters, waft out revolting odors to attract flies, carrion beetles, and other pollinators. Some of the

William A. Haber

Author standing next to *Dracontium gigas.*

anthuriums and spathiphyllums, on the other hand, emit pleasant perfumes that draw in male euglossine (orchid) bees.

Most Araceae contain calcium oxalate crystals. These crystals have barbs that can stick in the tongue of an animal that tries to eat the plant, creating an extremely unpleasant sensation. Other toxins produced by the plant may even be delivered into the body via these piercing crystals. Species used for food, such as taro, have been cultivated to be less irritating. The swollen underground stems, or corms, are prepared by removing their outer skin and cooking them.

spathe and spadix

Philodendron habit

Philodendron spathe opened, showing spadix, with female flowers below and male above.

Following are some of the distinguishing features of the larger Araceae family genera that are found in Costa Rica:

Anthurium: Many are epiphytic. Many sizes and shapes of leaves; often a collective vein that runs very close to the leaf margin, with some netted veins present. Uniform spadix (perfect flowers), spathe off to the side. *Anthurium* is the largest genus in the Aroid family.

Monstera: Usually begins on ground and then climbs. Juvenile and adult leaves differ—the former sometimes tightly appressed on a tree trunk, the latter often deeply lobed or with holes. Uniform spadix (perfect flowers); spathe erect and concave, forming an open hood around flowering spadix and later falling.

Philodendron: Often climbing or epiphytic. Usually with heart- or arrowhead-shaped leaves, venation parallel and much finer than in *Anthurium*. Spadix with female flowers below and male above, spathe constricted in middle, bulging below, partly open above when flowering and closing post-flowering.

Syngonium: Usually begins on ground and then climbs; milky sap. Leaves often with 3 or more lobes, collective veins near leaf margin; venation not parallel. Spadix with female flowers below, male above, spathe constricted in middle; loses upper part of spathe after flowering.

Dieffenbachia: Terrestrial, often erect with robust stem. Cut tissue of some species smells like skunk; lateral veins in leaf blade prominent and often ascending, no collective vein. Spadix with female flowers below, male above, upper part of spathe open when in flower, closing post-flowering and breaking open when fruit matures (orange-red).

Anthurium obtusilobum

Anthurium, *Anturio*

Anthurium obtusilobum; developing fruit on left, flowering spadix on right.

Description: Epiphytic or terrestrial, erect; size of plants varies from about 20 cm tall to more than 60 cm. Heart-shaped leaf blades to ca. 45 cm on long petioles; distinctive collective vein runs along most of blade, just in from leaf margin. Inflorescence on long stalk; white to light green spathe to ca. 12 cm long; white to cream (tinged pink) spadix to 10 cm, with a spicy toothpaste scent. Spadix turns purple-red after flowering; juicy red berries develop very gradually.

Distribution: Costa Rica to Colombia, 600–1,550 m. In Costa Rica, in forest on both slopes, widespread.

Comments: An individual spadix of *Anthurium obtusilobum* flowers for two or three weeks. It goes through a female phase (stigmas and stigmatic droplets present), a rest phase, and then a male phase (anthers dehiscing). Several

Male orchid bees visiting *Anthurium hoffmannii.*

anthurium species resemble *A. obtusilobum*. *A. hoffmannii* (photo opposite page) has a similar scent, but in both it and *A. monteverdense*, the collective vein is only in the distal portion of the leaf. The spadices are more cream-yellow in *A. monteverdense* (photo right) and emit an exquisite gardenia-like scent. Another species, *A. formosum*, tends to be larger, with blades to 80 cm long, and it has color variation (white to purplish) in both the spathe and the spadix. Male orchid bees (e.g., *Euglossa* and *Eulaema* spp.) that visit the flowers scrape off the scent compounds, which they convert into pheromones to attract female bees. Distinct scent compounds often attract particular bee species. Little is known about the pollination process for the hundreds of other *Anthurium* species that have very different colors and scents from those described here.

Anthurium monteverdense, late flowering stage.

Anthurium salvinii

Bird's-nest anthurium, *Tabacón*

Description: Epiphyte, or growing on steep, rocky slopes; roots thick and crowded. Leaves in rosette; leaf blade 1-1.8 m long, with relatively short petiole; thick, raised midrib; new leaves very plasticlike, with odd-looking, hooded cataphylls that resemble cabbage leaves curling around base of petiole; these cataphylls dry and shrivel with age. Spathe to 45 cm long, purple or green-tinged purple; pendant, lavender spadix is on a long (ca. 50+ cm) stalk; spadix is ca. 40 cm long when in flower, but up to 60 cm when in fruit. Red berries to 1.5 cm long.
Distribution: Mexico, parts of Central America, and Colombia, 0–1,500 m. In Costa Rica, often at ca. 1,000 m in moist to wet forest from north to south, both slopes.
Comments: This is one of the largest anthuriums in Costa Rica. The Spanish

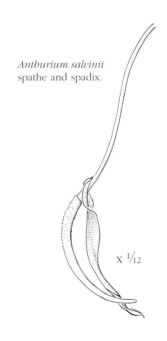

Anthurium salvinii spathe and spadix.

x ¹/₁₂

Anthurium salvinii

name, *tabacón*, is given to a variety of plant species whose leaves resemble those of tobacco. The rosette of leaves collects precipitation and leaf litter and directs it to the center of the plant.

Tabacón does best in partial shade and makes a great patio plant, but it eventually requires an incredible amount of space; the author has one obstructing the entryway to her house.

Dieffenbachia **spp.**

Some common names: Dumb cane, *lotería, sainillo, rayo de luna.*

Description: In general, terrestrial, fleshy plants, often ca. 1 m tall but to more than 2 m; milky sap; in some species, crushed plant gives off skunk-like odor; stems sometimes reclining. Oblong leaves sometimes mottled or with distinct white or yellow stripe along midvein; with long, sheathing petioles; no collective vein in leaves. Separate male (upper portion) and female flowers (lower) on spadix; spathe green, closed and overlapping around lower portion, open and hood-like above when flowering; female

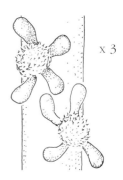

x 3

Dieffenbachia oerstedii female flowers with staminodia.

flowers spaced out and surrounded by 3–5 staminodia; male flowers tightly packed in upper part of the club-shaped spadix; spathe and spadix often curving when in fruit. Ripe, orange to red fruits, ca. 1.5 cm, make a showy display when the spathe splits open.

Distribution: Various species occur in wet or moist forest from Mexico to South America; often in lowlands, widespread. *D. oerstedii* (photo below left) reaches 1,500 m in Costa Rica.

Comments: Dumb cane species comprise a well-known group of houseplants that are frequently seen in shopping malls and offices. There are many variegated cultivars. There are a dozen or more species in Costa Rica—the exact number is not clear; at least four new species have been described from La Selva in recent years. The name dumb cane comes from the fact that ingesting the plant impairs speech, not to mention breathing and swallowing; calcium oxalate crystals and other unknown substances (possibly toxic proteins) produce burning and swelling of the mouth and throat. Although death in adults is unlikely, resultant fatal choking has been reported. The sap may be irritating to the skin. Researchers H. Young and D. Beath have studied the pollination of *D. nitidipetiolata*, a large species at La Selva that flowers from March to September. At dusk on the second day of being open, the spadix warms up, a foul odor is given off, and the female flowers are receptive. Large scarab beetles, some dusted with pollen, then arrive to mate and feed in the lower chamber of the spathe; the beetles munch on sterile stamens that are essentially protein food bodies. The next night the beetles get a dusting of pollen as they exit by climbing along the upper part of the spadix, where the male flowers are located. *D. oerstedii* has a similar pollination system, but produces no odor (Valerio 1984).

Dieffenbachia oerstedii

Dieffenbachia cultivar

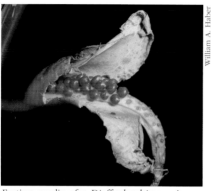

William A. Haber

The spathe tightens up and stays closed until the fruits are ripe (red-orange); it then splits apart to reveal the berries. The fruits are probably eaten by both birds and mammals. Dumb cane is used in homemade concoctions used to kill cockroaches and rodents. In South America, people take advantage of the nasty qualities of dieffenbachia and add it to curare preparations for poison arrows. Dumb cane sap is used externally to get rid of warts, and it is purported to cause sterility if ingested or injected.

Fruting spadix of a *Dieffenbachia* species.

Monstera deliciosa   Swiss cheese plant, *Costilla de Adán*

Other common names: Split-leaf philodendron, ceriman, *piñanona, mano de tigre.*

Description: Thick-stemmed climber to more than 8 m long; roots along stem,

Climbing shingle leaves of a *Monstera* species.

some dangling. Mature leaves huge, dark green, rounded in overall shape, but with deep cuts and holes of various sizes and shapes; leaf blade ca. 80 cm with equally long petiole. Bathtub-shaped, whitish spathe ca. 30–35 cm long; cream spadix, ca. 25 cm long, turning juicy in fruit, with individual hexagonal berries evident.

Distribution: Mexico to Panama. In Costa Rica, fairly widespread in forest, both slopes, from ca. 500 to 2,000+ m.

Comments: This well-known ornamental is seen indoors as well as outside. In nature, monsteras begin as vines on the forest floor that head toward a dark place (skototropic); this is usually the base of a tree, where the plants change their habit and begin to climb toward the sun-filled canopy. Most species of *Monstera* go through changes in leaf shape and size as they ascend, and many have juvenile shingle leaves that cling tightly to the tree trunk. The edible fruit of *Monstera deliciosa*, which takes about a year to mature, has a pineapple/banana flavor. Once the caplike hexagonal green pieces start falling off, it's ready to eat; pick it just as it is beginning to ripen and wrap it in a bag for a

Monstera deliciosa

few days. Washing the fruit before eating it seems to cut down on any calcium oxalate crystals that are prevalent in the green fruit and vegetative parts (see Araceae introduction, p. 354). Some people appear to be allergic to the fruit. The adaptive value of having holes in the leaves may be to allow light to filter down, to mimic insect damage, to shed excess water, and/or to prevent wind damage. The flowers appear to be bee-pollinated, possibly by *Trigona* spp. The fruits are probably eaten by birds and mammals.

Philodendron cultivar — Tree philodendron, *Filodendron*

Philodendron cultivar

Description: Stem 1.5–2 m tall, with many roots coming off of stem; large leaf scars. Petioles 1+ m long, with equally long leaf blades, heart-shaped in general outline, pinnately lobed with veins pinkish beneath. Spathe and spadix each 25 cm long; spathe lime green outside, pink inside; spadix white.
Distribution: Various hybrids and cultivars planted extensively as ornamentals throughout the world, doing best in warm, wet areas.
Comments: More than 500 species of *Philodendron* exist in the neotropics,

and a wide array are used in tropical landscaping. Many of the treelike philodendrons with deeply dissected leaves are cultivars or hybrids involving the Brazilian species *P. bipinnatifidum* (also called *P. selloum*). The one depicted here may be *P.* 'Evansii', a cross between *P. bipinnatifidum* and *P. speciosum*. A study of *P. bipinnatifidum* showed that the sterile florets in the spadix use fat to generate heat, a system comparable to that found in mammals (i.e., brown fat). (For information on pollination in philodendrons, see *P. radiatum*, p. 248)

Spathiphyllum friedrichsthalii   Peace lily, *Calita*

Other common name: *Anturio blanco.*
Description: Terrestrial plant typically less than 1 m tall. Elliptical leaf blades 30 cm or longer, with close parallel venation and sheathing petiole. Inflorescence on long stalk, spathe 13 cm or longer, white with some green extending into base on

back; spathe forms an open hood around cream to yellow, fat, cylindrical spadix, ca. 5 cm long; uniform with bisexual flowers with scent somewhat like a pleasant bathroom disinfectant. Fruits green. The size of plant parts can vary from one population to another.

Spathiphyllum friedrichsthalii

Orchid bees visiting flowers.

Distribution: Southern Mexico to Colombia. In Costa Rica, Atlantic lowland, partly open swamp forest to wet roadside and pasture areas; planted in gardens in southern Caribbean area.

Comments: The eight species of *Spathiphyllum* that occur in Costa Rica and the ca. 50 species in the rest of the world are nearly all similar in appearance.

Although a few are found in the Old World, the majority are New World tropical species that prefer shady forest or open, wet areas. Male orchid bees, which visit the flowers for their scent compound, are the main pollinators. The spathe turns green when the plants are in fruit. Several species and cultivars are widely used as shopping-mall ornamentals.

Xanthosoma undipes

Elephant ear, *Oreja de elefante*

Description: Large plant with thick, 1.5+-m-tall fissured trunk, tan and spongy/corky, with tubercles. Sheathing petiole ca. 1 m; arrow-shaped blade 1+ m long, with prominent lobes, intricate venation pattern. Inflorescences on long stalks; spathe, to 30 cm, surrounds shorter spadix; lower maroon-green part of spathe surrounds female and sterile flowers; upper white portion, which forms hood around male flowers, wilts and rots after flowering. Hundreds of many-seeded, fleshy, green fruits form on spadix.

Distribution: Central America and South America. In Costa Rica, both slopes; common in certain wet regions above 1,000 m and seen occasionally at lower wet sites such as La Selva and the Osa Peninsula; invades old pastures and natural openings in forest such as river edges.

Comments: This is another heat-producing aroid. At around 6:00 p.m., the spadix heats up to 42° C, wafts out a menthol-like scent and attracts pollinating scarab beetles (*Cyclocephala* spp.), which spend 24 hours inside of the base

Xanthosoma undipes

Developing fruit and flowers.

of the spadix. The temperature drops as evening progresses, but it rises again just before dark of the second day. Pollen is released and adheres to the beetles as they leave. There are many other insects associated with the inflorescence: mites that hitchhike on beetles from plant to plant take nectar from *Xanthosoma*; flies and sucking bugs reproduce in the inflorescences; and spiders and predatory and detritus-feeding insects are sometimes found in the spathe. The ripe fruit, which is exposed as the spathe breaks up and peels back, is eaten by bats. Five other species occur in Costa Rica, one of which, *X. wendlandii,* has compound leaves and mottled petioles. Tannia, or tiquisque (*Xanthosoma sagittifolium*, p. 167), is also in this genus.

Zantedeschia aethiopica   Arum lily, *Cala*

Other common names: Calla lily, *cartucho.*

Description: Clumped herbaceous plant, less than 1 m. Entire, arrow-shaped leaves, 20–40 cm long; white funnel-like spathe, ca. 15 cm long, enveloping yellow-orange spadix with female flowers below and male above.

Distribution: Originally from South Africa, but naturalized in Costa Rica; damp or wet areas, mid- to high elevations.

Comments: Common in spots on the route over Cerro de la Muerte and toward the high volcanoes Poás and Irazú, these ornamentals have escaped into moist, cool pastures. In Costa Rica, as elsewhere, they are a favorite in cut-flower arrangements for funerals. Many an artist (including Matisse, O'Keefe, and Rivera) have chosen calla lilies as subjects for their paintings. Although some sources list this as a poisonous plant, people use the plant topically to soothe burns and wounds, and others eat the cooked leaves and rhizomes; there are irritating calcium oxalate crystals present in fresh material. *Cartucho* means paper cone in Spanish.

Zantedeschia aethiopica

Bromeliad and Pineapple Family (Bromeliaceae)

This is another of the tropical families that is easy to recognize. Bromeliads look like pineapple tops perched on tree trunks and branches, where they collect water and organic debris in their rosette of straplike leaves. Not all bromeliads are epiphytes, however; some inhabit rocky ridges in dry forest, while others flourish on the ground in wet or dry forest. A few, such as Spanish moss (*Tillandsia usneoides*), have atypical growth forms. Members of the family exhibit an array of adaptations for amazingly different climates, ranging from hot, dry deserts to wet, cool cloud forest and even to paramo above tree line. Species vary in size from a few centimeters to 10 m.

The flowering stalk of a bromeliad, which forms in the center of the leaf cluster, may be a compact or tall spike, or a branching inflorescence. The perfect flowers, with three sepals and three petals, may be purple-black, white, yellow, orange, violet, or pink. Hummingbirds visit many species, but bats or insects pollinate others. Bracts and regular leaves often add bright red to the display. Epiphytic species use the prolific roots at their base for clinging to tree trunks; specialized hairs on the leaf surface, which have the form of stalked scales, play an active role in water and nutrient absorption. The leaves may be narrow and covered with grayish scales, or they may be wider, less scaly, and channeled.

Fruits are either berries eaten by birds and mammals or capsules with tufted or winged seeds that are wind-dispersed; the pineapple, with its multiple fruit, is an oddity. Side shoots—pups in gardeners' language—are typical; they eventually take over after the main plant has flowered and died. Estimates of the total number of species in the family vary, up to about 2,400, with many ornamental cultivars. With one exception (*Pitcairnia feliciana* of West Africa), this is a Neotropical family. Ca. 200 species occur in Costa Rica, the genus *Werauhia* (which now includes many species formerly in *Vriesea*), making up nearly a third.

Bromeliad-laden tree, Cerro de la Muerte.

Bromeliads and other epiphytes on a tree in Monteverde.

Bromeliads are popular with tropical gardeners because of their combination of continuously handsome foliage and interesting, often brightly colored, inflorescences. The plants also have a predictable compact form and are fairly easy to grow. Many do quite well as houseplants, even when neglected.

Besides pineapple, which is a source of food and proteolytic enzymes, humans use a few other bromeliad species—Spanish moss for stuffing furniture, and some *Aechmea* and *Ananas* species for fiber. In Costa Rica, people eat parts of the wild pineapple or piñuela (*Bromelia pinguin*).

A tank bromeliad.

Bromeliads that accumulate water and leaf detritus in the tanks formed by their tightly overlapping leaves provide habitat and/or a breeding ground for protozoa, aquatic insects, and frogs. Some species that live in drier climates have sheathing leaf bases that bulge out, forming hollows where ants can nest. These spaces, which are not vaselike, do not collect water as the tank bromeliads do, and the plants do not always grow upright. The feces, food waste, or organic garbage of these insect and other inhabitants probably contribute to nitrogen needs of the plants.

The genera *Ananas*, *Aechmea*, *Bromelia*, and *Ronnbergia* have fleshy fruits and leaves with toothed margins. *Catopsis*, *Guzmania*, *Racinaea*, *Tillandsia*, *Vriesea*, and *Werauhia* are usually epiphytes; they have entire leaf margins and seeds topped by a tuft of hairs. *Pitcairnia* and *Puya* species have a spiny leaf margin and a capsular fruit with appendaged seeds.

Aechmea mariae-reginae   Queen aechmea, *Corpus*

Aechmea mariae-reginae

Aechmea fasciata

Aechmea penduliflora

Other common name: *Espíritu santo.*
Description: Large epiphytic tank bromeliad, ca. 1 m across, to 1+ m tall in flower. Leaves with wide, sheathing bases; edge of leaves serrated/spiny. Erect flower stalk has skirt of ca. 10-cm-long pink bracts below the inflorescence of 50+ flowers; separate male and female plants (dioecious); flowers purple-blue with white at base, turning pinkish with age. Fleshy fruits.
Distribution: Found in Nicaragua and Costa Rica; in Costa Rica, lowland wet forest on Atlantic side, usually below 1,200 m; rare on Pacific slope.

Comments: The size and colorful display of this bromeliad make it a spectacular species. The bunch of ripe (supposedly edible) fruit may weigh more than 2 kilograms. The flowers are hummingbird-pollinated. This is the only dioecious *Aechmea.* A related terrestrial species, *A. magdalenae*, which can grow to 2.5 m tall, has a large cluster of reddish bracts in the center. There are many ornamental *Aechmea* species, including the Brazilian *A. fasciata* (photo above left), whose flowers and foliage make it popular with tropical gardeners, and the native *A. penduliflora* (photo above right).

Ananas comosus   Pineapple, *Piña*

Description: Terrestrial rosette of rigid, narrow leaves to 1 m long. Leaves glaucous beneath, margin with or without spines. Inflorescence ca. 10 cm tall; center made up of red bracts and up to 200 purple to blue flowers. Aggregate green-yellow to golden fruit, ca. 20 cm long, topped by cluster of green bracts.
Distribution: Probably originally from lowlands of Brazil and Paraguay; no

Pineapples for sale at a fruit market.

longer found in the wild. In Costa Rica, cultivated mostly in lower elevations (to 1,500 m); large plantations south of San Isidro de El General.

Comments: This bromeliad occurs on most continents since its delicious fruit has led to its cultivation around the world. Columbus had a taste of it on the island of Guadeloupe during his second voyage (1493), and by the mid-1500s, European traders had transported it as far as India and the Philippines. By the 1600s, it was growing in European greenhouses. It made its way to Hawaii, via England, in 1885; most of today's pineapple production is in Malaysia, the Philippines, and Hawaii. Besides being a wonderful fruit that contains B vitamins as well as vitamins A and C, pineapple yields bromelain, a proteolytic enzyme that alleviates inflammation (and possibly pain) and is used as a meat tenderizer. The leaves

Ananas comosus flowering.

produce usable fibers. The immature fruit has been used medicinally as an abortifacient and against internal worms; the rind and leaves for the latter, also. Teas of ripe fruit or rind are taken as a diuretic, and various parts are used externally on hemorrhoids, sprains, and wounds. Bromelain appears to inhibit *E. coli* bacteria from lodging in the intestinal wall. Although pineapple was probably once bird-pol-linated, today's cultivars form fruit without pollination. Growers apply hormones to manipulate fruit production. New plants can be grown from plants that form at both the base of the fruit and plant, or by planting the leafy top of a pineapple. Colorful cultivars are grown in tropical gardens. The many uses of pineapple in the kitchen range from baking it with ham to making jam and the famous piña colada.

Bromelia pinguin

Wild pineapple, *Piñuela*

Description: Terrestrial pineapple-like plant, ca. 1 m, growing in large, dense colonies. Long spiny-edged leaves yellow-green above, suffused with scarlet and orange; glaucous on underside. Pinkish inflorescence, ca. 50 cm; usually appears in early rainy season. Yellow fruits ca. 4 cm long. Sends offshoots out via umbilical-cord-like runners.

Distribution: Mexico and West Indies to northern South America. In Costa Rica, from sea level to 800 m; common along Pacific beaches and other dry areas.

Comments: The young flowers are tasty in stews with onion and bell pepper, and the peeled fruit makes a good fruit drink that has been taken both for intestinal parasites and as a diuretic. Wild pineapple is a host plant for satyrid caterpillars (*Dynastor darius*). Hummingbirds visit the flowers. A study done in Veracruz, Mexico, found that land crabs (*Gecarcinus lateralis*) are serious predators on the seeds and seedlings. This species and *B. karatas* (= *B. plumieri*), which has a sessile, compact inflorescence, are useful as living fences.

Bromelia pinguin

Catopsis paniculata

Catopsis paniculata

Description: Epiphyte 60+ cm tall, in tight rosette. Yellow-green leaves to 4 cm wide (to 6 cm at sheathing base) and 40 cm long; pointed at tip. Numerous white flowers on 20-cm-long, nodding branches of inflorescence. The seed has a plume of silky hairs that are 2 cm when stretched out; the hairs are folded up in the capsule.

Distribution: Mexico to Costa Rica. In Costa Rica, 700–2,000+ m, seen in Central Valley region, Monteverde, and close to the Panama border.

Comments: *Catopsis* is a small genus, with a total of 12 species in Costa Rica. They tend to be less colorful than many bromeliads, but their form can be attractive, as is seen in *C. paniculata*. The seed capsules open in the dry season; the manner in which the plume is folded up inside the seed capsule is characteristic of the genus *Catopsis*.

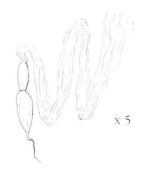

x 5

Catopsis paniculata seed recently removed from a seed capsule.

Guzmania monostachia   Striped torch

Description: Epiphyte ca. 30 cm tall in flower, with light green leaves, ca. 30 cm, in rosette. Cylindrical flowering spike with upper bracts red, lower bracts green with purple longitudinal stripes, surrounding 2.5-cm–long white flowers much of year. Capsular 3-cm fruit with plumed seeds.

Distribution: Southern Florida, West Indies, and Nicaragua to Bolivia. Widespread in Costa Rica, in humid forest and on pasture trees; from sea level to 1,600+ m.

Comments: *Guzmania monostachia* is one of the most geographically widespread members of its genus. You can see this colorful bromeliad not only in forests, but also being sold on the streets of San José. It is a popular ornamental, along with scarlet star (*G. lingulata*, photo below), which has a more compact flower head subtended by large, spreading red bracts.

William A. Haber

Guzmania monostachia

Willow Zuchowski

Guzmania lingulata

Pitcairnia brittoniana

Description: Epiphyte, often grows on side of tree trunk, or is a terrestrial plant on embankments. Somewhat wavy, 1-m-long leaves, extending laterally to one side. Spines, 0.5 cm long, along margin near leaf base and on the narrow petiole that widens into an orange-tan-sheathing base. Horizontal inflorescence to 1.5 m, including stalk; red sepals, orange petals. Dehiscent capsular fruit; seed tiny and narrow, with a thin bristle at each end; total length 1 cm.

Distribution: Nicaragua to Bolivia and Peru. In Costa Rica, in forest or edge 700–1,900 m; around Central Valley and all major mountains.

Comments: This species differs from many bromeliads in having a horizontal inflorescence and less rigid, laterally arranged leaves. Most *Pitcairnia* species are terrestrial, often growing on rocky slopes; and some, such as *P. heterophylla*, have two types of leaves. The genus *Pitcairnia* contains the only bromeliad that is native to an area outside of the neotropics—*P. feliciana* of West Africa.

Pitcairnia brittoniana flowers.

Pitcairnia brittoniana growing on an embankment in the cloud forest.

Tillandsia caput-medusae   Medusa's head

Description: Epiphyte to 32 cm tall when in flower; base of plant bulbous due to wide-sheathed, bulging leaf bases that create dark, hollow spaces. Silvery gray-green leaves with tones of purple-pink, to 30 cm, covered with scales giving them a suede feel; leaves channeled, and curving in all directions; drier leaves sometimes forming spiraling ringlets. Flower spike 13–20 cm, including stalk; simple to branching, with overlapping bracts green, tinged with rose; flowers violet. Valves of old seed capsule 4 cm long, dark brown, twisted.

Distribution: Mexico to Panama. In Costa Rica, from lowland deciduous forest to around 1,000 m; Guanacaste, Carara, and Central Valley region.

Comments: The special scale hairs that cover the leaves create a large surface for taking in water and nutrients. This is very important for plants growing in windy, dry areas. The hollows created by the bulging leaf bases form cavelike nesting places for ants (*Crematogaster*, *Camponotus*, and *Solenopsis*). Decaying debris that accumulates in the ant nest provides nutri-

William A. Haber

Tillandsia caput-medusae growing on a dry-forest tree.

ents to the bromeliad; the ants may protect the plant against herbivores (for more information on ant-plant interactions, see p. 47 and p. 270).

Tillandsia punctulata   Fairy queen

Description: Epiphyte to 30+ cm tall in flower. Many crowded leaves narrowing to tip of less than 1 mm wide. Broad spoon-shaped bases are deep purple, some purple marking on leaves; many small scales on the outer surface giving a whitish cast to the plant. Erect inflorescence with overlapping flower bracts; lower ones red, upper ones green; purple flower has white tip. Capsules to ca. 3 cm long.

Distribution: Mexico to Panama. In Costa Rica, on pasture trees and in for-

est, in all of the major mountain ranges, at 1,000–2,300 m, sometimes lower or higher.

Comments: The structure of this tillandsia appears to allow it to take advantage of varying weather conditions. During relatively dry periods, the narrow, scale-covered leaves cut down on water loss and collect whatever mist is in the air; but during heavy rains, the wide leaf bases, which form cups, can collect and store water.

Tillandsia punctulata

A tillandsia species at Lankester Gardens.

Tillandsia usneoides

Spanish moss, *Barba de viejo*

Tillandsia usneoides

Other common name: *Musgo de pino.*

Description: Prolific epiphyte on tree branches; dangling gray tapering clumps, to 4+ m, appear wispy in the wind. Narrow 2.5- to 5-cm-long leaves. Small greenish flowers, supposedly fragrant at night. Dehiscent capsules, ca. 2.5 cm long.

Distribution: Antilles; southern United States to Patagonia, sea level–3,000+ m. In Costa Rica, close to sea level in Osa region, to 2,000 m around central region; also in Lake Arenal and San Carlos areas.

Comments: This species, which has the widest distribution of any bromeliad, resembles lichen of the genus *Usnea*. The scales covering the plant absorb water and nutrients. Pieces of the plant may be dispersed by wind or by birds and other animals that carry it away for nesting material. It has been used as packing and stuffing material.

Vriesea incurva

Description: Epiphyte with rosette of leaves to 25 cm tall. Leaves grayish green or yellow-green with some maroon blotching, rather thick and leathery, triangular with wide bases 5.5 cm across and tapering tips. Ca. 5 pendant inflorescences, to 30 cm, in cluster; green (to purple) flowers, green and pink bracts covered with scales that mute color. Capsular fruit.

Distribution: Antilles; Costa Rica to South America. In Costa Rica, ca. 1,000–2,000 m, from Monteverde south to Talamancas, in forests or in pastures.

Comments: A number of species in the genus *Vriesea* serve as ornamentals and house plants. The pink bracts and narrow-tubed flowers of *V. incurva* indicate that it is probably hummingbird-pollinated. Many species of the closely related genus *Werauhia* (illus. and photo below) have cream to green flowers and are bat-pollinated.

Vriesea incurva

x 2

Plumed seed of *Werauhia tonduziana*.

Werauhia tonduziana in fruit.

Palm Family (Arecaceae)

There are few tropical landscapes worldwide that do not include palm trees. Costa Rican rainforests, beaches, gardens, and pastures all feature one kind of palm or another, with their graceful, multipurpose foliage that provides ornament, shelter, and shade. Some species can tolerate arid conditions, but palms typically prefer a wet climate and grow naturally in tropical or subtropical regions. The palm family, Arecaceae, plays an important economic role in the world, serving as a source of oil and starch as well as weaving and construction materials. Palms are a mainstay in the lives of some indigenous South American groups; people living around Iquitos, Peru, for example, use more than twenty species for food.

In Costa Rica, the greatest diversity of palms occurs in lowland rainforests, but some species grow at elevations as high as 2,500 m. In lowland forests, short trunkless species may be seen in the understory beside the towering large-crowned palms of the canopy.

Palms are generally easy to recognize since they do not branch as typical trees do, and because their compound or simple leaves have a characteristic feather- or fanlike pattern. The leaf scars that encircle the trunks of taller palms often make them appear bamboolike. Certain species have a noticeable smooth, cylindrical crown shaft just below the leaves that is made up of long, tubular leaf sheaths that wrap around the trunk. Palms have conspicuous bracts and inflorescences that are held among or below the leaves. The small unisexual or bisexual flowers have three sepals and three petals. The largest palm fruits in Costa Rica are coconuts, but most fruits in the family range from pea to golf-ball size.

The Arecaceae family contains around 2,500 species, with 109 species in Costa Rica. Interestingly, the La Selva region alone has more than 30 species. The genus *Chamaedorea*, which includes the ornamental parlor palms, is the most numerous genus in Costa Rica, where 31 species occur. Other large genera are *Geonoma* and *Bactris*. Spines occur on *Bactris* as well as on various parts of other palms (*Acrocomia*, the rare *Acoelorraphe* and *Aiphanes*, *Astrocaryum*, *Cryosophila*, *Desmoncus*, *Elaeis*, and *Raphia*). *Socratea* and *Iriartea* have spines on their stilt roots.

The flowers of most species are insect- or wind-pollinated, although *Calyptrogyne ghiesbreghtiana*, which occurs at La Selva, is bat-pollinated. Coconuts and a few other palm fruits and seeds can float and may be transported some distance in moving water; in general, however, palm seeds are dispersed by mammals or birds that eat the fruits.

Because they can be so striking, palms have great potential as components of tropical landscaping. A few of the species planted most often in Costa Rica are the royal palm (*Roystonea regia*, p. 391), the butterfly palm (*Dypsis lutescens*, p. 119), and the native pacaya (*Chamaedorea costaricana*, p. 383).

Palms at Braulio Carillo National Park.

Acrocomia aculeata   *Coyol*

William A. Haber

Acrocomia aculeata habit.

Description: To 10 m tall with spiny trunk, leaves, and bracts; often has skirt of old leaves. Pinnately compound leaves, ca. 3 m long; the narrow leaf segments and also the long black spines on the trunk and leaves project out at varying angles. Brown bract to 1 m. Large, light yellow inflorescences are a mix of 50,000+ male and zero–100s female flowers. Pendant clusters of round fruits, each ca. 4 cm; bony endocarp with 3 equidistant holes around midsection.

Distribution: Mexico to Central America and parts of South America and Antilles. In Costa Rica, mostly Pacific slope, abundant in some pastures; dry to moist sites to 500 m.

Comments: Coyol is a common pasture palm (compare with palma real, *Attalea rostrata*, p. 381) that can survive fires and cattle. It flowers and fruits in the late dry to early wet season. The

Willow Zuchowski

Acrocomia aculeata in flower.

flowers are visited by beetles and *Trigona* bees. Rodents eat the flesh of fallen fruits. Collared peccaries appear to be seed predators since they split open the hard nuts; cows disperse the seeds in pastures. Paper wasps (*Polybia* spp.) choose *Acrocomia*, as well as other spiny plants, as nesting sites. The hard, bony shell (endocarp) that encloses the seed can stay afloat for at least two years, and it sometimes washes up on distant beaches. The heart of the palm trunk, the fruit, and the seed are all edible. Wine and vinegar can be made from the sap. To make coyol wine, the palm is cut down, a rectangular trough is made in the trunk near the crown, and sap that oozes into it is collected over several weeks and then fermented. In some areas of South America, oil is extracted from the seed and used in cooking and for soap. The name *A. vinifera* is a synonym. There is only one other species in the genus, *A. hassleri*, a short grasslike palm of Brazil and Paraguay.

Attalea rostrata — *Palma real*

Description: To 25 m with nonspiny trunk. Fanlike arrangement of arching, vertically held, 5-m-long, pinnately compound leaves with 1-m-long leaf segments. Pendant, furrowed bract, to 3 m, with long pointed tip, encloses ca. 1-m-long branched inflorescence of either male or mix of female (majority) and male flowers. Oblong, beaked, yellow to orange fruit, ca. 6 cm, with 1–3 seeds. **Distribution:** Guatemala to Panama. In Costa Rica, Pacific slope and some Atlantic lowland sites. Needs some moisture, so occurs in swampy habitats, near streams, or on hills, depending on the region's climate; seen in pastures on the

Attalea rostrata in a pasture on the Nicoya Peninsula.

southern Nicoya Peninsula and in the Valle de El General.

Comments: Coatis, squirrels, and white-faced capuchins eat palma real fruits. Scarlet macaws, rodents, and bruchid beetles are seed predators. Small insects visit the flowers. Holes in the trunk are sometimes the nest of the orange-chinned parakeet (*Brotogeris jugularis*).

The leaves are durable and are frequently used for roof thatching, notably on the Nicoya Peninsula. The oily fruit is called *corozo*. The genus *Attalea* is more diverse in South America. *Attalea rostrata* was formerly called *Scheelea rostrata*. Some taxonomists lump *Attalea rostrata* under the South American species *A. butyracea*.

Bactris gasipaes — Peach palm, *Pejibaye*

Description: Spiny palm to 20 m, trunks often clustered; bands of long, black spines around trunk. Spines on leaf midrib and on bract covering flowers. Leaves to 3+ m, leaflets in groups sticking out in different directions, drooping and shaggy–looking. Monoecious; many-branched inflorescence a mix of male and female flowers, white to yellow. One-seeded fruit, ca. 5 cm long, usually orange-red, but may be yellow or green, in clusters of 200+.

Distribution: In cultivation from Honduras to northern Bolivia. Origin unclear; somewhere in South America and then introduced, traveling with migrating indigenous groups. In Costa Rica planted sea level–1,000+ m, but does best in low, wet, warm regions; especially noticeable around Puerto Viejo de Sarapiquí and San Isidro de El General.

Bactris gasipaes

Pejibaye fruits for sale.

Comments: Peach palm is associated with people throughout its range today, and there is debate as to whether wild plants exist. It has been in cultivation for a long time, and various indigenous groups have selected plants for certain traits, such as a lack of spines. Gene banks of the many Central American and South American forms have been created through efforts at CATIE and the University of Costa Rica. The true number of species of *Bactris* is unclear and needs more study. About 16 native species of the genus *Bactris* occur in Costa Rica, most of them below 1,000 m, although *B. dianeura* occurs to 1,600 m. Pejibaye fruits are a common site on street corners in San José, where vendors simmer them in metal boxes. Once peeled, they can be eaten as is, but because the nutty-starchy orange flesh is rather dry, a dab of mayonnaise is recommended. Cooks add them to stews, soups, dips, bread, and cakes. The fruits are a good source of vitamin A, starch, and protein. The seed is also edible once its bony shell is removed. Historically, the indigenous people of Costa Rica had peach palm plantations; one very large one existed in the Sixaola Valley. As with other Amerindians living in wet lowlands, the fruit, in a variety of forms, was a staple in the Costa Rican diet in the same way that various types of grain have been in other cultures. This palm has a number of other useful products, including needles made from the spines, roots taken medicinally against worms, palm wine made from the sap, and wood used for construction and bows and arrows. In areas of the Atlantic slope, growing *Bactris gasipaes* for its heart of palm for local use and export appears to be an economically promising activity. The palm grows rapidly and resprouts when cut, and there is great potential for improving this species as a crop and developing other products, such as oil and meal made from the fruit. Weevils and scarab beetles have been suggested as pollinators. Rodents and livestock eat fallen fruit. Although the spineless varieties are easier to harvest, they are more vulnerable to climbing rodents, which are naturally deterred by spines.

Chamaedorea costaricana Costa Rican bamboo palm

Other common names: *Pacaya*, parlor palm.
Description: Trunks, slender and green, to 6+ m, clumped. Smooth, pinnately compound leaves, ca. 1 m long, with segments that are narrow and slightly S shaped; distinctive flaps projecting from top of petiole sheath. Dioecious; branching clusters of yellowish flowers. Fruiting stem orange, with 1-cm fruits dark olive, black when ripe. *C. tepejilote*, another common species, is usually not colonial, has more conspicuous prop roots, a prominent yellow line along underside of leaf midrib, and a stockier, denser cluster of fruits.
Distribution: Mexico to Panama. In Costa Rica, both slopes from ca. 600 to more than 2,000 m; more widespread on Pacific slope.
Comments: *Chamaedorea*, with ca. 95 species, is the largest palm genus in the neotropics; 31 species are found in Costa Rica. They tend to grow in the understory, and some are trunkless. *C. costaricana* is the only species in the

Chamaedorea costaricana

genus native to Costa Rica that forms conspicuous colonies. In the forest understory, the lanky stems, with their prominent nodes reaching toward the canopy, may look like bamboo to the casual hiker. This is a variable species, with some individuals, especially those on the Atlantic slope of the Talamanca mountains, looking more robust than others. Bees, beetles, and flies visit the flowers; other species are wind-pollinated. The name *pacaya* is given to many chamaedoreas. *Chamaedorea costaricana* is attractive as an ornamental and is especially popular in the Costa Rican town Pacayas, east of San José. Another favorite ornamental is the Mexican species, *C. elegans*, the typical parlor palm.

Cocos nucifera

Coconut, *Coco*

Other common name: *Cocotero.*
Description: To 30 m, with grayish, often arching, trunk with swelling at base, where thousands of adventitious roots emerge. Pinnately compound leaves, to 6+ m long, in a spiral, with ca. 100 narrow, rather stiff segments in a plane. Monoecious; branched inflorescence with many more male than female flowers. Fruit smooth; green, orange, or yellow; ca. 30 cm long, sometimes slightly triangular in cross section. Nut inside fruit ca. 15 cm in diameter, with 3 indents at one end. This shell (endocarp) contains the endosperm, which is the meat and the liquid; the endocarp is covered by both a thick, fibrous layer (mesocarp) and a final smooth layer (exocarp).
Distribution: Origin unclear, but most likely on Pacific or Indian Ocean islands;

now cultivated in many tropical regions and planted to 1,000 m. In Costa Rica, occurs naturally along the shores at the highest tide line, but also planted in yards; dwarf varieties are fairly common.
Comments: The quintessential tropical palm, the coconut plays an important economic role throughout the world. In Costa Rica, it has many uses in Caribbean communities. It formed an important industry in the past, when people would sell whole coconuts or the oil made from them; the remains of the oil-making process were fed to pigs and chickens. The southern Caribbean coast of Costa Rica has coconut walks, which are paths bordered by a mix of planted and ocean-borne coconuts. In the rest of Costa Rica, the coconut is

Cocos nucifera growing on Pacific coast.

A dwarf variety of *Cocos nucifera*.

Coconuts germinating.

known mainly as an ornamental and as a source of nuts and *pipas*, the immature fruits that provide a refreshing drink; the green ones are sweeter than the yellow. The major production of coconut oil and fiber today is in the Old World tropics, particularly in the Philippines and Indonesia. Some small-scale plantations exist in Costa Rica, and the coconut could work well in agro-forestry systems. Insects visit the sweet-scented flowers, which are produced year-round. It takes about a year for fruit to mature. The fruits can travel thousands of kilometers in the ocean, and they can stay viable for at least several months while they are afloat. The thought of getting conked on the head by a falling coconut may seem comical, however some such incidents have resulted in death in Costa Rica. Scientists believe some wild coconut palms found in the Philippines and in Australia are the ancestors of the varieties seen today. In Costa Rica, the Atlantic coast variety, which is of Indian Ocean origin, was introduced via West Africa and the Caribbean; the variety on the Pacific coast is from the central and western Pacific islands and either floated or was carried there. Since this is a pantropical species, people from around the world have accumulated many uses for different parts of *C. nucifera*. The bony shell is used for utensils or fuel, and the leaves are woven into hats and baskets and used for thatch. The husk fiber, called coir, serves for stuffing, rope, and mats; it is not the ideal plant fiber, but it is abundant and rope and nets made of coir hold up well in saltwater. Wood from the trunk may be used in construction as well as in smaller carved articles. Palm heart, from the center of the crown, can be eaten as a vegetable. The coconut meat itself is shredded and incorporated into pies, cookies, and candies; the dried flesh—copra—is processed for its highly saturated oil, which is an ingredient of margarine, candles, soaps, cosmetics, and suntan lotion. *Pipa* water is taken as a diuretic and as a source of potassium; root decoctions are taken for fever, diarrhea, venereal diseases, and toothache. The oil functions as an emollient. Scientific studies have shown that the shell has antifungal activity and that a coconut-oil diet suppresses mammary tumors in mice. Coco-de-mer, or double coconut (*Lodoicea maldivica*), of the Seychelles in the western Indian Ocean, beats out *C. nucifera* for having the largest seed in the world; it's encodarp is 30+ cm long. Although the two species share a similar common name, they are not closely related.

Cryosophila warscewiczii
Rootspine palm, *Guágara*

Other common name: *Palma de escoba.*
Description: Solitary palm, to ca. 10 m tall; stem has branching spiny roots that are denser at the base, where they form a cone. Rounded, palmately lobed leaves, with white-gray coating beneath, new ones conspicuously pleated; leaf to nearly 2 m across, ca. 1 m long, on a petiole to 2+ m; fibrous, pubescent leaf sheath; leaf blade deeply split into two sections, each section divided into 6–11 segments that are also split. Pendant, branched, inflorescence, ca. 60 cm long, with many bracts that are not persistent; whitish flowers with 4-mm petals. White fruit to ca. 2.5 cm with 1 seed.
Distribution: Southern Nicaragua to Panama. In Costa Rica, wet forest to ca. 900 m, occurring at La Selva, Tortuguero, Braulio Carillo, and other Atlantic slope sites.
Comments: This is the most common native palm with palmately divided leaves; other genera in Costa Rica also referred to as fan palms are *Acoelorraphe, Colpothrinax,* and *Sabal,* however none are broadly distributed. The Spanish name for *Cryosophila, palma de escoba,* comes from using the leaves to make homemade brooms (*escoba* means broom). The flowers are probably beetle–pollinated. *C. guagara,* a Pacific slope species that is found mainly on the Osa Peninsula, has many persistent inflorescence bracts. The two other species that occur in Costa Rica, *C. grayumii* and *C. cookii,* are rare endemics.

Cryosophila warscewiczii

Elaeis guineensis   African oil palm, *Palma Africana*

Description: Trunk 10–20 m, covered with bases of old leaves. Leaves to 6 m with more than 100 pairs of pinnae grouped in bunches and sticking out in different planes; toothy margin on the petiole. Male and female flowers in separate clusters, but often occur on same plant (at different times); branching inflorescence, to 40 cm, on stalk of equal length. Orange to red fruits, 2–5 cm long, in clusters to 300; yellow-orange pulp surrounds hard stone in center containing 1 seed.

Distribution: Origin is West and Central Africa; now in cultivation in Southeast Asia, Central and South America. In Costa Rica, one of the most conspicuous growing areas is near Quepos, where rows of oil palms line the road for miles. Other plantations are in the southwest, toward Golfito, and there are a few in the Atlantic lowlands.

Comments: Large green blocks of African oil palms in neat rows are a dominant feature of the landscape as one approaches Quepos and Manuel Antonio by land or by air. In the occasional patch that has been burned and cut, you may see black-bellied whistling ducks (*Dendrocygna autumnalis*) sitting atop, and possibly nesting in, old palm snags. In Africa, this palm grows naturally in second-growth habitat. The natural pollinators are weevils, which breed in the old male inflorescences. In plantations in Costa Rica, the fruits are sometimes eaten by birds and rodents. Ferns and other epiphytes take up residence on the trunks. African oil palm is the source of two major oils in the world market—palm oil, from the flesh of the fruit, and palm kernel oil, from the seed. It remains an important source of calories for people in developing countries. These oils are outstanding

Adrian Hepworth

Plantation of *Elaeis guineensis.*

sources of vitamin A and do not contain cholesterol, but they have a highly saturated fat (higher than butter) that may play a role in increasing blood cholesterol levels. In Costa Rica, margarine and *manteca*, a vegetable lard, are staples in most households. Some country people throw a stick of Numar margarine in their saddle bag for sustenance on the trail. Palm oil is also added to baked products, candles, artificial coffee creamer, and cosmetics. In Costa Rica, large factories process the fruits. Steaming arrests the natural breakdown of oil after fruits are picked, and pressing extracts oil. To obtain palm kernel oil, for use in soap and chocolate, the hard pits must be cracked. African oil palm arrived in the New World tropics during the slave-trade era; the American oil palm (*Elaeis oleifera*, photo right), which is known as coquito or palmiche, has its origin along the Pacific slope of Central America and was probably spread by humans. It is squatter then *E. guineensis* and somewhat reclining, has leaflets in a plane, and is seen in wet, some-

Elaeis oleifera

times inundated, areas; it may be a good source of genetic material for increasing resistance to diseases (e.g., spear rot) found in African oil palm.

Geonoma congesta

Caña de danta

Description: Colonial palm with 20 or more stems to 5+ m, often leaning. Leaves to ca. 1.5 m long, with sharp edge on petiole; up to 12 leaflets of various widths. Inflorescence fairly short and sturdy, with few to 14 branches, turning orange in fruit. Fruit ca. 1.5 cm in diameter, rough and purple-black.

Distribution: Honduras to Colombia. In Costa Rica, Caribbean slope forest to 800 m, common at La Selva; also southern Pacific slope and Osa Peninsula.

Geonoma congesta fruits.

M. & P. Fogden

Geonoma congesta

Comments: The genus *Geonoma* is a conspicuous group of palms in Costa Rica, with a variety of species, some short and trunkless, that are found in a wide range of elevations. *G. congesta* is an obvious clonal species, typically growing in clumps of around 7 stems (ramets). The advantages of this system include less chance of the entire plant being killed when, say, a tree falls on it, increased access to patchy resources, and the ability to channel resources from older stems of the plant to newly forming parts. These clones may live for 100 years or more, and the estimated lifespan of individual stems is 60–70 years (Chazdon 1992). The flowers are pollinated by small bees. People use the leaves for roof thatch. There are 14 other species of *Geonoma* in Costa Rica.

Prestoea acuminata   *Palmito morado*

Other common names: *Chonta, palmito mantequilla.*

Description: Usually stems in a cluster, to 12 m tall, ca. 8 cm in diameter. Leaves to 3+ m long, with up to 50 pairs of leaflets; from the middle of the leaf toward each of the two ends, the leaflets become progressively shorter; crownshaft of ca. 1 m formed by sheathing, green to maroon leaf bases; canoe-shaped spathe, 1+ m, around glabrous inflorescence that has many branches. Monoecious; small, white to pinkish flowers arranged in groups of threes—2 males and 1 female. Black fruit ca. 1 cm.

Distribution: Widespread; in parts of the Antilles as well as from Nicaragua to Bolivia. In Costa Rica, ca. 1,000–2,450 m,

wet forest in major mountain ranges; commonly on Atlantic slope, but some populations occur on Pacific slope near Continental Divide (e.g., Monteverde).

Comments: Four smaller prestoeas also occur in Costa Rica, none of them having the distinctive crownshaft seen in this species. Parrots feed on the fruit of *P. acuminata* in some areas. Today, most of the heart of palm produced in Costa Rica comes from the commercially farmed *Bactris gasipaes* (p. 382), however *P. acuminata* and species of *Euterpe* were traditionally sought out in the wild for their particularly delicious hearts. The very similar species *Euterpe precatoria* occurs in areas of the Osa Peninsula and wet Caribbean slopes

Prestoea acuminata

below 1,000 m. It usually grows as a solitary stem, unlike *P. acuminata*, whose stems are clustered. The inflorescence branches are covered with woolly hairs in *E. precatoria*, while those of *P. acuminata* are nearly smooth. *P. allenii* is a synonym of *P. acuminata*.

Roystonea regia   Royal palm, *Palma real de Cuba*

Description: To more than 25 m; light gray, smooth trunk that looks like a concrete pillar, bulging out about halfway up; conspicuous smooth, green crownshaft. Pinnately compound, arching leaves 3–4 m long—the many 1-m-long leaflets sticking out in various directions. Long, erect inflorescence bud encloses not only the cream to white flowers, but fluff made up of millions of tiny hairs that shower down to the ground

Roystonea regia

beneath the palm. Purplish fruit to 1.5 cm long, in large, heavy clusters.

Distribution: Various sources say origin is Cuba, but *R. regia* also appears to be native to adjacent islands, parts of the Yucatan coast of Mexico, northern Central America, and southwest Florida; elsewhere planted as ornamental. In Costa Rica, grows best at low to midelevations, in fairly wet climates.

Comments: The royal palms one is likely to find in Costa Rica are *Roystonea regia* and *R. oleracea*, the former being the more common. The lower leaves of *R. oleracea* are held more erect or horizontally than those of *R. regia*, and the leaflets are in a plane; the trunk also does not have irregular bulges. The flower/fruit stems are wavy in *R. oleracea*, but straight in *R. regia*. Like other palms, royal palms have served many uses in their native land. People eat the heart of palm and use the fruit as food for pigs. The leaves can be used for thatching, and the sheath at the base of the leaf to cover harvested tobacco. The trunks are used in furniture-making or split into boards for other construction. The genus is named in honor of Roy Stone, a United States Army general. There are 10 species of royal palm in the world, all of them in the Caribbean and neighboring countries.

Socratea exorrhiza

Stilt palm, *Maquenque*

Description: Single trunk to 30 m tall; its base is a 2-m-high, open, cone-shaped cluster of brown, spiny, stilt roots; noticeable crown shaft to 1.5 m. Ca. 7 pinnately compound fronds to more than 3 m long; wedge-shaped leaflets (pinnae) with ragged tips, split into segments that stick out in different directions. Inflorescence to 45+ cm long including stalk, with branches ca. 50 cm long; thousands of female and male flowers mixed in same inflorescence; buds erect; once the buds open, branched, flowering stalks hang below leaves. Yellowish fruit, usually oblong, ca. 3 cm. Similar *Iriartea deltoidea* has pendant, horn-shaped buds (illus., opposite page).

Distribution: Nicaragua to Bolivia, sea level–1,000 m. In Costa Rica, in wet, lowland areas such as La Selva, where it is one of the most common trees in the subcanopy of primary forest; also Osa Peninsula.

Comments: While walking in lowland wet forest, you cannot miss the impressive prickly stilt roots of the *Socratea* and *Iriartea* palms. The leaves of the two species are quite similar, but the prop roots differ: those of *Socratea* are brown and loosely arranged, while those of *Iriartea* are black and densely packed. In young individuals of *Socratea exorrhiza*, you can see the original basal part of the stem, but over time this dies and the spreading roots take over. The leaves of stilt palms change as the plants get larger; leaflets in older trees may be straplike or wedge-shaped, and they change from being in one plane to being in many. Palms are usually not thought of as a source of wood, yet the stilt palm has a very hard outer trunk that, once split, can be used in floors and walls, as well as for smaller items such as bows and harpoons. The spiny roots are too rough for back-scratchers, but they make good graters for manioc. The palm heart is edible, although a little bitter. Weevils, other beetles, and possibly bees pollinate the flowers. Mammals, including bats, and large birds such as toucans eat the fruit; spiny pocket mice (*Heteromys desmarestianus*) eat the seeds. *S. exorrhiza* was formerly known as *S. durissima*.

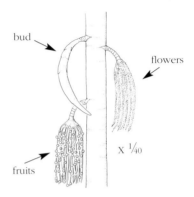

Iriartea deltoidea

Socratea exorrhiza juvenile.

Socratea exorrhiza stilt roots.

Heliconia and Bird of Paradise Families
(Heliconiaceae and Strelitziaceae)

Heliconias are the quintessential tropical plants. Striking and substantial, they are showiest in the neotropics; some less-colorful species occur in Indonesia and the South Pacific. There are between 200 and 250 species in the Heliconiaceae family, with perhaps as many, if not more, forms or cultivars. There are seven species in the related Strelitziaceae family. This section focuses on some of the more than 35 native Costa Rican heliconia species. Among the nonnative species included here are the parrot's flower (*Heliconia psittacorum*), a South American species, and the bird of paradise (*Strelitzia reginae*), which is from South Africa. Both are frequently seen as ornamentals and cut flowers.

Although their general form and leaf shape is similar to bird of paradise and banana, members of the Heliconiaceae family have a distinctive flower display. Many bracts of brilliant colors enclose the much smaller green, yellow, or orange flowers, while in the bird of paradise it is the flowers, not the single bracts, that attract attention. The flashy colors function as flags that attract pollinating birds—hummingbirds in the case of heliconias, and sunbirds for birds of paradise.

The sheathing leaf bases of heliconias form what is called a pseudostem, with the inflorescence developing at the top. While the paddlelike leaves are

Strelitzia reginae (Strelitziaceae)

typical, various species have smaller leaves that come off of the stem either at a 45-degree angle or horizontally as in some species of the ginger family. The plants can spread vegetatively via underground stems called rhizomes.

The flowers have three sepals and three petals that are held together to form a tube. Although just one or two flowers open at a time in a bract and only last a day, the buds are packed into each bract like sardines so flowering often goes on for long periods. The fruit, which has three seeds, turns blue on the day of ripening.

A close look at the flowers reveals a variety of lengths and shapes, demonstrating how heliconias have coevolved with hummingbirds. The curved flowers are adapted for pollination by hum-

Heliconia latispatha (Heliconiaceae)

mingbirds such as the hermits, which have long, curved bills; the shorter straight flowers are suitable for visits from a variety of non-hermits. Different flower species dab pollen onto distinct parts of a bird's bill or head, making transfer of the pollen to the stigma of the next individual of that species likely to occur. Hermit hummingbirds are trapliners, which means that they visit a few flowers in a small clump of heliconias, then fly several hundred meters and visit a few more flowers, and so on. Some of the non-hermit species are territorial in their behavior, staying close to, and defending, a single extensive heliconia patch.

Heliconia leaves are perfect nest sites for hermit hummingbirds. The birds attach their elongated nests to a strip of leaf that has torn away from the main leaf structure; the nest hangs on the underside where it is hidden and has a natural roof. In addition to hummingbirds, numerous other organisms, ranging from protozoa to bats, interact with heliconia plants. Furled leaves create narrow funnels where spiders, frogs and disk-winged bats (*Thyroptera*) hide and roost. The flat-bodied larvae and adults of rolled-leaf hispine beetles eat the topmost layer of the leaf. If you see an unfurled leaf that droops down on both sides of the midvein, you may discover white bats (*Ectophylla alba*) roosting underneath; they create a tent by chewing holes along the vein. Watch for female owl butterflies (*Caligo* spp.)—which are larger than, but not as resplendent as, morpho butterflies—flying around and laying eggs on heliconia foliage, which becomes food for their caterpillars.

Aggressive ants nest in the inflorescence bracts of some species (e.g. *H. osaënsis*). Their presence may protect fruits against herbivores, but the ants also chew into flowers and take nectar, although this most likely happens after early-morning pollination by hummingbirds has occurred. Mites are common flower inhabitants and hitch rides to other plants on hummingbird bills. Many heliconias that have erect boat-shaped bracts contain liquid that houses protozoa and insect larvae, especially those of flies.

When the fruits are ripe, other players come on the scene. Manakins, flycatchers, and motmots are among the 28 bird species known to seek out fruits of one species, *H. latispatha*, at La Selva. Birds are the main agents in seed dispersal, but a seed-hoarding squirrel or a stream may occasionally carry a fruit away.

Heliconias may be seen in most natural areas in Costa Rica as well as in botanical gardens. The Wilson Botanical Garden in San Vito, which is one of the Heliconia Society International Plant Conservation Centers, has a superb collection. The greatest number of heliconia species occur at midelevations; except for a few species such as *H. lankesteri*, they normally do not occur above 2,000 m. In natural settings, sun-loving species grow along streams or in gaps in the forest where there has been a tree fall or landslide. Some of these are almost weedy, growing along roadsides and in second-growth situations. The less abundant semi-shade or forest species become rare as older forest disappears. Wet season is generally the best time for flowering, but some species, like *H. wagneriana*, peak in dry season.

Indigenous and country people use heliconia leaves as food wrappers and for thatching. In Costa Rica, where the plants are known as *platanilla*, Stiles (1979) reported country people using rhizomes of certain species to treat cancer. Some South American indigenous groups eat the rhizomes as food and use them internally and externally for a variety of medicinal purposes.

Heliconias may have some utilitarian value in the lives of humans, but they are most widely known for their ornamental qualities. Species from the New World now appear in cultivation in Asia, Africa, and Hawaii. Commercial growing for cut flowers, which began in Hawaii and in greenhouses in Holland and Germany, has spread to other regions, including Costa Rica, where growers mainly export heliconias to the United States market. A cut inflorescence will last about two weeks. The best approach to growing heliconias in gardens is to plant small plants or rhizomes, because the seeds may take many months to germinate.

Heliconia clinophila

Heliconia clinophila

Description: 0.5–3 m tall. Leaf blade ca. 80 cm; 6–9 yellow bracts erect and distichous, with green-yellow flowers.
Distribution: Costa Rica and Panama, cultivated in Florida. In Costa Rica, near streams and on steep slopes, from 1,000–1,800 m on Caribbean side in areas such as Braulio Carillo National Park.
Comments: This is one of about a half dozen naturally occurring Costa Rican heliconia species that have yellowish or greenish bracts. It has been hybridized with *H. secunda.*

Heliconia imbricata

Description: 2–5 m, in clumps. Long-stemmed leaf blade ca. 1- to 2-m long, sometimes with red-purple coloration in midvein and underside of leaf. Upright inflorescence to more than 50 cm tall, with 15–40 deep, tightly overlapping, distichous bracts that are mostly dark red fading into yellow or green toward center of inflorescence and along base of bracts; rotting and blackish along edges. White S-shaped flower, with only the greenish tip sticking out of bract.
Distribution: Costa Rica to Colombia. In Costa Rica, wet lowlands, both slopes, sea level–700 m elevation; common on forest edge and in second growth at Carara, Osa Peninsula, and La Selva.
Comments: The deep bracts of this plant can hold a lot of liquid. An experimental study done at La Selva showed that the plant secretes liquid into bracts—the bract liquid was drained

Heliconia imbricata

out, the inflorescence was covered, and liquid was found to have accumulated by the next day (Bronstein 1986). The fruits are taken by many species of birds. Natural hybrids of *H. imbricata* form by crosses with *H. latispatha* or *H. sarapiquensis*; the cross with the former is a frequently encountered hybrid that is often seen as a distinct patch in a more uniform population.

Heliconia lankesteri var. rubra

Francis X. Faigal

Heliconia lankesteri var. *rubra*

Description: 2–4 m tall. Leaf blade ca. 1 m; 10–17 red (or yellow in var. *lankesteri*) bracts in a plane. Flowers yellow with green tinge.
Distribution: Costa Rica and Panama. Red variety in Central Mountains and Atlantic slope of Talamancas, with yellow variety mostly on Pacific slope of the Talamancas; 1,300–2,000+ m; in forests, second growth, and along edges of streams.
Comments: This species reaches the highest elevations of any heliconia in Costa Rica. Two varieties exist: var. *lankesteri*, with yellow bracts, and var. *rubra*, with red bracts. Natural hybrids occur where they meet.

Heliconia latispatha

Description: 2–4 m, in clumps; spreading by rhizomes. Leaf blades to more than 1.5 m long, often with narrow maroon border. Erect inflorescence to 60 cm, with 8–17 spiraling bracts that decrease in size toward tip; color of bracts variable—green-yellow, orange, or red; flower tubes are nearly straight and may be yellow-green to orange with dark green edges.
Distribution: Mexico to northern South America and cultivated in various parts of the world. In Costa Rica, occurs naturally in sunny second growth along rivers; also in man-made disturbances; sea level–1,300 m, both slopes and Isla de Caño.
Comments: This is probably the most common heliconia in Costa Rica and one of the most visible species to travelers

Heliconia latispatha

since it frequently occurs on roadsides. It tolerates a range of climates, including that of the dry forest, where it often grows near rivers or other moist pockets. Besides the variability in bract color, there is also a range in the size of the plants, with those of the upper Atlantic slope generally being taller than individuals from dry lowlands. Although an occasional hermit hummingbird may be seen visiting the flowers, territorial nonhermit species such as the rufous-tailed hummingbird (*Amazilia tzacatl*) are the principal pollinators. The fruits are popular—data from La Selva show that 28 species of birds have been seen eating them. *H. latispatha* is called caliguate by some Guanacaste residents. It sometimes hybridizes with *H. imbricata*, and various horticultural cultivars exist.

Heliconia longa

Description: Plants 5–7 m tall. Blades ca. 1–2.5 m long, underside with white, waxy layer and midvein sometimes maroon. Pendant inflorescence, ca.1.5 m long, distichous to somewhat spiraling; rachis (stem where bracts attached) sinuous; 12–40 10-cm-long red bracts with or without fine pubescence. Flower white-yellow at base and deeper yellow toward tip, to 6.5 cm long, curving sharply downward.
Distribution: Southern Nicaragua to Ecuador. In Costa Rica, to 900 m, in disturbed areas, forest edge, and old fields; Caribbean slope and Osa Peninsula.
Comments: This is one of about a dozen Costa Rican heliconias with pendant flower stalks. Some recent publications have used the name *H. curtispatha* for this species, but that applies to a species further south (Panama to Ecuador). *H. longa* can be seen in flower during much of year. *H. stilesii* is similar but its bracts are always in a plane and are much more crowded.

Heliconia longa

Heliconia monteverdensis

Heliconia monteverdensis

Description: 1–2 m tall. Leaf blade ca. 60 cm long, often with maroon on stem and dark margin on leaf. Erect, spiral inflorescence with 5–9 red bracts; flowers cream or light yellow.

Distribution: Costa Rican endemic. Generally on cloud forest ridges ca. 1,500–1,700 m in Tilarán and Guanacaste mountain ranges.

Comments: This is one of several Costa Rican heliconias named after specific places, in this case Monteverde, where the species was first encountered. It resembles *H. tortuosa*, which is also seen in the Monteverde region, but it is smaller, with paler flowers, and it is able to grow in cooler, windier habitats.

Heliconia pogonantha var. pogonantha

Heliconia pogonantha var. pogonantha

Description: 4–8 m, growing in clumps. Leaf blade ca. 2 m long. Pendant 1- to 2-m-long inflorescence with 20–50 bracts, distichous but in loose spiral, red with yellow near attachment to rachis; flower yellow, white toward base.

Distribution: Nicaragua to northern South America. In various parts of Costa Rica, mostly below 700 m; frequent on forest edge and in second growth in Atlantic lowland sites such as La Selva Biological Station; also on the Osa Peninsula.

Comments: The plant's overall size and the color and pendant form of the inflorescence combine to make this a very impressive heliconia. The upside-down bracts of *H. pogonantha* lack the bract liquid that accumulates in many heliconias and that helps protect flowers from nectar robbers. However, this species' tightly packed flowers, hairs in the inflorescence, and tougher flower tissue all deter would-be robbers. The fruit stalk quickly elongates when the fruit is ripe so as to expose it to avian seed dispersers. Another variety, var. *bolerythra*, lacks the yellow on the bracts.

Heliconia wagneriana

Description: Plant to more than 4 m tall. Leaf blade 1.5 m long with wavy edge. Erect inflorescence shorter than leaves; 6–15 overlapping bracts in a plane; pink-red-orange in center of bract, surrounded by yellow, then green toward edge; flower green with white base.

Distribution: Through Central America to Colombia, commonly cultivated. In Costa Rica, in forest, on forest edge, and along streams; especially common in wet, sunny second growth on Osa Peninsula and in wet Atlantic lowland sites such as Cahuita and Punta Uva; occasional on Pacific slope, usually below 300 m.

Comments: Rat-tailed maggots, the larvae of a syrphid fly, live in the bract water of *H. wagneriana* and eat detritus and flower parts. In general, the liquid that the plant secretes into the bracts serves as defense against most of the insects that rob the nectar and damage the flower ovaries, where the seeds form (Wootton and Sun 1990). Large, boatlike bracts thus not only attract hummingbird pollinators, but they create protective moats for

Heliconia wagneriana

flowers and developing fruit. The ripe fruits are pushed out from the bract by the rapid lengthening of the stalks. Pollinated by hermit hummingbirds.

SEXY PINK, LOBSTER CLAW, AND BIRD OF PARADISE

In the tropical forest, the common name for a *Heliconia* species is heliconia, but when one steps into a garden or a florist shop, the names for heliconias proliferate into a variety of common names and cultivar epithets as colorful and unusual as the plants themselves. Heliconias that have originated through lengthy selection and garden cultivation have fanciful cultivar names such as 'Andromeda', 'Sexy Pink', 'Chocolate Dancer', 'Maya Blood', and 'Orange Christmas'. These may be used alone as common names at times, but in formal scientific contexts they are always attached to species names—*Heliconia chartacea* 'Sexy Pink', for example (photo opposite page). Some species, such as *H. rostrata* (photo opposite page), which is variously called lobster claw and beaked, or hanging, heliconia, have not been tampered with much by gardeners and thus do not have many cultivars.

Many of these extremely showy plants have become popular as garden ornamentals or for cut flowers. A number of companies in Costa Rica, like Costa Flores of Guácimo, for example, produce heliconias, various gingers, and other flowers and foliage for export to the cold countries of the north, where a flourish of tropical blossoms is welcome.

Parrot's flower
(*Heliconia psittacorum*)

Bird of paradise
(*Strelitzia reginae*)

Small heliconia species can be kept as potted plants. Some Old World species, such as cultivars of *H. indica* of the South Pacific islands, have bronze and reddish foliage that is more striking than their flowering display.

A few exotics are shown here. Perhaps the most widespread in Costa Rica is parrot's flower (*Heliconia psittacorum*, photo opposite page), a South American species that is often planted around hotels and in parks. It is usually 0.5–2 m tall, with bracts of various shades of green and red and banana-shaped flowers of yellow, orange, or red that have a distinctive dark green or black band below a light (usually white or yellow) tip. Parrot's flower is daintier than many other heliconia species; dwarf forms, which are quite slender, are frequently used as garden or container plants. 'Andromeda', 'Strawberries and Cream', and 'Lady Di' are a few of the many cultivars.

Another horticultural favorite, a cousin of the heliconias, is the well-known bird of paradise (*Strelitzia reginae*, photo opposite page) of the family Strelitziaceae. Originally from South Africa, it grows in clumps 1 to l.5 m tall and has gray-green leaves and boat-shaped green bracts with reddish borders. The showy orange parts of the flower are sepals; the blue parts are petals that form a keel that envelops the long anthers and style. A shorter third petal covers the nectary. In its natural habitat, bird of paradise is pollinated by sunbirds that land on the sturdy blue petals, which form a perch that allows a bird to reach the nectar at the base of the flower. The weight of the bird causes the petal lobes to spread apart, exposing the pollen-bearing anthers, which contact the bird's feet. This family contains some very large plants, some taller than 10 m, including white bird of paradise (*S. nicolai*), traveler's tree (*Ravenala madagascariensis*), and a South American species, *Phenakospermum guyannense*. The family name, Strelitziaceae, honors Charlotte von Mecklenburg-Strelitz, the queen to England's George III.

Heliconia chartacea 'Sexy Pink'

Lobster claw
(*Heliconia rostrata*)

Orchid Family (Orchidaceae)

Orchids are at once the most beautiful and the most bizarre plants, exhibiting a variety in flower size, color, and shape that is astounding. They are found throughout the world, in habitats that range from cold to very hot, in both wet and dry regions. Most orchids grow as epiphytes, and many are not very conspicuous since hundreds of species have flowers that are just 1 cm or smaller. Those that do not live in the forest canopy grow on the ground or on rocks. Vanilla is an orchid that grows as a high-climbing vine. The Orchidaceae is one of the largest families of plants in the world with over 20,000 species; only the Asteraceae may be a little larger. An even greater number of ornamental hybrid orchids now exist in gardens, florist shops, and live collections around the world. In the tropics the diversity of orchids is phenomenal; in Costa Rica alone there are roughly 1,400 species, and some parts of the country are particularly rich. There are about 500 species in the Monteverde region, for example.

One might logically wonder how so many species that look so different can be in the same family. What holds them all together is a distinctive and unique set of flower characteristics. The three sepals and three petals are often similar but one of the petals is showier than the rest. This labellum, or lip, may be adorned with spots of color, wartlike projections, and frills. The lip is often what attracts the pollinator to the flower and/or serves as a landing platform. In addition to this labellum, orchid flowers have a structure called a column, which bears both pollen-bearing and receptive parts. Pollen is in two to eight packets or masses called pollinia, and it is sometimes attached to a stalk with a sticky pad. The mature seed capsules split along three or six lines; inside are tens of thousands of tiny, dustlike seeds that lack food reserves and need fungal mycorrhizae in order to germinate. The seeds are usually wind-dispersed. Other characteristics of some, but not all, orchids are alternate, fleshy, parallel-veined leaves; pseudobulbs, which are thickened parts of the stem that store water; and velamen, a spongy layer of dead cells on the roots that absorbs water and nutrients.

Pollination systems in orchids are fascinating. Hummingbirds visit a few of the species with red, purple, or orange flowers to feed on nectar, but other orchids lure insects in with various rewards, or use trickery to attract them. Many orchids have no nectar, but offer scent compounds, wax, or oil to their pollinators. Some species have fake pollen, or they mimic other flowers, female insects, rotten meat, or mushrooms. Others have traplike flowers or structures that force the pollinator into certain positions that assure pollination will take place. Insects from large hawkmoths to tiny flies are among known pollinators; male euglossine bees, also called orchid bees, collect the scent compounds of certain orchids in order to make their pheromones (see *Stanhopea wardii*, p. 431).

There is no particular orchid-flowering season in Costa Rica, but the seasonal transitions from wet to dry (and vice versa) sometimes bring about more activity. Since these plants are difficult to see in the wild, one way to view many orchids in a small amount of time and space is to visit gardens or flower shows. An impressive orchid exhibition presented in San José each March features displays by various garden and educational organizations, including Lankester Garden, which specializes in orchids. The Orchid Garden in Monteverde is another wonderful place to get a close look at orchids. Keep an eye out for them on fallen branches in the cloud forest, also.

Many orchid species are small, rare, and epiphytic, all characteristics that make them challenging to study. In fact, much about the life of orchids, their role in the forest, and their interactions with other organisms remains a mystery. As forest destruction and illegal collection and trade continue, some important information will be lost forever.

The species presented here are few in comparison to the size of the family, but the descriptions help demonstrate the amazing diversity and beauty of this family.

William A. Haber

A *Sobralia* orchid.

Arundina graminifolia

Bamboo orchid

Arundina graminifolia

Description: To more than 2 m tall; reedlike stems with many leaves in ladderlike pattern. Leaves narrow, to 30 cm long, wider at sheathing base, tapering toward tip. Flowers terminal, 1 to a few open at once; lilac-tinged sepals and petals 4 cm long, petals broader than sepals; deeper rose-purple 5-cm-long lip (rolled into a tube toward base), touch of yellow in throat.

Distribution: Native to Asia, has become a common tropical ornamental. In Costa Rica, in yards, especially at mid- to lower elevations.

Comments: The flowers, which superficially look similar to *Cattleya*, *Guarianthe*, and *Laelia*, are short-lived. They are probably bee-pollinated.

Close-up of *Arundina graminifolia* flowers.

Brassavola nodosa — Lady of the night, *Huele de noche*

Brassavola nodosa

Description: Epiphyte with narrow, but succulent, leaves ca. 15 cm long, folded lengthwise. Up to 8 flowers in inflorescence; sepals and petals similar—narrow, light green to near yellow-green, 6–7 cm long; the rolled, green tube section (2.5 cm long) of lip leads into a 3.5-cm-long flaring white lobe that comes to a point; some maroon spotting may be seen in throat.

Distribution: Chiapas, Mexico to northern South America. In Costa Rica, lowlands below 500 m, in Pacific dry and wet areas and southern Caribbean.

Comments: This is one of the most common epiphytic orchids in the dry Guanacaste region, but this species also occurs in wetter areas along both Pacific and southern Atlantic coasts, so you may see it growing on enormous rain trees (*Albizia saman*, p. 271) in Guanacaste, on coconut palms near the beach, or even on mangrove trees. The nocturnal perfume of the flowers attracts hawkmoths.

Brassia gireoudiana — Spider orchid, *Arañas*

Other common name: *Chapulín.*

Description: Epiphyte; pseudobulbs to 14 cm tall, topped by 2 straplike leaves, ca. 30 cm long, one leaf emerging above the other leaf. Up to 15 flowers in lateral inflorescence to 1 m; flowers to 25 cm long; greenish petals, to 10 cm, have bands of brown on lower third; sepals to 20 cm; some spotting on sepals and lip; lacks green warts on lip that

Brassia gireoudiana

characterize some species of *Brassia*. Fruit a capsule.

Distribution: Costa Rica, Panama, ca. 500–1,000+ m.

Comments: All Costa Rican species of *Brassia* have long, thin sepals. *B. caudata* has a smooth lip like *B. gireoudiana*, but it lacks solid bands of dark color on the petals; it may have spots, however. Each pseudobulb of *B. arcuigera* has sharp edges and a single leaf. *B. verrucosa* has wartlike bumps on the lip. A recent revision places *B. gireoudiana* as a subspecies of *B. verrucosa* (Grayum et al. 2004).

Catasetum maculatum

Catasetum, *Zapatico*

Description: Large epiphytic orchid with dense roots, some of which point upward; thin, dry sheaths surround furrowed pseudobulbs; pseudobulbs to 30 cm, with spines at top where the deciduous leaves fall off. Pleated leaves to 50 cm long. Lateral inflorescence, to 40 cm, of up to 14 fragrant flowers that are green with purple/maroon markings; separate inflorescences of male (staminate) and female (pistillate) flowers. Male flowers with sepals and petals 4–5 cm long, fleshy, hoodlike lip ca. 3 cm; a 3-cm-long column; antennae project from the column into lip area; pollinia are on a stalk that is curved and under tension. Less common female flowers, which lack pollinia and antennae, have shorter petals, sepals, and column; shape and color pattern is different also.

Distribution: Central America and northern South America, dry and wet environments. In Costa Rica, both slopes, below 1,000 m; occurs at Santa Rosa, Manuel Antonio, and Carara.

Catasetum maculatum

Comments: This is one of the few orchid species that has separate male and female flowers. Male *Eulaema* bees come to the flowers to scrape scent compounds from the inner back wall of the hoodlike lip. When visiting a male flower, the bee disturbs the antennae and triggers the release of the pollinarium—pollinia with a stem and a sticky pad—which gets shot, and glued, onto the bee's back. When the bee visits a female flower, the pollinia hanging off its back get caught in the groove of the stigma as the bee backs out. The flowers loose their scent after the pollinia are removed or after being pollinated. The male bees make a pheromone out of the perfume they collect. Of the 75–100 species of *Catasetum*, only three occur in Costa Rica. *C. maculatum* flowers mostly in the rainy season.

Encyclia cordigera

Easter orchid, *Semana Santa*

Description: Large swollen pseudo-bulbs, ca. 6 cm long; 2–3 strap-shaped, leathery leaves to ca. 30 cm long. Flowering stalk to 40 cm long, with 3 to 16 flowers; flowers variable, ca. 6 cm across; sepals to 3.5 cm long, curving at tips; petals and sepals green with brown toward tips; large lobe of lip to 3 cm wide, white with 3 purple lines (or one blotch); delicate perfume; old flowers yellow and maroon.

Distribution: Mexico to northern South America. In Costa Rica, to 900 m; most common in dry forest.

Comments: The English and Spanish common names refer to the flowering time, which is around Easter. This species is common on deciduous trees in Santa Rosa. Large carpenter bees (*Xylocopa* spp.) visit the flowers in the late afternoon, and nectaries at the base of the plant attract ants. Panamanian plants, which have a completely rose-purple lip, are considered a variety (var. *rosea*). *Encyclia cordigera* is easy to care for in cultivation.

Encyclia cordigera

Epidendrum ciliare   *Plumilla*

Other common name: *Lengua de gallina.*
Description: Epiphytic or growing on
rocks; pseudobulbs to 30+ cm with 1–2
leathery leaves, to 30 cm long. Terminal
inflorescence of a few to ca. 12 showy
flowers; narrow, greenish-white to pale
yellow sepals and petals to 7 cm long;
3-lobed white lip with side lobes deeply
cut and fringed; delicate tonguelike
midlobe, 5-6 cm long; fragrant at night.
As with other epidendrums, the lip is
fused with the column.
Distribution: Mexico to South America,
and West Indies. In Costa Rica, found in
a range of elevations (700–1,900 m) and
in all of the major mountain ranges; also
on the Nicoya Peninsula.
Comments: The size of the pseudobulbs and the size and shape of the
leaves varies considerably in this
species. Without flowers, the plants
look somewhat like those of

Epidendrum ciliare

Guarianthe skinneri, the showy national flower; thus orchid sellers sometimes
try to pass this and other similar species
off as *Guarianthe skinneri.* The heavy
nocturnal perfume suggests that moths
are the likely pollinators. In the 1700s,
Linnaeus purportedly named the genus
Epidendrum (Greek for upon a tree),
including within it all epiphytic orchids

of the tropics with which he was familiar. Needless to say, the definition of
the genus has been refined over time.
About 159 species of *Epidendrum*
occur in Costa Rica. In addition to
Epidendrum ciliare, photographs for
two other species are included here to
illustrate the diversity within this genus.

Epidendrum congestoides

Epidendrum barbeyanum

William A. Haber

Eriopsis rutidobulbon

Description: Large, rough pseudobulbs with 2–3 leaves. Many flowers on long erect spikes; flower ca. 4 cm in diameter; sepals and petals, to ca. 2 cm long, burnt gold with maroon edging; yellow and green column; very dark throat below base of column; lip dull gold with purplish or red markings, with purple-speckled white midlobe.

Distribution: Parts of Central America to South America; mountains and some lower-elevation sites.

Comments: This is a small genus of just 4 to 6 species, only one of which is in Central America. While much orchid literature notes *E. biloba* as the name for Costa Rican specimens, the correct name is *E. rutidobulbon*.

Eriopsis rutidobulbon

Guarianthe skinneri

Guaria morada

Adrian Hepworth

Guarianthe skinneri

Description: To more than 40 cm tall, epiphytic or sometimes growing on rocky cliffs; creeping rhizome. Stalked, furrowed, spindle-shaped pseudobulbs, 15–30 cm by 3 cm, topped with two leaves 10–20 cm long. Spectacular flowers in clusters of 4–15, pinkish to purple; other color forms exist, one all white; flowers 6.5–11 cm in diameter; showy lip ca. 6 cm long by 4.5 cm, tubular at base, surrounding the column; whitish throat. Sepals and petals ca. 5.5 cm long; petals wider than sepals and with wavy margin; light fragrance. Ribbed capsule 4–5 cm long.

Distribution: Southern Mexico to Costa Rica. In Costa Rica, seasonally dry forest 800–1,400 m.

Comments: Although this is the national flower of Costa Rica, it is difficult to find in the wild because it has been heavily collected. You may see *Guarianthe skinneri* decorating patios and yards, however. It is probably bee-pollinated. *G. skinneri* generally flowers January to April, with a peak in March. *Guarianthe patinii*, which is similar but generally smaller than *G. skinneri*, grows in southeastern Costa Rica; it flowers late in the year and has a dark throat. Up until recently, these species were considered part of the genus *Cattleya*. *Cattleya dowiana*, a rare, showy, Atlantic lowland-forest species, has yellowish petals and sepals and a deep purple lip with fine yellow lines. Members of the genus *Cattleya* are known around the world for the extreme showiness of some hybrids and for decorative use in corsages. Hybrids are sometimes formed by crossing species of *Guarianthe* and *Cattleya* with other genera.

Habenaria monorrhiza   Rein orchid

Habenaria monorrhiza

Description: Terrestrial, single stemmed, 0.3–1 m tall, with underground tuber. Up to 16 sheathing leaves, to ca. 10 cm long, spiraling up stem. Long terminal inflorescence with many white flowers; sepals less than 1 cm long, uppermost sepal concave; petals have lobes that vary in length from flower to flower; lip with three 1-cm linear lobes; spur, to 2.5 cm, at base of lip; petals and lobes of lip curved back. Ribbed capsule to 2 cm long.

Distribution: Antilles, Mexico to South America. In Costa Rica, usually at 500–1,600 m, but occasionally lower; in old pastures and other disturbed sites in foothills and mountains of both slopes.

Comments: Habenarias, in general, are called rein orchids. They often grow in wet meadows or forest. There are ca. 600 species around the world, with ca.

17 in Costa Rica. Many are moth-pollinated. *H. monorrhiza* produces a slight fragrance; its pollinator is unknown. Mosquitoes pollinate a North American species, *H. obtusata*. Toward the end of the rainy season, the leaves of *Habenaria monorrhiza* begin appearing on cut banks and in pastures; if they are spared by the machete, they'll produce flowers from September to November.

Huntleya burtii   *Escudo*

Other common name: *Cara de gato.*

Description: Epiphyte without pseudobulbs. Leaves in fanlike arrangement, straplike, of medium thickness and to 40 cm long. Flower ca. 10 cm across; pimply textured sepals and petals in a plane, yellow and toasty golden-brown and maroon, with white at base; lip white with brown to red-brown toward tip; hairlike appendages on lower part of lip.

Distribution: Parts of Central America and South America. In Costa Rica, generally from ca. 400–1,200 m, Atlantic slope.

Comments: *Huntleya burtii* flowers in the mid- to late dry season. Usually there are just 1 or 2 flowers open at once. Male orchid bees of the genus *Eulaema* visit the flowers. This is the only species of *Huntleya* in Costa Rica.

Huntleya burtii

Maxillaria inaudita

Maxillaria inaudita

Description: Epiphyte, usually ca. 50 cm tall; but up to 1 m; stem erect to somewhat arching. Straplike leaves to 25 cm long, in two opposite rows along stem; leaf bases overlapping. White to cream-colored flowers in leaf axils; wavy sepals 4–5 cm long, petals shorter; yellow on 1-cm-wide lip; pungent, honeylike scent.

Distribution: Costa Rica, Panama, and Ecuador. In Costa Rica, 900–1,800 m, in forest and on trees left in pastures.

Comments: An identifying characteristic of this stocky plant is its arrangement of the many leaves in a plane, with their bases overlapping. It usually flowers in the dry season, but is variable.

Maxillaria sanguinea

Maxillaria sanguinea

William A. Haber

Maxillaria umbratilis

Maxillaria biolleyi

Description: Epiphyte; in dense bunches; small spindle-shaped pseudobulbs to 2.5 cm long. Very narrow, grasslike, leaves, 2–3 mm wide, ca. 30 cm long. Flowers yellow-green permeated with red-brown; lip rosy purple, with white edge and a dark purple callus; sepals and petals ca. 2 cm long; mild scent.

Distribution: Southern Nicaragua to western Panama, in forest from lowlands to ca. 1,000 m.

Comments: A similar but wider-leaved species, *M. tenuifolia*, which has leaves wider than 3 mm and more cylindrical pseudobulbs, occurs from northern Costa Rica to Mexico. In addition to *Maxillaria sanguinea*, photos for two other species are included (previous page) to illustrate the diversity within this genus, which contains ca. 110 species in Costa Rica.

Oerstedella centradenia

Oerstedella centradenia

Description: Epiphyte, with stems to ca. 50 cm, branching and bushy; roots and plantlets emerging from stems. Very narrow, 5-cm-long leaves. Flowers rosy magenta with petals and sepals just less than 1 cm long; lip, 1–2 cm long, is white at base; although the lip appears to be 4-parted, it is actually 3-lobed, with the midlobe deeply indented. Capsule ca. 2 cm long.

Distribution: Nicaragua to Panama. In Costa Rica, ca. 1,000–1,500 m.

Comments: This species grows in full to partial sun, and is often seen as an epiphyte in yards. It flowers in the dry season, from November to March. *O. centropetala*, a species of wetter habitats, is somewhat similar but less bushy in appearance. The midlobe of its lip has two narrow fingerlike sections instead of being deeply cleft. Recent molecular studies show *Oerstedella* to be part of *Epidendrum*.

Oerstedella exasperata

People orchid

Description: Terrestrial or epiphytic to 3 m, sometimes branching. Leaves, ca. 10 cm long, in ladderlike pattern along stout and canelike stem; base of leaf clasps the stem. Large inflorescences; flowers to ca. 2 cm across; honey-scented; sepals brown with dull yellow edging, petals white or yellow suffused with brown, green at base; 3-lobed lip, to 1 cm across, white with purple markings, midlobe split, side lobes fringed; lip overall darker with age. **Distribution:** Costa Rica and Panama; 1,000–2,200+ m, in all major mountain ranges. Common along Braulio Carillo Highway near toll booth. **Comments:** This is among the tallest native orchids in Costa Rica. The size of its leaves and flowers is variable. The shape of the lip suggests the arms and legs of a human figure, with the head formed by the anther cap and the opening of the flower. The people orchid flowers during various times of the year. Also called *Epidendrum exasperatum*.

Oerstedella exasperata

Oncidium spp.

Some common names: Dancing ladies, golden rain, *lluvia de oro*. **Description:** Most members of this genus are epiphytic; variable in many aspects, but often with medium to large showy flowers, yellow and reddish brown (some with pink or white). Many have pseudobulbs topped by one or two leaves. Leaf size and texture vary with species. Arching, branching inflorescence, usually many-flowered, arising from base of pseudobulb or from leaf axil, ranges from a few centimeters to more than 2 m long; sepals and petals usually fairly similar; lip often with 3 lobes, the middle one (forming "skirt" of dancing lady) larger and sometimes cleft; projections (calli or papillae) common on lip; 2 waxy pollinia have stem with sticky pad (viscidium) at base. Capsular fruit. **Distribution:** Florida, Caribbean, and Mexico to Argentina. In Costa Rica, various species can be found, sea level–2,500+ m. **Comments:** The broad, often undulating, lower lobe of the flower's lip, along with smaller lateral lobes, look like the skirt and arms of an elegantly dressed woman, thus its common name, dancing ladies. With some imagination, one can make out a head, perhaps a headdress, and breasts. The

Oncidium polycladium

Oncidium isthmii

genus name, *Oncidium*, comes from the Greek word for swelling (*onkos*) and pertains to the protuberances at the base of the lip. Different species attract pollinators (mostly bees) in various ways; they mimic bees, produce oil, and mimic oil-producing flowers such as those of the Malpighiaceae family. Some species are particularly favored by orchid growers because of their extremely long and full flower clusters. There are ca. 30 *Oncidium* species in Costa Rica, and ca. 450 in the world. *Oncidium* contains some species that may be more appropriately placed in other genera. Pictured here are: *O. isthmii* (photo above), which has a narrow isthmus on the lip, and *O. polycladium* (photo, p. 421), with 1-m-long inflorescences.

Phalaenopsis spp. and cultivars — Moth orchids

Description: Members of this genus are epiphytes that lack pseudobulbs. Overlapping leaf bases cover short stem. Leathery leaves, sometimes mottled, to 20 cm wide and to 60 cm long. Inflorescence lateral and often 0.5 to 1 m long; flower 4–10 cm in diameter; side petals may be broader than (or equal in size to) sepals; color ranges from white, rose, purple, to light yellow; some with red, purple, or bronze markings; lip 3-lobed, with mid-lobe (or its tip) sometimes anchor-shaped and/or with appendages that look like horns or antennae. Some species have fragrant flowers.

Distribution: Taiwan and south to Indonesia, the Philippines, and Australia; also in India.

Comments: Most of the species in this section are orchids that are native to

Costa Rica or to the New World tropics. The moth orchids, however, are from the Old World tropics. They are frequently seen in the Costa Rican florist trade, and some growers produce them (along with dendrobiums, vanda, and cymbidiums) for local sales as well as for export. There are ca. 45 species of *Phalaenopsis*, as well as a myriad of hybrids and cultivars.

Phalaenopsis cultivar

Phragmipedium spp.

Some common names: Slipper orchid, tropical American lady's slipper, *zapatillas, chinelas.*

Description: Members of this genus may be terrestrial or epiphytic; without pseudobulbs. Leaves keeled, long (to 90 cm in *P. longifolium*), and leathery. Flower with bucket-shaped lip and long, narrow petals; lateral sepals fused; short column between 2 fertile anthers; flowers green, dull yellow or gold, pink, maroon. The pendant, spiraling petals in *P. caudatum* complex to 60+ cm long.

Distribution: The genus *Phragmipedium* occurs from Mexico to South America. In Costa Rica, *P. longifolium* is mostly on rocks along Caribbean-slope streams at ca. 700 m, but reported from Pacific slope also. The rarer *P. warscewiczii* subsp. *warscewiczii,* which is usually epiphytic, occurs in wet forest of the Talamanca mountains.

Comments: Lady's slippers, known to orchid fanciers as cyps, occur in different regions

Phragmipedium Grande, a hybrid between *P. longifolium* and a member of the *P. caudatum* complex.

of the world, and this common name is applied to members of several genera. In Europe, Asia, and North America, one finds species of *Cypripedium,* some with pink or yellow, bee-pollinated flowers. *Paphiopedilum* of Asia and *Phragmipedium* of the neotropics often have yellow-green slipper flowers marked with browns, pink, or maroon. Some members of the genus *Phragmipedium* give off unpleasant scents and attract flies and/or small bees as pollinators. Flies head toward the upper back wall of the sac-like lip, slip down, and then find a passageway behind the lip wall that leads them up past the column and anthers. Parts of the lip have an oily surface and purple spots. Slipper orchids have sticky pollen instead of the pollinia typical of most orchids. Recent taxonomic studies have split *P. caudatum* into various entities, and names are in flux; *P. warscewiczii* subsp. *warscewiczii* is the latest name given to the Costa Rican member of that group.

MINIATURE ORCHIDS

Of the approximately 1,400 species of orchids that occur in Costa Rica, a large percentage of them are in the subtribe Pleurothallidinae. Particularly abundant in cloud forests, these orchids are often inconspicuous not only due to their diminutive size but because many grow high in the forest canopy. While there is much variation in flower form, all of these miniature orchids lack pseudobulbs and have sheaths on the stems, which bear just a single leaf. In some species, the flowers are borne on the leaf;

Dracula vespertilio

the old flower stalks persist in all species. Most pleurothallids appear to be insect-pollinated, especially by small flies, although relatively few have been studied extensively. The flowers have various odors, colors, and odd-looking appendages to attract the flies, which pollinate them in the process of taking nectar. Minute details of the flowers are not included here, although those characteristics often help in identifying genera and species. The descriptions here each discuss a different genus of the miniature orchids found in Costa Rica. The photographs illustrate some of the variety and elegance in these often overlooked, small to wee members of the orchid family.

Dracula flowers are pendant, with inconspicuous petals. The tips of the fused sepals form long, thin tails. The form of the flower is similar to that of *Masdevallia*, but the outer part of the lip resembles a tiny white mushroom (and in some species smells like one!) that attracts fungus-seeking gnats as pollinators. There are seven species in Costa Rica; the species pictured, *D. vespertilio*, has a white, mobile lip and hairy sepals, 3–8 cm long, that are light yellow-green with purple blotches. Its range spans from Central America to the Andes.

A typical *Lepanthes* has a single tiny flower in the middle of a stiff, small, round leaf; it may be on either the upper or the lower surface. The inflores-

Lepanthes turialvae

Lepanthes helleri

MINIATURE ORCHIDS

Adrian Hepworth

Masdevallia zahlbruckneri

Platystele jungermannioides

cence, which originates at the leaf base, has buds that develop over time. The showiest parts of the flower are the petals, which are often 2-parted and glandular or pubescent. The sepals lay beneath, and the lip is small, with winglike lobes. Since the flower resembles a fly on a leaf, pollination may take place by pseudocopulation. More than 90 species of *Lepanthes* occur in Costa Rica.

Masdevallia are small to rather large orchids, with leaves often narrowing toward the base. The sepals are partly fused, forming a tube or cup, but then narrowing abruptly into tail-like tips. The petals and lip are smaller and less conspicuous. Some species have exquisite colors and markings. There are approximately 31 species in Costa Rica; *M. zahlbruckneri* has leaves ca. 6 cm long. The flowers are borne near the substrate. The upper sepal and the tails are yellow, with lower two sepals, ca. 2 cm long, fused and mottled purple.

With its extremely small flowers, the genus *Platystele* includes some of the smallest orchids in the world; there are ca. 12 species in Costa Rica. Plants of *P. jungermannioides* stand just under 2 cm tall. The flowers are translucent pale green with pink shading. The sepals are 1 mm long; and there is a fleshy protuberance, a callus, at the base of the lip.

The ca. 150 *Pleurothallis* species in Costa Rica are being split up into a variety of genera, leaving only 60 true *Pleurothallis*. The genus has always been difficult to define because there is quite a bit of variation in leaf shape and inflorescence size, and it ranges from miniatures to the 1 m tall *P. colossus*. The sepals, which are often longer than the petals, are free or somewhat fused; the two lateral ones are sometimes completely joined. The inflorescences, which are terminal and originate from the leaf base, sometimes rest on the leaf. The column usually has a projecting foot at the base to which the small lip is attached.

Restrepia trichoglossa is about 8 cm tall and grows in clumps. It often has papery bracts along the stem. The leaf blades are 3–4 cm long. This orchid has one to a few yellow and wine-red flowers, with the upper sepal and the petals, to ca. 2 cm long, very thin, ending in clublike thickenings. The side sepals join together to form a concave platform beneath the small lip. Only two other species of *Restrepia* occur in Costa Rica: *R. lankesteri* and *R. muscifera*. In some species of *Restrepia*, the clublike tips produce an odor that may attract flies.

Stelis species have a long stalk with many flowers. The fleshy leaves are elliptical or wider toward the tip. The three relatively broad sepals, sometimes fused toward their bases, are the most prominent aspect of the flowers, the petals and lip being much smaller. These orchids are probably fly-pollinated. *Stelis* is a large genus, with more than 90 described species (and counting) in Costa Rica. It can be difficult to identify an individual plant to the species level. The species pictured, *Stelis microchila*, has leaves ca. 2 cm long; its inflorescence is ca. 9 cm long, with tiny fuzzy flowers less than 0.5 cm in diameter.

There are several other species of small orchids that occur in Costa Rica that are not in the Pleurothallidinae subtribe. Both of the following genera are in the subtribe Maxillarieae.

Dichaea are epiphytic orchids that lack pseudobulbs and have stems that are often pendant. The alternate leaves are in a plane, with overlapping bases that enclose the stem. Some species are deciduous. The single flowers, borne in leaf axils near the end of the stem, are often tinged with blue violet. The 3-lobed lip is typically shaped like an anchor. The seed capsule may be smooth, rough, or bristly, depending on the species. There are ca. 29 species in Costa Rica.

Stellilabium (now part of *Telipogon*) has tiny flowers and is often leafless when flowering—this orchid basically looks like a green stem. A series of winged bracts sheathing the inflorescence gives it a flattened appearance. It lacks pseudobulbs, and the leaves are very small and, when present, near the base of the plant. The flowers are ca. 6 mm in diameter. This group has recently attracted a lot of attention from orchidologists—six new species have been discovered and described during the past ten years in Monteverde alone.

Acronia palliolata (=*Pleurothallis palliolata*)

William A. Haber

Restrepia trichoglossa

Stelis microchila

Dichaea species

William A. Haber

Stellilabium boylei

Ponthieva formosa

Description: An erect epiphyte, sometimes on base of tree trunks close to ground; 15 cm tall; entire plant with long whitish hairs. Soft leaves ca. 8 cm

long (possibly longer in shade). Up to 10 flowers per plant; flower ca. 1.5 cm across; sepals a bit more than 1 cm long; petals less than 1 cm long, partially united and attached to column, forming a false lip; true lip small and like a hood above column; sepals dull, yellow-green to brownish, with glandular dots; petals brownish-green and white.

Distribution: Mexico to Panama; in Costa Rica 1,350-2,100 m; in cloud forest, from Tilarán range to Talamancas.

Comments: The arrangement of the petals in this species is unlike many orchids since the true lip is not conspicuous and is located above the column. Another hairy species, *P. brenesii*, with reddish flowers, occurs in Costa Rica, along with two glabrous ones, *P. racemosa* and *P. tuerckheimii*.

Ponthieva formosa

Prosthechea prismatocarpa

Description: Epiphyte with large, thick, smooth pseudobulbs, 5–15+ cm long; 2-3 fairly thick, straplike leaves, 25 cm long. Erect inflorescence to 30 cm long, with many fragrant flowers, each ca. 4 cm across; sepals ca. 3 cm long (petals a bit shorter), very pale green—almost white—with dark purple blotches; 2-cm-long lip, partially fused to col-

umn, white with lilac midlobe that comes to a point.

Distribution: Costa Rica and Panama, forest 1,200–2,000+ m.

Comments: Some populations have papillose sepals that are visited by ants. *Prosthechea prismatocarpa* is widely cultivated. The similar *P. ionocentra* lacks large spots. Both were formerly considered to be in the genus *Encyclia*.

Prosthechea prismatocarpa

Psychopsis krameriana

Butterfly orchid, *Mariposa*

Description: Epiphyte with clustered, flattened, wrinkly pseudobulbs, ca. 3 cm across, each with one
thick, oblong leaf, ca. 22 cm long,
somewhat mottled or spotted with
red or purple. Flowering stalk, to 80
cm, topped with a number of
flower buds; only one opens at a
time; large flowers ca. 7 cm in
diameter, 10+ cm tall; 2 deep brown
(or red-brown) petals and the
uppermost sepal are each ca. 5–8
cm long and very thin with crisped
edge toward tips; the two broader,
sickle-shaped side sepals, 4–5 cm
long, with wavy margin, yellow
with brown blotching; lip ca. 3 cm
wide, with 3 lobes, the lower one—
largest and deeply notched—has
yellow center and reddish-brown
edge. Two appendages projecting
from the column end in knobs and
look like antennae.

Psychopsis krameriana

Distribution: Costa Rica to South America. In Costa Rica, rainforests of lowlands and foothills, up to ca. 1,200 m. Found at La Selva and Caribbean sites to the south.

Comments: This lovely species, a relative of the genus *Oncidium*, has practically been wiped out by commercial collectors. The long-lived flowers, long flowering period, and its general popularity make it a prime choice for experimentation in laboratories and for greenhouse propagation. The genus name comes from Greek—like (*opsis*) a butterfly (*psyche*).

Schomburgkia lueddemannii   *Guarión rojo*

Description: Epiphytic; furrowed pseudobulbs, to 30 cm long, topped by 2–3 leaves to 30 cm. Erect, terminal flower stalk to 1+ m tall. Ca. 15 maroon

Schomburgkia lueddemannii

flowers; sepals and petals similar, ca. 4 cm long with wavy margins; 3-lobed lip rose to purple, with yellow or white markings. Dressler distinguished between 2 species in Costa Rica: *S. lueddemannii*, which has 3 keels and flowers with 4- to 5-cm sepals, and *S. undulata*, with 5 keels on the lip and sepals 3.3–3.5 cm.

Distribution: Costa Rica and Panama. In Costa Rica, 200–800 m; mostly in central Pacific and Guanacaste regions.

Comments: This is an odd-looking orchid that perches on tree trunks or on large branches exposed to the sun. Its sepals and petals look somewhat like crispy bacon strips. It is closely related to the genus *Laelia*; hybrids have been produced with other genera (e.g., *Cattleya*, *Laelia*, *Epidendrum*). The related *Myrmecophila tibicinis* (formerly *Schomburgkia tibicinis*), a rare species of northern Guanacaste, houses ants in its older, hollow pseudobulbs. Bees are reported to be the pollinators for that and for various species of *Schomburgkia* (Rico-Gray and Thien 1987).

Sobralia spp.

Some common names: Sobralia, *flor de un día*.

Description: Members of this genus are epiphytic, or often terrestrial on embankments; with thick roots; most species 0.5–1 m tall, lacking pseudobulbs; with canelike stems. Alternate leaves, ca. 12 cm long, strongly veined, pleated, and rather stiff.

Sobralia species

Inflorescence terminal; showy, delicate flowers may be white, yellow, lavender, some with yellow in throat or reddish markings on lip; the 3 sepals similar, lip tubular, and lateral petals often wider than sepals. The sobralia shown above has sepals ca. 7 cm long by 2 cm, petals 6 cm by 3 cm; tube to 6 cm; lip notched, ruffled; yellow in throat; some flowers with touch of lavender. Ribbed fruit 15 cm long.

Distribution: Central and South America. In Costa Rica, various species occur from lowlands to ca. 3,000 m.

Comments: The delicate flowers of this genus, often large and fragrant, are known for their short lives; as their Spanish name, *flor de un día*, suggests, they last only a day—or mere hours. Some *Sobralia* species exhibit synchronized flowering with an ephemeral burst of blossoms of all the individuals within a population. Species may be very distinctive in color and form, but there are a number of similar species that are difficult to identify. Carpenter (*Xylocopa* spp.) and euglossine bees are major pollinators; hummingbirds visit the darker-colored species, such as the widespread magenta *S. amabilis*. There are ca. 30 species in Costa Rica.

Stanhopea wardii　Stanhopea, *Torito*

Description: Epiphyte. Pseudobulb ca. 5 cm long. Elliptical erect leaves to ca. 45 cm long including petiole; several veins especially prominent on underside. Pendant inflorescence of ca. 6 flowers arising from base of pseudobulb; flower ca. 9 cm long; sturdy, waxy lip ca. 4.5 cm long, golden yellow with large, purple-brown eye spots near base; distal portion, which is yellow-green with maroon speckling, has hornlike projections; the 3 large, maroon-speckled sepals, 5 cm long by 3 cm wide; the petals 4 cm long, narrower, curled and

Stanhopea wardii

twisted; long column faces the ornate lip. Flowers very fragrant.

Distribution: Mexico to Panama; in Costa Rica, 400-1500 m, moist forest to rainforest.

Comments: The form of these odd flowers has evolved in response to the anatomy and behavior of orchid bees. A male orchid bee, seeking scent compounds to use as building blocks for pheromones, heads straight to the indented part of the lip between the eye spots, and scrapes it to gather scent. The bee then slides down the slippery lip, hitting his back on the long, narrow structure opposite the lip—the column, where pollen packets are either picked up or deposited. Orchid bees of the genus *Eulaema* have been seen at this species. The flowers last several days. The Spanish common name means little bull and refers to the curved, hornlike lobes sticking out from the midsection of the lip. There are about eight species of *Stanhopea* in Costa Rica. *S. costaricensis* flowers are distinctive, with O patterns on the petals and sepals rather than the solid dots of other species. *S. ecornuta* lacks horns; its inflorescence has only two flowers, which are suffused with orange and pink, marked with brown. Most stanhopeas grow in foothills and mountains, although *S. cirrhata*, of the Pacific slope, occurs lower (0-700 m).

Stenorrhynchos albidomaculatum

Description: Epiphytic or terrestrial; thick, white, fingerlike roots; to 35 cm tall when in flower, the erect inflorescence arising from rosette of leaves blotched with light green. Leaves to 20 cm long. Flowering stalk has bracts to 5 cm long; flowers 1.5 cm long, red-pink with white lip, tinged with rose.

Distribution: Mexico to northern South America. In Costa Rica, in mountains 1,200–1,600 m.

Comments: This orchid is part of what is known as the *Stenorrhynchos speciosum* complex. It flowers December to February and is most likely hummingbird-pollinated. The similar, *S. glicensteinii*, is smaller and epiphytic. *S. lanceolatum*, a species seen at lower elevations, is leafless when flowering. Various *Coccineorchis* species that occur in mountainous habitats and are usually terrestrial are somewhat similar to *S. albidomaculatum*, but overall they are larger and several have yellow or yellow and red flowers.

Stenorrhynchos albidomaculatum

Trichopilia suavis *Tricopilia*

Description: Epiphyte; pseudobulb to ca. 7 cm tall (to nearly as wide), somewhat flattened, with 1 leaf. Inflorescence of ca. 3–6 flowers starts at base of pseudobulb. Broad leaf to 20+ cm long. Petals and sepals, ca. 4.5 cm long with wavy margin, white with washed-out wine-colored spots; long tubular lip, ca. 6 cm, with bottom lobe crisped and wavy, cleft, and ca. 4 cm broad, with wine-pink blotches and spots; golden-orange in outer throat, with smaller wine dots deep in throat; sweet scent. Fringe at tip of column is a characteristic of the genus *Trichopilia*.

Distribution: Costa Rica to Colombia. In Costa Rica, usually in wet forest habitat, 500–1,500 m, both slopes.

Comments: Populations of *Trichopilia suavis* are dwindling in Costa Rica due to habitat destruction and overcollecting. It usually flowers in the dry season. The flowers' pleasant perfume most likely attracts male orchid bees.

Trichopilia suavis

Vanilla spp.

Some common names: Vanilla, *vainilla.*

Description: Members of this genus are large vines climbing by aid of roots on stem. Alternate leaves often thick and fleshy, although small and scalelike in some species. Greenish or cream-yellow flowers; lip, with wavy margin, rolled into tube; pollen mealy, not in typical pollinia as in other orchids. Fruit are fleshy pods looking like green beans or scrawny bananas; thousands of tiny seeds embedded in pulp. *V. pompona* has broad leaves to 20+ cm long, usually wider at base. Cream-yellow flowers with ca. 7-cm-long petals and sepals; tube, ca. 7 cm long, formed by lip. Fruit ca. 15 cm long.

Distribution: Vanilla species found in Old and New World tropics. *V. pompona* occurs from Mexico to Bolivia. In Costa Rica, collected at around 500 m on Pacific slope.

Comments: Rarely do people think about eating orchids, but when you indulge in ice cream made from real vanilla, you experience an exquisite flavor that comes not from the flower, but from the specially cured fruit, or bean, of this orchid vine. Five species occur in Costa Rica. In cultivation, *V. planifolia*, the use of which dates back to Aztec chocolate drinks, is preferred over *V. pompona.* Vanilla is cultivated in Sarapiquí, Quepos, and other hot, humid areas of Costa Rica. Most vanilla is produced in Madagascar and Indonesia. Dutch and French travelers carried vanilla to islands in the Old World for planting. In wild populations, bees (*Melipona* or larger species) are likely pollinators, but in plantations, hand-pollination leads to a successful crop. The traditional fermentation process for vanilla requires a series of treatments that take months. This involves picking pods before they are ripe and arresting their development by putting them in the sun, although sometimes a boiling water bath is used. For a month, sun baths are alternated with a process called "sweating," in which pods are bundled in blankets and boxed up to exclude air and light. After another month of air drying out of the sun, the pods are usually stored for a few more months. This procedure creates the fine flavor of real vanilla. Artificial vanillin, a poor substitute made from wood pulp or from eugenol, found in clove oil, lacks many of the compounds found in vanilla beans. Tissue culture research may result in a capability to produce real vanilla in the lab. Seed dispersal in wild vanilla may be by animals (perhaps ants) since the seeds are in a sweet pulp that is atypical for orchids. Another name used for *V. planifolia* is *V. fragrans.* The Spanish word *vaina* means sheath or pod.

Vanilla pompona

Ferns and Lycophytes

Ferns are not exclusively tropical, but in the wet tropics they grow exuberantly. The humid environment of a rainforest canopy is an ideal home for ferns, and many species have evolved characteristics for clinging to tree trunks and branches. Fern species have a wide range of climatic adaptations; some species grow in hot, dry conditions, while others thrive in cold and snowy regions, and they can be found from sea level (and almost in the sea) to elevations above 4,000 m.

Although this book focuses on flowering plants, it would seem incomplete to omit the pteridophytes, as this large body of vascular plants is called, whose ancestors comprised most of the vegetation on earth during the Carboniferous period 300 million years ago. This section provides a glimpse of the beauty and diversity of the ferns and lycophytes of Costa Rica. Its more than 1,100 species, out of about 12,000 species worldwide, range in size from tiny, 2-cm water ferns (*Azolla* spp.) to 20-m-tall tree ferns.

Pteridophytes do not bear flowers, fruits, or seeds, but reproduce by minute spores that are produced on the mature plant, which is called a sporophyte. These spores germinate and develop into gametophytes, which are plantlets, typically flattened and heart-shaped, that are from 2 mm to 2 cm across. Structures on the underside produce eggs and sperm. Moisture facilitates fertilization since it enables the sperm to swim to and join either an egg on the same gametophyte, or, more likely, one on another gametophyte. Often the sperm is produced later than eggs on any individual gametophyte, and, in some cases, the eggs produce hormones that induce sperm production in nearby gametophytes. Once fertilization occurs, a tiny fern begins to grow and a new sporophyte phase, or generation, begins. Besides this form of sexual reproduction, many pteridophytes can also reproduce asexually by means of tubers, plantlets at tips of fronds, or root buds.

The study of ferns involves a technical vocabulary that has been simplified here—fern-related terms are illustrated and defined in the glossary. Scales and hairs on the rhizome (stem), venation in the leaves, arrangement of the sori (groups of spore-producing bodies), as well as the form and texture of the frond, are all important characteristics to look at when identifying ferns.

The Pteridophytes are grouped into two classes: the Lycopodiopsida (lycophytes) and Polypodiopsida (ferns). Until recently, the horsetails (*Equisetum*) and whisk ferns (*Psilotum*), were classified apart, in their own groups, but it has now been shown from DNA studies that these unusual plants are ferns, "nested" among other ferns in the fern tree-of-life. Ferns typically have large leaves with several to many veins and numerous spore cases borne in clusters (sori) along the margin or lower surface of the leaf. The lycophytes typically have small, entire, single-veined leaves, with a single spore case on the upper surface.

Fern classification is an active field, and taxonomists do not all agree on how to group the different taxa. Differences occur at all levels of classification, in both popular and technical references. And adding to the confusion, hybrids are common, and cultivars abound in horticulture.

Ferns' soft and lacy decorative foliage lends itself nicely to the art of horticulturists and florists. A more practical benefit for humans, which is equally appreciated but much less obvious, are coal reserves. These are a result of the build-up, over eons, of lycophytes' ancestors. In some cultures, especially those of Asia and Pacific Islands, the fiddleheads, rhizomes, or fronds of various species are eaten, and a few have medicinal uses.

The ferns in this section are alphabetized according to their species names.

A *Cyathea* tree fern.

Acrostichum danaeifolium
Family: Pteridaceae

Giant leather fern, *Negra forra*

Other common name: *Helecho mangle.*
Description: Immense terrestrial ferns, more than 3 m tall, growing in dense stands in wet areas. Leaf underside has short hairs and some channeling in the rachis; once pinnate, ca. 30 pairs of entire, thick pinnae, plus one terminal

pinna similar to others; pinnae ca. 30 cm long; netted venation. In fertile fronds, majority of pinnae with carpet of sori on underside.
Distribution: Florida, Mexico, Central America, South America, and West Indies. In Costa Rica, swamps near Atlantic coast and inland wetlands, to ca. 1,000 m.
Comments: Another common fern in this genus is the pantropical *A. aureum*, which lacks hair on the leaf underside, has fewer pinnae, and whose fertile fronds have sori on 7 or fewer of its pinnae. It often occurs in more brackish areas than does *A. danaeifolium*, and it is also seen in Pacific coastal areas. Both species can tolerate salty sea spray, heat, and sunlight, along with fluctuating water levels. Some Central American indigenous people, and early settlers, found the fire-resistant salty fronds suitable for use as thatch above cooking areas. In some parts of world, young fronds of *A. aureum* are eaten. The genus name derives from the botanical term *acrostichoid*, describing the uniform covering of sori on the underside of the frond. A third species, *A. speciosum*, occurs in Australia and tropical Asia.

Acrostichum danaeifolium

Adiantum concinnum
Family: Pteridaceae

Brittle maidenhair fern, *Alientos*

Other common names: Dwarf maidenhair fern, *culantrillo.*
Description: Terrestrial, often on vertical banks of roads and streams; sometimes pendant, in clumps. Fronds to more than 50 cm long, overall triangular shape; shiny, smooth, black midribs; 2–3

pinnate, 10–25 pairs of pinnae, divisions tapering toward tip and somewhat toward base; pinnules like little lobed fans, ca. 1 cm by 1 cm; pinnule at the base of each division 2-parted and overlapping the rachis (this is a distinguishing characteristic of the species). Spores

Adiantum concinnum

borne on kidney-shaped patches under curved-over pinnule edges; the patches are in notches between lobes. Pleasantly scented.

Distribution: Mexico to South America, and Antilles; widespread in Central America on rocks, riversides, and road banks; from sunny and dry to shady and moist areas; sometimes cultivated. In Costa Rica, low to high elevations, to 2,000 m, widespread, especially along Pacific slope.

Adiantum macrophyllum

Comments: These are delicate and attractive ferns, with fanlike pinnules (leaflets) and black wiry stems. Several of the 30 Costa Rican species of *Adiantum*, including the elegant diamond, or giant, maidenhair fern (*A. trapeziforme*) and the broad-leaved maidenhair (*A. macrophyllum*, photo above right) are popular in horticul-

ture. Two common maidenhair ferns that grow in disturbed areas near La Selva Biological Station are *A. latifolium* and *A. petiolatum*. In Belize, *A. tenerum* is used as an antiparasite medicine, to treat coughs, and externally for dandruff. The black stems of various species are incorporated into baskets.

Asplenium auritum
Family: Aspleniaceae

Auricled spleenwort

Description: Epiphytic; a variable species with small to medium bunched fronds. Often forms loose colonies by root proliferation. One-pinnate pinnatifid; 10–20 pairs stalked pinnae; frond with tapering tip; bases of pinnae (at least lower ones) have earlike lobes on the margin; the lobes point toward tip of frond; veins ending in tiny white streaks—salt-encrusted hydathodes (water-secreting pores) on upper surface of leaf; clathrate (latticelike) rhizome scales. Elongated sori along veins. Plant becomes fragrant as it dries.

Distribution: Central and south Florida, southern Mexico to Argentina, and Madagascar; 100–2,200 m. Widespread in Costa Rica, in fairly wet habitats, from lowlands to more than 2,000 m.

Comments: *Asplenium* is a large, mostly tropical, genus with ca. 700 species; there are ca. 60 species in Costa Rica and many are popular in horticulture. Some *Asplenium* species reproduce by buds. Species like *A. bulbiferum*, of the Old World, produce buds on the surface of their fronds; other species, like *A. maxonii* and *A. cirrhatum* of Costa Rica, produce a bud on the whiplike extension (cirrus) of the rachis. *Asplenium* species are recognized by the latticelike scales on their stems and single linear sori along the veins below. *Diplazium* (Woodsiaceae), a similar

x ⅓

tip

Plantlets start at the tip of a long cirrus in some *Asplenium* species.

Asplenium auritum

genus, has at least a few paired, back-to-back sori and is always terrestrial. Many *Asplenium* are one to three pinnate pinnatifid, but long, simple leaves occur in a group known as bird's-nest ferns, which are mostly found in the Old World, although one species, American bird's-nest fern (*A. serratum*), occurs in Costa Rica (La Selva, Tortuguero, Osa Peninsula).

Azolla spp.
Family: Salviniaceae

Some common names: Mosquito fern, water fern, fairy moss.

Description: Members of this genus are aquatic plants that float on the water surface; plants often 1–2 cm long, with branching stems and unbranched, thin roots. Scalelike, overlapping, alternate leaves yellow-green to purple-red, to 2 mm, in 2 rows along stem. Sori usually paired on the underside of leaf.

Distribution: Various species are found from the United States to South America; also in Asia.

Comments: These aquatic plants, the smallest species of ferns, form velvety, mosslike mats on calm water in temperate to tropical regions of the world. The name mosquito fern comes from azolla's ability to choke out mosquito larvae by covering the water surface. In some regions they are cultivated as ornamentals, for mosquito control, or for their nitrogen-fixing qualities. The fern leaves have cavities in their upper lobes that house the nitrogen-fixing cyanobacterium *Anabaena azollae*. Capitalizing on this characteristic, rice farmers in Asia and the Philippines grow azolla in rice paddies and use it as an inexpensive green manure. Experimental duck-azolla-rice systems are being developed in Japan to cut

Azolla cristata

down on the use of pesticides and artificial fertilizer (Watanabe 2003). *Azolla* serves as food for insects and other invertebrates, as well as water fowl and livestock. The ferns are also edible to humans. The plants turn purplish-red with more sun exposure, and although they are attractive, azolla species can be noxious weeds because they can reproduce vegetatively and grow rapidly. Introducing the Asian species *A. pinna-* *ta* into the United States is against the law. There are ca. seven species worldwide. *Azolla cristata* may be the only species that occurs in Costa Rica. Azollas are sometimes classified in their own family, Azollaceae. Identification to the species level is difficult because the best distinguishing characteristics, which are on the megaspores, can only be seen using scanning electron microscopy (Armstrong 1998).

Blechnum occidentale
Family: Blechnaceae

Hammock fern

Description: Terrestrial; new growth reddish; shiny and leathery when mature. Blades, tapering toward tip, are generally ca. 30 cm long on a ca. 15-cm smooth stalk, but some plants are up to twice that size; 1-pinnate, except for area toward tip, which is lobed (pinnatisect); ca. 20 pairs pinnae (and lobes), but variable; pinnae widest at their bases, tapering toward tip, lower ones with slightly cordate base, upper ones with base attached along rachis. Fertile and sterile fronds look similar; on fertile fronds, sori are in two lines on the sides of the midvein of each pinna and form velvety, chocolate-brown strips when mature.

Distribution: Southern United States to Argentina, and Antilles. In Costa Rica, roadside and forest, sea level–2,000 m.

Comments: Hammock fern can be planted as a ground cover out of full sun; it spreads by running rhizomes. Indigenous people of northwest Venezuela bathe in an infusion they consider a cure-all. A leaf decoction is used in Argentina to dissolve kidney or bladder stones. Some individuals of the variable *B. occidentale* are easy to confuse with *B. glandulosum*, which has hairs on the

Blechnum occidentale

Blechnum proliferum

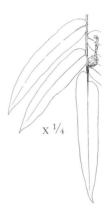

Plantlet sprouting from a
Blechnum proliferum frond.

underside of the rachis, has more pinnae (23–40 pairs), darker rhizome scales, and generally grows at higher elevations, 1,200–3,000 m. In Costa Rica, however, a few collections of B. occidentale have been made up to ca. 2,000 m. The two species sometimes hybridize. The new

leaves of many blechnums, including B. proliferum, are red and showy (photo above left). B. proliferum can reproduce in two ways; it produces spores (sexual reproduction) and it also grows buds and plantlets at the tips of its fronds (vegetative reproduction).

Elaphoglossum peltatum   Parsley fern
Family: Lomariopsidaceae (*Elaphoglossum* is now in Dryopteridaceae)

Description: Epiphyte with creeping rhizomes. Small fertile and sterile fronds, the latter undivided or with up to 6 dichotomous divisions ending in slender, finger-like lobes. The fertile fronds, to 2 cm in diameter, rounded or 2-lobed, with covering of spores on one side; sterile blades ca. 2.5 cm long and often wider across, with short or relatively long petioles to 7 cm.
Distribution: Mexico to South America, and Antilles. In Costa Rica, in mature, evergreen forest, sea level to more than 2,500 m.

Comments: The genus *Elaphoglossum* has over 600 species, more than 100 of them in Costa Rica. Although *Elaphoglossum peltatum* (photo following page) shares many characteristics with other members of its genus, it does not have the tongue-shaped leaf that is typical of most species (see *Elaphoglossum lingua*, p. 313). Its unusual leaf form, which resembles parsley, has led some experts to classify it as *Peltapteris peltata*, with four forms defined according to the degree to which the leaves are split.

Elaphoglossum peltatum

Equisetum spp.
Family: Equisetaceae

Some common names: Horsetail, *cola de caballo*.

Description: Members of the genus *Equisetum* are terrestrial and generally grow in wet areas; branching underground rhizomes are present. Vertical, jointed, hollow green stems; longitudinally grooved; some species with whorls of horizontal branches. Minute, toothlike leaves fused into sheaths that encircle joints; rough texture is from silica in cell walls. Fertile and sterile stems differ in some species; spores in strobili (conelike structures at tip of stem or branches). *E.* x *schaffneri* (photo right) is a hybrid between *E. giganteum* and *E. myriochaetum*. It is 3+ m tall; with 10–19 branches at a node, each ca. 45 cm long; 8-cm internodes; stem 0.7 cm in diameter, with black sheaths formed by 1-cm-long leaves that have 0.6-cm-long gray teeth.

Equisetum x *schaffneri*, a giant horsetail.

Equisetum x *schaffneri* at Wilson Botanical Garden, San Vito.

Distribution: Members of the genus *Equisetum* occur around the world except in New Zealand and Australia. In Costa Rica, giant horsetails are rare and difficult to find. *E.* x *schaffneri* has been found in Mexico and parts of Central and South America; it grows in marshy places. The smaller, more common *E. bogotense* occurs in many parts of Costa Rica, from foothills to mountains, in wet ditches and along rivers.

Comments: Horsetails have been called fern allies, thought to be different from but related to ferns, but it is now known that horsetails *are* ferns, perhaps most closely related to marattias or osmundas. Relatives of the horsetails comprised an important component of earth's flora during the Carboniferous period. The most common horsetail in Costa Rica is *E. bogotense*, which is less than 1 meter tall, a typical size for most temperate-zone horsetails. In the tropics, however, some species can reach more than 8 m. Another common name for horsetails is scouring rush; because of their high silica content, some species are used to clean pots and pans. Various medicinal uses include treatments for dysentery and urinary tract infections; vendors in Costa Rica sell horsetail in the streets and at markets. Some North American species have been found to contain a suite of alkaloids; they also contain thiaminase (see *Pteridium* species, p. 452). This is a difficult group of plants to identify to the species level both due to the existence of hybrids and because the identifying characteristics, such as the shape of the tiny tubercles on the ridges of the branches and the arrangement of stomata, are not evident to the naked eye. Three species, plus this hybrid, occur in Costa Rica.

Hymenophyllum spp.
Family: Hymenophyllaceae

Filmy fern

Hymenophyllum species

Description: Members of this genus are epiphytic, translucent ferns with long, slender rhizomes with hairs. Leaf size is generally 1–30 cm long (although the South American species *H. speciosum* reaches 2 m); frond simple to divided several times, with toothed or entire margin; midvein sometimes winged. Sori in 2-lobed, clamlike structures (involucres) along leaf margin.

Distribution: Pantropical, also in some wet temperate regions; most diverse in midelevation cloud forest, such as those of the Tilarán, Talamanca, and central volcanic mountains in Costa Rica.

Comments: Filmy ferns (*Hymenophyllum*) are closely related to the bristle ferns (*Trichomanes*)—both genera have translucent leaves that are mostly one cell thick,

Hymenophyllum   *Trichomanes*

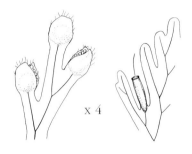

X 4

Comparison of reproductive structures along edge of fronds of *Hymenophyllum* and *Trichomanes*.

except around the veins. And, both filmy and bristle ferns grow low down on tree trunks in wet, tropical forests. Although both groups occur at a range of elevations, filmy ferns are more common in the mountains, whereas bristle ferns are more abundant in wet lowlands. *Hymenophyllum* have 2-valved, clamlike structures with no filament. In *Trichomanes*, the tips of the segments on spore-producing leaves have a tubular structure with a thin filament (bristle) sticking out. Two rare species of *Hymenophyllum* and five species of *Trichomanes* exist in the United States and Canada. There are approximately 37 *Hymenophyllum* species and 47 *Trichomanes* species in Costa Rica.

Lycopodium clavatum   Running club-moss, *Cipresillo*
Family: Lycopodiaceae (Class Lycopodiopsida)

Other common names: Ground pine, staghorn clubmoss.

Description: Trailing plant, sometimes dense, with stems forking, growing along ground or as an epiphyte; running stems as well as erect stems have small, narrow, 1-nerved leaves ca. 0.5 cm long. Leaves are green with a tan-white, hairlike tip. Occasional roots, often exposed, may be 12 cm long. Sporangia, borne on sporophylls, arranged in strobili (cones), 1–3 in a group, terminal, distinct from leafy stems and often stalked.

Distribution: One of the most widespread plants in the world: Europe, Africa, Southeast Asia, Pacific Islands; and Canada south to Paraguay. In Costa Rica, 1,200–3,500+ m, in mountains from Tilarán range south through Talamancas, often on damp embankments.

Comments: Club mosses are in a separate class (Lycopodiopsida)

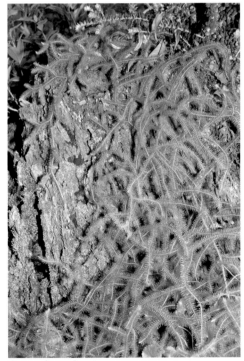

Lycopodium clavatum

from typical ferns (Polypodiopsida) and, along with selaginellas, are commonly referred to as fern allies. The family Lycopodiaceae comprises ca. 450 species found around the world from the Arctic to Antarctica. The three genera seen in Costa Rica are *Lycopodium*, *Lycopodiella*, and *Huperzia*. *Lycopodium* has erect strobili, while some species of *Lycopodiella* have nodding strobili. In both, forking branches differ in length, whereas in *Huperzia* they are equal. Also, species of *Huperzia* do not have a principal running stem from which branches arise. The gametophyte stage of lycopodium's reproductive cycle is underground and dependent on fungi (i.e., mycorrhizal). Oils in the spores make the plants flammable, a trait that has led to their use in fireworks and various applications where a flash of light is needed. They've also been used in homeopathy, in hair powders, for polishing furniture, and as a coating on latex gloves and condoms, although some people are allergic. Taxonomists divide *L. clavatum* into two subspecies or even two species: Plants in which the branches diverge and the strobili are long-stalked are placed in the subspecies *clavatum*, which occurs at 1,000–3,300 m; if the branches are parallel and the strobili short-stalked or sessile, the plant is placed in the subspecies *contiguum*, which grows at elevations above 2,600 m.

Niphidium nidulare
Family: Polypodiaceae

Bird's-nest fern

Description: Epiphyte with long, entire, straplike leaves, to ca. 1 m long, in a dense clump, forming a basket. Fertile and sterile leaves similar; base of blades either narrowing or spreading and rounded; petiole to 1 cm long, with

Niphidium nidulare

Underside of *Niphidium nidulare* frond showing arrangement of sori.

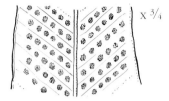

x ³⁄₄

Niphidium nidulare

blackish scale mass at base; cells close to central vein are 1.5–3 times longer than wide (in *N. crassifolium*, they are usually 3–5 times longer than wide). The sori, or groupings of spore-bearing structures, are roundish and occur on the frond underside in single lines between the lateral veins.

Distribution: Costa Rica and Panama, 1,300–2,000 m. In Costa Rica, occurs in all major mountain ranges.

Comments: The basket of leaves catches falling tree leaves and other organic matter. Three species, of an estimated ten in the genus, occur in Costa Rica. The leaf of *N. crassifolium*, a similar, more widespread species, has a narrowing, wedge-shaped base, grayish scales, a definite petiole 5–15+ cm, and more elongate cells near the midvein. A newly described species, *N. oblanceolatum*, is similar to *N. crassifolium*, but differs in range and in the details of the rhizome scales. *Niphidium* is close to the genus *Campyloneurum*, which has more than one row of sori between veins. The name bird's-nest fern is also used for some species of *Asplenium* (American bird's-nest fern, *A. serratum*, occurs at La Selva).

Odontosoria gymnogrammoides
Family: Dennstaedtiaceae

Bramble fern

Description: Scandent fern lime green, with lacy appearance, several meters or more in length. Four to five pinnate; often growing on forest edge and road banks; copper to green rachises with small prickles, especially on underside of young stems; main portion of rachis rather wiry, 3-parted at top—central fiddlehead and 2 side branches with curling tips that expand more rapidly; ca. 17-cm long internodes along rachis. Midribs of the pinnae zigzag; final segments fingerlike, notched at tips. The sori are at tips of—and span the width of—ultimate segments of the fronds.

Distribution: Costa Rica, Panama, and Colombia. In Costa Rica, on both slopes, ca. 500–2,000 m; common in cloud forest; on road banks, landslides, and forest edge.

Odontosoria gymnogrammoides habit.

Close-up of Odontosoria gymnogrammoides.

Comments: This common scrambling fern with tiny prickles on its stems often clings to other vegetation. A non-spiny species, *O. schlechtendalii*, also occurs in Costa Rica. Two species of *Eriosorus* (Pteridaceae) resemble *O. gymnogrammoides*, but they lack prickles and have sori that extend along the veins, not just at the tips. The Asian *O. chinensis*, also called *Sphenomeris chinensis*, yields a red dye. *Lygodium venustum* (Schizaeaceae) is another common climber, often seen below 500 m in Costa Rica's dry lowlands (illus. right).

X ⅓

Lygodium venustum (Schizaeaceae), another climbing fern.

Pleopeltis polypodioides
Family: Polypodiaceae

Resurrection fern

Other common name: *Helecho que resucita*.

Description: Fern with similar fertile and sterile fronds spaced out along rhizome that creeps over rocks, tree trunks, and branches. Scales on rhizome blackish, with light margin. Pinnatifid, deeply divided, with ca. 10–15 pairs of pinnae;

grooved petiole ca. 4.5 cm, blades often 8–10 cm long with a frosting of silvery scales on underside. Line of round sori along each side of midveins of pinnae.

Distribution: Depending on variety, southern United States, Central America, South America, West Indies, and Africa; sea level–3,000 m. In Costa Rica, widespread; found in major mountain ranges, Pacific slope, and the Nicoya Peninsula.

Comments: This is a fascinating little fern that dries and curls up when the humidity in the air drops, but it revives after the next light rain; the scales on the lower surface of the leaf help revive the fern after rains by funneling water into the middle layer of the leaf. There are five distinct varieties, three of which occur in Costa Rica. *P. polypodioides* var. *aciculare* has a dense covering of scales on the underside that are rust-colored with a light margin; some scales have pointed tips. It has a creeping rhizome that travels over branches and rocks. In Cuba and Puerto Rico, resurrection fern has medicinal uses as a heart tonic and in treating liver problems and high blood pressure. This species, along with some other scaly-leaved relatives, was formerly in the genus *Polypodium*. *Polypodium* and related genera are generally nonfrilly

In this arranged photograph, note revived frond of *Pleopeltis polypodioides* at left and dessicated fronds at right.

ferns with running rhizomes; the fronds are pinnatisect or 1-pinnate and have a comblike appearance, round sori, and yellow spores; an indusium (protective covering over the sorus) is always lacking in the genus.

Pteridium spp.
Family: Dennstaedtiaceae

Note: For many years, the name *Pteridium aquilinum* has been applied to bracken ferns worldwide. Fern specialists now recognize three separate species: *Pteridium arachnoideum*, *Pteridium caudatum*, and *Pteridium feei*, all of which occur in Costa Rica. Unless otherwise noted, the information given here is for *Pteridium aquilinum* in the broad sense.

Some common names: Bracken fern, brake, *helechón*, *helecho alambre*.

Description: The various bracken fern species are large, terrestrial, coarse ferns, 1–3 m tall, with deep, creeping rhizomes; stiff, wiry texture. Sterile and fertile leaves similar; fronds about 1/2 petiole and 1/2 blade; 3–4 pinnate blade forms a broad triangle; internodes between primary pinnae long; pinnae dark green, smooth above, lighter below, edges inrolled. Sori on underside along edge of pinnae; nectaries

Pteridium arachnoideum

present near where main pinnae connect to rachis. *P. arachnoideum* (photo above) has lobes between the ultimate segments, wide grooves in midveins on the hairy underside, and an orange-tinged rachis.

Distribution: Bracken ferns are worldwide, sea level–3,000 m elevation; most grow in burned or cut-over areas. *P. arachnoideum* occurs from Mexico to South America, and the Antilles, in open sites as well as cloud forest, in foothills and mountains. In Costa Rica, often in old pastures, to 2,500 m.

Comments: Bracken fern, or brake, is considered a weed in most parts of the world. As humans created large gaps in once-forested landscapes, bracken extended its cover. The underground stems are aggressive and hard to eradicate. Clones of bracken often form in old fields—one clone in Finland is estimated to be 1,500 years old. During the seventeenth and eighteenth centuries, in Scotland and elsewhere, bracken was harvested for potash, an ingredient of soap and glass. Its springy texture makes it satisfactory bedding or packing material. Pigs relish bracken rhizomes. Ingestion of large quantities of the fronds by some domestic animals affects the nervous system; it can result in death because thiaminase, which degrades thiamine (vitamin B-1), is present in the raw plants. Bracken also contains carcinogens such as ptaquiloside; cows that eat bracken have cancer-causing substances in their milk. Fiddleheads of braken ferns are eaten in Japan, which has the world's highest rate of stomach cancer; Costa Rica has the second-highest rate. People in various cultures have eaten the rhizome, sometimes made into a flour, and it has been used medicinally against worms. The nectaries at the base of the pinnae attract ants, but a typical ant-plant mutualism (i.e., food for ants and protection against herbivory for the plant) is not clear.

Selaginella pallescens
Family: Selaginellaceae (Class: Lycopodiopsida)

Moss fern

Other common name: Sweat plant.

Description: Growing as a rosette, on rocks or ground; spreading to ca. 15 cm across. Main stems ca. 10 cm long, branching, covered with tiny, one-veined, scalelike leaves of two sizes; green above, silky and silvery below, with fine white hairs along leaf margins. Plants curl up into loose lettucelike heads when dry and expand (resurrect) soon after a soaking rain. Spores arranged in conelike structures (strobili) at the tips of the branches.

Distribution: Mexico to Colombia. In Costa Rica, 100–2,000 m; sunny to shaded, dry to moist environments; central region north to Tempisque, Nicoya, and Arenal areas.

Comments: Resurrection plant is another common name applied to this species and to similar species of the American Southwest, such as *S. pilifera*. It may be grown as a house plant or in rock gardens. Some selaginellas (known as doradillas in Costa Rica) are used medicinally; for example, the Kuna indigenous people of Panama make a tea from the roots of

Selaginella pallescens; dessicated fronds at left, revived frond at right.

S. exaltata to treat stomach ache. There are around 700 species of *Selaginella* worldwide, and ca. 40 in Costa Rica. This family is in the same class as the club mosses (Lycopodiopsida).

Sphaeropteris brunei
Family: Cyatheaceae

Monkey tail tree fern, *Rabo de mico*

Description: Stout trunk to 10+ m; black, adventitious roots near base; old, tan, scale-covered bases of leaf stems covering much of trunk. Leaves to 4+ m; 3-pinnate with first segments ca. 80 cm long; gray-green below; no spines.

Overall shaggy appearance is from scales covering fiddleheads and bases of old fronds; scales of varying width and length (some 7 cm long) glistening, tan, almost transparent with minute, dark teeth along edge (use hand lens to see). Sori are round.

Distribution: Costa Rica to Colombia. In Costa Rica, 800–2,000 m, from Tilarán Mountains south to the Talamancas; riverside and forest gaps and edges; common along the El Camino trail in Monteverde Cloud Forest Preserve.

Comments: Several of the major tree fern groups are easy to identify if one looks at the spines and scales on the trunks and leaf stems. *Sphaeropteris brunei*, for instance, has shaggy/woolly scales that look like fur. The *Alsophila* tree ferns, by comparison, have glossy, black, often curved, spines, ca. 1 cm long, on the trunks, curled-up fronds (fiddleheads), and the bases of leaf stalks; the crooked scales have ragged, irregular edges that are best seen under a dissecting scope. *Cyathea* (photo, p. 437) has some spinelike projections, but

they are not black. All three of these genera belong to the Cyatheaceae, the largest of four families of tree ferns. Some taxonomists lump the three genera into the genus *Cyathea*. Tree ferns are basically like other ferns, but instead of the rhizome running horizontally, it stands vertically, supported by either the remnant bases of fallen fronds, which appear as patterned scars on the trunk in some species, or by a mantle of dense, hard roots. The lacy fronds and the tree ferns' pleasing form make them attractive to landscapers. They can be grown from spores; some tree fern species produce billions of spores on each plant. Some species are known to live for at least 130 years. The fibrous root mats are often collected as material for orchid growing, and certain species of fern epiphytes (e.g., *Blechnum fragile* and *Trichomanes capillaceum*) are largely restricted in occurrence to the fibrous root mantle on tree fern trunks (Moran et al. 2003). Majestic tree ferns flourished in the era of the dinosaurs. Fifty species of tree ferns occur in Costa Rica.

Sphaeropteris brunei; note how unfurling frond resembles a monkey's tail.

Sphaeropteris brunei

Sticherus bifidus
Family: Gleicheniaceae

Fan fern

Description: Terrestrial; stiff, erect, and clambering to 1 m or more; often in large clumps. Dark maroon stem with light-colored scales at base. Central rachis with tiers of pinnae laterally forking a number of times; horizontal, penultimate segments to 40+ cm long; last segments comblike, with edges turned under, smooth above and orange-tan woolly below. Orange-tan, scale-covered buds in axils of forks. Sporangia in groups of 3–5 (illus. right).

Distribution: Mexico to South America, and Antilles. In Costa Rica, 200–2,000 m, widespread from north to south. Occurs in disturbed habitats: steep, sunny road banks, riversides, old pastures.

Comments: This group of ferns is conspicuous because of its large colonies of forking fronds; a resting bud occurs at the base of each fork. *S. bifidus* is one of about 17 species in this family (Gleicheniaceae) in Costa Rica. A similar genus, *Dicranopteris*, has axillary buds with hairs instead of scales, sporangia in groups of 8–15, and veins that are 2–4 forked, as opposed to 1-forked as in *Sticherus*. Some botanists consider the genus *Sticherus* to be part of *Gleichenia*, but others see enough distinct characteristics in New World species to separate *Sticherus* from the Old World *Gleichenia*.

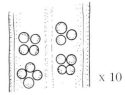

x 10

Sporangia on the underside of a *Sticherus bifidus* frond.

Sticherus bifidus

Thelypteris rudis
Family: Thelypteridaceae

Thelypteris rudis

Description: Terrestrial, in clumps; 2 vascular bundles in base of petiole. Fronds to ca. 1.5 m long; 1 pinnate pinnatifid, the pinnae divided almost to midrib; size of pinnae tapering gradually toward tip and abruptly toward base, where they are reduced to vestigial pinnae (small, leaflike projections). Leaves smooth above, fine hairs below and on stem; pleasant, fresh scent.

Distribution: Mexico to South America, and greater Antilles; 1,000–3,000 m. In Costa Rica, along forest edge and roadsides, in mountains from north to south, often near cloud forest (e.g., bordering the El Camino trail in the Monteverde Cloud Forest Preserve).

Comments: *Thelypteris* species are usually terrestrial and pinnate pinnatifid, with pubescent leaves that have two vascular bundles, which can be seen by looking at a cross section of the petiole with a hand lens. There are more than 90 species of *Thelypteris* in Costa Rica, most of them native; two have been introduced (*T. dentata* and *T. opulenta*). The downy maiden fern (*T. dentata*), a weedy species introduced from the Old World, has a flannel-like surface and does not have vestigial pinnae as in *T. rudis*. *T. opulenta*, which occurs in the lowlands, generally below 500 m, has yellow glands along the veins on the underside of the pinnae. The maiden and marsh ferns of the United States are in this genus.

8. Conspicuous Grasses

The grass family, Poaceae, is ubiquitous. Grasses emerge from cracks in city sidewalks, cover expanses of Arctic tundra, and grow in the understory of wet, tropical forests. They appear in the daily diet of people around the world, either in the form of sugar, which comes from the stems of sugar cane, or as seeds from wheat, rice, or maize plants. Those three grains provide more than 50% of the calories people ingest. Other grasses whose grains are eaten less often but that still have economic value are barley, oats, sorghum, and millet. Some species that are not ingested directly by people frequently serve as food for domesticated grazing animals. Grasses impact us in other ways: they are used as construction, thatch, or weaving material; they are planted as lawns or ornamentals; and some of them are invasive weeds. Bamboos, which are indispensable in many Asian communities, are also members of the family Poaceae.

Grasses are herbaceous—or, in the case of bamboos, woody—plants that often spread by runners. They have jointed stems; alternate, sheathing, linear leaves in a plane; and inflorescences of spikelets. These spikelets consist of simple flowers and tiny, specialized bracts.

Certain natural habitats of the world are dominated by grasses that provide food for grazing animals. In Costa Rica, however, these types of habitats were never common. The extensive fields that support cattle today have taken the place of what was once tropical forest. Nevertheless, there are almost 400 native grasses in Costa Rica and some 100 introduced species, many of which came from Africa. Others, such as rice and sugarcane, originated in Asia or the Pacific Islands. Worldwide, the Poaceae comprises nearly 10,000 species.

The grasses described here are abundant; they are also conspicuous since they grow in disturbed areas such as alongside roads and rivers or in pastures and old fields.

Left, *Rhynchelytrum repens*

Andropogon bicornis Broom sedge, *Cola de venado*

Andropogon bicornis

Description: Usually 1 m (but up to 2.5 m), clumped; straight vertical stems are green, yellow, and pinkish tan. Leaves to 50 cm long. Conspicuous, silky, plumose head of flowers and seeds; bracts (lemmas) lack bristles (awns).

Distribution: Mexico to South America, also Antilles. In Costa Rica, common along roadsides and in old pastures; both slopes, to ca. 1,800 m.

Comments: Deer tail, the translation of the Spanish common name, is an apt description of the fluffy inflorescence of *A. bicornis. A. glomeratus* is similar but smaller, and it has flower bracts that end in 1-cm-long bristles in the flower/fruit head. It lacks the enlarged terminal spikelets, which look like 0.5-cm-long, thin brown seeds, that are seen in *A. bicornis.*

Cynodon nlemfuensis   African star grass, *Estrella africana*

Description: Stems to 1 m tall, but plants can climb several meters into surrounding vegetation; runners occur along the ground. Narrow leaves 5–16 cm long, scratchy if one runs a finger down toward base of leaf; wine coloration where leaves attach to stem and sometimes on runners. Inflorescences composed of 4–9 (usually 5) spikes, each 5–10 cm long, radiating horizontally from tip of stalk, sometimes with one branch sticking straight up; often purplish.

Distribution: Originally from Africa, now naturalized in many parts of the tropics, sea level–1,700 m. In Costa Rica, open sunny fields, roadsides; dominant pasture grass in certain areas (e.g., Monteverde).

Comments: Unlike most of the other grasses included in this chapter, this species is not particularly showy, however it is abundant. Star grass is basically a hefty Bermuda grass (*C. dactylon*); it grows taller and has larger flowering branches, and is planted more for forage than as a lawn grass. Its stems and runners are very tough. *C. dactylon* has horizontal, leafy stems while *C. nlemfuensis*

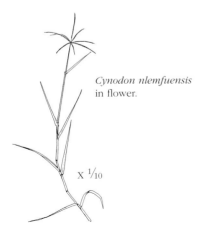

Cynodon nlemfuensis in flower.

X ¹⁄₁₀

Cynodon nlemfuensis in foreground.

spreads by long, smooth runners (stolons). In some areas of the world, when it has been introduced as a forage grass for cattle and dairy cows, it has taken over aggressively. In Costa Rica, star grass pervades in pastures of middle and upper elevations in the same fashion that jaragua (*Hyparrhenia rufa*, p. 277) does in the Guanacaste lowlands. To get rid of it in areas where it is unwelcome, planting shrubs and trees that create shade is an alternative to using herbicides.

Gynerium sagittatum   Giant cane, *Caña brava*

Description: Thick, semiwoody stems to 10 m. Leaves 0.5–2 m long, toward top of stem, fanned out in a plane; leaf margins rough. Stem topped by long-stalked, erect, feathery inflorescence, to 1.5+ m, with pendent branches. Male and female plants often in separate groups; female flower/seed heads showier than males.

Distribution: Native to New World tropics, southern Mexico to Paraguay; also in West Indies. In Costa Rica, it occurs along river edges and wet slopes, sea level–1,500 m; noticeable along sections of the Braulio Carillo highway, disturbed streams on Osa Peninsula, and in Arenal region.

Comments: There are no other species in the genus *Gynerium*, but several related genera have showy, plumelike inflorescences. *Phragmites australis*, a reed abundant in certain wetlands in the United States is occasionally seen in Costa Rica. Giant reed (*Arundo donax*), of the Old World tropics, and pampas grass (*Cortaderia selloana*), from South America, are sometimes cultivated in Costa Rica. There are a few paramo (very high elevation) *Cortaderia* species native to

Gynerium sagittatum

Costa Rica. The sturdy stems of giant cane have been used for propping up banana plants, creating simple dwellings and fences, and as lath in house construction. The leaves can be woven into mats, and the stems serve as arrow shafts. A less utilitarian, but aesthetic, use is made of the fluffy seed heads in dry flower arrangements. Some peoples of South America consider the plant a diuretic and an asthma remedy.

Melinis minutiflora — Molasses grass, *Calinguero*

Description: Usually less than 1 m (to more than 1.5 m), growing in dense, uniform stands; sprawling stems that root at the lower nodes. Leaves 5–15 cm long; sheaths, as well as the leaves, have soft flannel pubescence and a sticky, scented resin. Inflorescence, 10–23 cm long, pyramid-shaped when in flower, narrower when in seed. Spike is a rich pinkish-purple in flower; even when not in flower, this grass is colorful because the stems and leaves are sometimes red-purple or pink-tan.

Distribution: Originally from Africa; now cultivated as forage grass, from Mexico and the Caribbean to South America; also in Hawaii. In Costa Rica, sea level–1,900 m, but most commonly found in midelevations, along roads and in pastures.

Comments: This grass creates a stunning display in November and December when it is in peak flower. Fields and hillsides turn pink-purple and shimmer in the wind; later, a characteristic golden tan sets in as the seed heads dry and fall off. It is a forage grass and is known for its distinctive sweet scent. A folk belief suggests that this sticky grass traps ticks and that the resin acts as a mosquito repellent. In Africa, pest management researchers have experimented with intercropping *Melinis* with maize; volatile chemicals in *Melinis* both repel egg-laying, female stem borers and attract insect parasites of the borers. Take a look at the sheaths on the stems with a hand lens to see the resin droplets on the hairs. *Gordura* and *capín* are two other Spanish names.

Melinis minutiflora

Paspalum saccharoides

Description: Creeping stems, 1–2+ m. Leaves to 40 cm long, by ca. 1 cm, with some long, silky hairs on leaf sheaths. Silvery white inflorescence to ca. 30 cm long, with many delicate, individual fingerlike, pendant branches; silky hairs border the spikelets.

Distribution: Native from West Indies and Nicaragua to South America. In Costa Rica, most common

Paspalum saccharoides

X ¹⁄₁₀

Typical inflorescence form of *Paspalum* species.

at 100–1,800 m; occasionally seen at lower or higher elevations; on streambanks and moist roadbanks; common around Lake Arenal.

Comments: While there are several hundred species of *Paspalum*, this species has no close relatives. It is distinctive from most members of the genus since a typical paspalum lacks the silky hairs and has oval, often paired spikelets that look like little grains lined up along the horizontal branches of the inflorescence (illus. left). The feathery plumes of *P. saccharoides* can be seen at various times of the year.

Pennisetum purpureum — Elephant grass, *Gigante*

Other common name: *Pasto elefante.*
Description: Very tall, thick, stems, to 8 m. Larger leaves to more than a meter long, by 4 cm wide, keeled and with a white stripe down center; somewhat rough when you run your finger down leaf; leafy side shoots sometimes form along the main stem. Cylindrical soft, bristly flower/seed spikes, to 30 cm long, at top of stems.
Distribution: Native to Africa, this grass is now found in the tropics and subtropics worldwide. In Costa Rica, cultivated in fields, but also seen as an escape along roads and river banks; to 1,800 m.
Comments: This tall forage grass looks superficially like sugar cane, but the tail-like purplish or yellowish spikes are very different than the plumelike heads on sugar cane. It is extremely persistent and difficult to eradicate from old fields. There are about a half dozen native and a few cultivated species and hybrids of *Pennisetum* in Costa Rica. Another related, but much shorter, African pasture grass that is planted in areas above 1,300 m is kikuyo (*P. clandestinum*).

Pennisetum purpureum

Rhynchelytrum repens Ruby grass, *Zacate ilusión*

Rhynchelytrum repens

Other common name: *Zacate de seda.*
Description: To 1+ m, stems clumped, somewhat creeping. Leaves to ca. 20 cm. Terminal, pyramid-shaped inflorescences, 14–23 cm, with long hairs on spikelets, appearing whitish or reddish.
Distribution: An African species that is naturalized in the southern United States, Caribbean, and Mexico to South America; from sea level to 1,700 m (to more than 2,000 m in some South American locales). In Costa Rica, grows well in sun, along roads and other disturbed areas.

Comments: This is one of the most colorful grasses along the Inter-American Highway and other major open roads in the country. The silky rose or white inflorescences glisten in sunlight, especially after a rain. Both ruby grass and molasses grass (*Melinis minutiflora*) produce a reddish carpet, but the former does not grow as thickly as the latter. Also, ruby grass may be seen flowering various months of the year, whereas molasses grass is most conspicuous from November to January. Ruby grass is sometimes classified in the genus *Melinis*.

Urochloa maxima Guinea grass, *Guinea*

Other common name: *Zacate de Guinea.*
Description: Bunched grass, 0.5–3 m. Light green leaves to ca. 60 cm by 2.5 cm; rough margin due to tiny teeth; last few leaf blades toward top often bent over; tufts of hair where leaves connect to stem. Flower/seed heads, to 60 cm tall, with long branches.
Distribution: After being introduced as a forage, this African grass is now naturalized in many tropical areas and in parts of the United States. In Costa Rica, one of the most common grasses along

Inter-American Highway as one travels north into Guanacaste. More noticeable in rainy season; lowlands to 1,100 m.

Comments: Pittier (1978) suggested that this species was brought in from Jamaica; herbarium specimens from Costa Rica date back to 1890. He also noted that this grass grows well even in drought conditions. It is nutritious, but with age it grows so tough that livestock cannot eat it. This may be just as well because plants grown in stressful condi-tions and in poor soil may contain hydrocyanic acid that could poison live-stock. Some indigenous South Americans use the root medicinally to remedy flu. *Panicum grande*, a very large colonial grass, 2–4 m, is somewhat similar and grows on muddy edges of marshes and along canals. *P. grande* has leaves that are longer and about twice as wide as those of *U. maxima*. *U. maxima* is sometimes called *Panicum maximum* or *Megathyrsus maximus*.

William A. Haber

Urochloa maxima

Glossary

aquatic. Lives in water.

adventitious. Produced from an unusual position (as in above-ground roots, for example).

agroforestry. A land-use practice that combines crops, trees, shrubs, and sometimes livestock. Pasture or crops are planted beneath, or interspersed with, harvestable fruit or timber trees (native or exotic) or native trees that will form the basis of future forest.

alternate. With one leaf (or flower, bract, etc.) per node. See fig. 6.

ament. A spike made up of tiny, inconspicuous, often unisexual flowers; also called *catkin*.

anther. The bulging apical part of the stamen that contains the pollen. See fig. 1.

appressed. Lies flat or nearly flat against a surface; also called *adpressed*.

aril. Fleshy, edible tissue, often brightly colored, attached to a seed. Usually found in capsular fruits. See fig. 2.

aroid. A member of the Araceae, the philodendron and arum family.

axil. The upper angle between the leaf and the stem, or between a secondary vein and the midvein.

ballistically dispersed. Refers to seeds that are shot out of a seed capsule.

Batesian mimicry. When an organism evolves to resemble another organism that has some advantageous characteristic (e.g., nectar for pollinators) that it lacks, thus gaining benefits (e.g., visitation by pollinators). In this type of mimicry, an animal is fooled to behave as if a plant or another animal is actually something else.

berry. A fleshy, indehiscent fruit containing one to numerous seeds. See fig. 2.

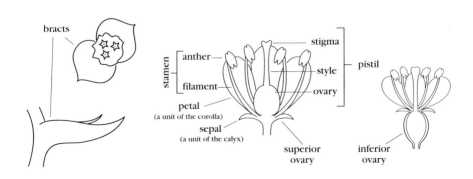

Figure 1. Bracts and Flower Parts

bipinnate. Twice pinnate (see **pinnate**). See fig. 3.

blade. The broad, flat part of a leaf or leaflet; the lamina.

bract. A modified leaf, sometimes scale-like, associated with inflorescences; often beneath or surrounding flowers, or at a branching point of an inflorescence. See fig. 1.

brown fat. A specialized type of fat found in mammals that generates heat to regulate body temperature, especially in hibernating animals.

bulbil. A small bulblike structure that originates on some aboveground part of a plant. It separates and serves as a vegetative propagule.

buttressed trunk. A type of trunk in certain tropical trees in which laterally flattened, somewhat triangular-shaped, aboveground extensions of the roots form perpendicular to the ground in the angle between the trunk and lateral roots.

buzz pollination. Vibratory pollen collection, often by bees, in which the wings are vibrated, setting off vibration in the anthers of a flower that causes the pollen to shake out.

callus. A hard, thickened part of some plant tissue.

calyx (PL. **calyces**). The outer whorl of floral parts below the corolla; a group of sepals. See fig. 1.

capsule. Usually dry fruit, composed of several cells, that opens into a number of valves when mature by dehiscing along regular suture lines. Usually contains several to many seeds, which are often winged or bear arils. See fig. 2.

cardenolide. A member of a group of cardiac glycosides that share the same general chemical structure. Especially abundant in Asclepiadaceae and Apocynaceae.

cardiac glycoside. A plant substance that consists of sugars combined with nonsugars (aglycones) and that affects the heartbeat.

cardiotonic. Having a stimulating effect on the heart.

cataphyll. A modified leaf, sometimes scalelike or not well developed, that protects newly expanding growth in certain plants (e.g., cycads, some aroids).

catkin. See **ament**.

cauliflorous. Having flowers and fruits that develop on the trunk or older branches.

cirrus. A long threadlike extension of the rachis by which some plants climb or send out propagules; also called *cirrhus*.

clonal. Producing genetically identical offspring via asexual reproduction by one individual.

colonial. Growing in large clumps or colonies.

column. A structure consisting of the fusion of stigma, style, and stamens; found in orchid flowers.

complex. In plant classification, a group of very similar taxa that have not been clearly delimited.

compound leaf. A leaf divided into distinct, leaflike segments called leaflets that are separated to the base. Compound leaves usually have a distinct petiole with a swollen base (pulvinus) and lack buds at the tip of the rachis and in the axils of the leaflets. See fig. 3.

cordate. Heart-shaped.

corm. A bulblike underground portion of stem.

corolla. An attractive whorl of flower parts above the calyx; a group of petals. See fig. 1.

corona. A structure, often showy, located between the petals and stamens of certain flowers (e.g., Passifloraceae, some Asclepiadaceae).

crenate. With rounded teeth, usually referring to a leaf margin.

cultivar. A cultivated variety; a distinct group of plants in cultivation that, when reproduced sexually or asexually, retains a particular set of characteristics.

cupule. The cuplike expansion of the receptacle surrounding the base of some fruits, such as many in the family Lauraceae.

cyanogenic glycoside. A substance that consists of sugars combined with nonsugars and that yields hydrogen cyanide when broken down by the addition of water.

cytotoxic. Destroys cells.

deciduous. Falling away at the end of a growth period, such as trees that drop their leaves in the dry season.

decoction. A liquid preparation made by boiling plant parts.

decurrent. With the base extending downward along the edge of a stalk, such as in some leaf blades.

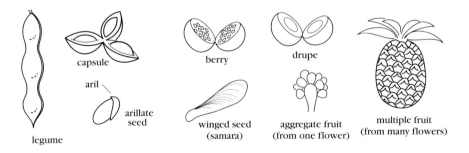

Figure 2. Fruit Types

dehiscent. Splits open naturally when mature; used to describe anthers and some fruits.

dichotomous. Branched into one or more pairs.

dioecious. Having staminate and pistillate flowers that are borne on separate plants of the same species. That is, the flowers of a given individual are unisexual and of only one kind—male or female. See **monoecious**.

disc flower. A small, tubular, radially symmetrical flower in many Asteraceae species; often many are grouped into a head.

distichous. Leaves, bracts, or fruits, alternately arranged on a stem and in the same plane.

domatium (PL. **domatia**). Tiny pouchlike cavity (or sometimes just a tuft of hair) found in the axil of a leaf vein, or at the base of a leaf blade, in a variety of plant species. Domatia are usually inhabited by tiny arthropods such as mites. See fig. 7.

dominant. A plant species that comprises a large percentage of individuals or biomass in a defined area.

drupelet. A small drupe; a drupe is a fleshy, indehiscent fruit with a soft layer enclosing a hard stone(s) with seeds inside.

elfin forest. A forest composed of short trees; often occurs in windswept areas or where other environmental stresses exist.

emergent. A tree that grows taller than most others in the surrounding forest; one that sticks out of the canopy. Also, a rooted aquatic plant that shoots up above the water surface.

endemic. Occurring naturally only in a specific area such as a valley, mountain range, or country.

endocarp. The innermost, often bony, layer in the wall of a ripened fruit.

entire. Referring to a margin that lacks teeth. See fig. 5.

epiphyte. A plant that grows rooted on another plant; many orchids and bromeliads are epiphytes. Note that epiphytic plants also occasionally grow on rocks, tree stumps, logs, and cut road banks, while terrestrial species sometimes grow like epiphytes on logs and tree trunks.

escaped. Growing and reproducing outside of cultivation after initially occurring only in cultivation. This refers to introduced species that begin appearing in nature outside of captivity or gardens.

exocarp. The outermost layer of a ripened fruit wall.

extrafloral nectary. A nectar-secreting gland occurring outside a flower; produces nectar as an attractant for ants that guard the plant from herbivores. Extrafloral nectaries range from a small spot on a leaf blade to a distinctive gland on a narrow stalk. See fig. 5.

fertile frond. In fern species that have two different kinds of fronds, the ones that bear spores.

fissured. Roughly and irregularly grooved.

fluted. With rounded furrows or grooves extending along the length of a stem or a trunk.

frugivorous. Fruit-eating.

gametophyte. In ferns, the plantlet that grows from a spore and produces eggs and sperm. The gametophyte is typically flattened and heart-shaped, and from 2 mm to 2 cm across. It contains half the number of chromosomes of the sporophyte, the fern plant that produces spores.

gap specialist. A plant species that specializes in growing in forest openings, such as where a tree has fallen or a landslide has occurred.

gland. A cell or group of cells that secretes a substance (e.g., oil, nectar).

glandular. Covered with tiny glands or with hairs tipped with a gland or gland-like structure.

glabrous. Lacking hairs or pubescence.

glaucous. 1. A pale blue-green, blue-gray, or gray color. 2. Covered with a whitish bloom that can be rubbed off (e.g., grapes and stems of some raspberries).

globose. More or less spherical.

gymnosperm. A vascular plant with seeds not enclosed in an ovary. Conifers and cycads are included in this group.

hemiparasitic. Partially parasitic; deriving some but not all of its nutrients from its host.

hemiepiphyte. A plant that begins life as an epiphyte, but sends roots to the ground as it grows in order to take up water and nutrients from the soil (e.g., *Clusia, Ficus*).

herbivory. The consumption of live plant material.

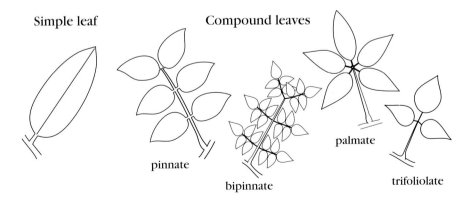

Figure 3. Types of Leaves

hilum. The scar on a seed marking its point of attachment to the stalk on the ovary wall during development; seen in seeds of legumes (Fabaceae) and zapotes (Sapotaceae).

hoary. Covered with a layer of white, or whitish, pubescence.

indehiscent. Not opening naturally at maturity, usually refers to fruits.

inferior ovary. An ovary submerged in the flower base below the attachment point of the sepals and petals (as in Rubiaceae, for example). See fig. 1.

inflorescence. The reproductive part of the plant bearing the flowers, often a stalked and branched structure with numerous flowers.

infructescence. The fruit-bearing structure of the plant.

invasive species. A nonnative species that disrupts and replaces native species.

labellum. A lip; in orchids, the distinctive, usually showy, petal that often serves as a landing platform for pollinating insects.

lanceolate. Lance-shaped; longer than broad, wider at the base, and tapering toward the tip.

leaflet. One of the individual leaflike units that make up a compound leaf. See fig. 5.

legume. 1) A type of fruit similar to a bean pod, with sutures along opposite edges. See fig. 2. 2) A member of the bean family (Fabaceae).

lenticel. Lens-shaped, wartlike patch on a stem or trunk that allows for movement of gases.

ligule. 1) In grasses and similar plants, an appendage on the upper edge of the leaf sheath. 2) In some Araliaceae, a tongue-shaped stipule fused to the petiole base. 3) In some Asteraceae flowers, an elongated, tonguelike part of the corolla.

membranaceous. Thin, flexible, and often translucent.

mesocarp. The usually fleshy or fibrous middle layer in the wall of a ripened fruit. See **endocarp** and **exocarp**.

midrib. The central, main vein of a leaf, leaflet, or bract; midvein.

midvein. See **midrib**. See fig. 5.

monoecious. A plant species in which each individual produces both staminate and pistillate flowers, that is, unisexual flowers of both sexes. See **dioecious**.

mucilaginous. Containing mucilage, a slimy substance.

Müllerian body. A glycogen-rich food packet produced by many species of *Cecropia* to feed *Azteca* ants. These food bodies are found on the base of the leaf petiole.

Müllerian mimicry. When a number of species with a similar characteristic (e.g., flowers with nectar, or insects with an unpleasant taste) evolve to look alike, the result being a larger advertisement that benefits all species involved.

mycorrhizal. Referring to an association between a plant and a fungus in which the fungal mycelium grows on or in the roots of the plant and both organisms benefit through an exchange of nutrients.

naturalized. Refers to an organism introduced and successfully established in an area other than where it naturally occurs; it then reproduces and spreads as if native.

nectar guide. A marking on a flower that guides a pollinator to the nectar.

nectar robber. An organism that takes nectar from a flower without entering it in the normal fashion. Often a bee or a hummingbird, for instance, pierces the corolla, and probes for nectar from the side of a flower instead of from the mouth.

nectary. A plant organ or tissue that produces or secretes nectar.

netted venation. A netlike pattern of veins in a leaf, where some veins run perpendicular to others.

node. The point on a stem where the leaf and bud emerge. See fig. 5.

obovate. Shaped somewhat like an upsidedown egg; in leaves, widest toward the tip.

odd-pinnate. Having an odd number of leaflets.

opposite. When each node has two leaves, one attached on each side of the stem (can also apply to bracts, flowers, etc.). See fig. 6.

ovary. The basal, usually rounded part of the pistil (female part of the flower), containing the ovules, or undeveloped seeds, that will develop into a fruit. See fig. 1.

palmate. Divided into several lobes or into separate leaflets that are attached at one point like the fingers on the palm of a hand. Also used to describe leaf venation of this pattern. See figures 3 and 6.

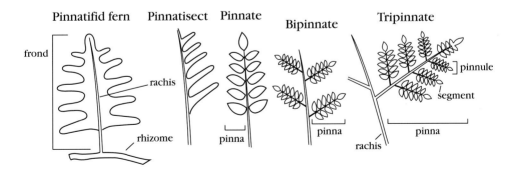

Figure 4. Fern Terminology

pantropical. Distributed in tropical regions throughout the world.

papillae. Tiny protuberances with rounded tips.

papillose. Covered with papillae.

paramo. The cold, wet habitat that occurs above timberline. This high elevation area is dominated by shrubs, herbaceous plants, and bamboo and other grasses.

peltate. With a stalk attached within the blade area of a leaf or scale. See fig. 6.

perfect flower. A flower that is bisexual.

persistent. Remains attached to; opposite of **deciduous**.

petiole. The leaf stalk that supports a leaf blade and connects it to the node. See fig. 5.

phytochemical. Any chemical compound produced by a plant. The term is often used when discussing the medicinal and health benefits of plant compounds.

pinna (PL. **pinnae**). A leaflet of a pinnately compound leaf (see **pinnate**).

pinnate. With four or more leaflets arranged in two rows along the rachis. Also applies to venation arranged in this pattern. See figures 3, 4, and 6.

pinnate pinnatifid. Divided into pinnae that are incised part way to the axis. Describes fern fronds that are not quite bipinnate.

pinnatifid. Divided, but not incised to the axis. Used to describe fern leaves that are pinnately lobed. See fig. 4.

pinnatisect. With lobes that divide all the way, or nearly all the way, to the axis. Each lobe does not have a stalk at the base, but is fused along its base to the axis. See fig. 4.

pinnule. A pinna, or segment, of the second order in bipinnate ferns. See fig. 4.

pistillate. Flowers that have pistils but lack functional stamens.

pith. The soft, innermost tissue of a stem.

plumose. 1) Similar to a feathery tuft. 2) With fine hairs branching from a central main stem.

pneumatophore. A root extension that grows vertically upward out of the soil; seen in certain tree species in inundated areas (e.g., mangrove swamps), where they aid in aeration.

pollinium (PL. **pollinia**). A mass of pollen grains.

precursor. A molecule that exists as an intermediate of another product.

propagule. A unit of dispersal of a plant; a seed, a germinated seed, a spore, or a vegetative structure capable of growth if detached from the parent.

prop root. An aboveground root that originates from the stem of a plant, roots in the ground, and supports the plant.

pseudobulb. A swollen bulblike part of an orchid stem where water and nutrients are stored.

pseudocopulation. The act of attempted mating by a male insect with a flower, as seen in some orchids in which the flower mimics a female of the insect species. Pollination of the flower takes place in the process.

pseudostem. An erect, false stem made up of overlapping leaf bases; common in Heliconiaeae, Musaceae, and Zingiberaceae.

pubescent. Coated with hairs or fuzz.

pulvinus. The swollen base or tip of a petiole. Characteristic of many compound leaves. See fig. 5.

rachis. The midvein of a pinnate leaf, attached at the tip of the petiole. See figures 4 and 5.

ramet. An individual member or stem of a clone.

rank. A vertical row, as leaves along a stem.

ray flower (or ray floret). A small flower composed of a corolla tube and petal-like limb or blade. Ray flowers create the outer showy part surrounding disc flowers, as in many Asteraceae. See **disc flower**.

receptacle. The flattened, concave, or convex top of a flowering stalk where the flower parts are attached.

recurved. Curved backward or downward.

relict tree. A tree that survives from an earlier period.

rhizome. A stem running on or beneath the ground that may bear rootlets, vertical stems, and leaves.

saponin. A certain type of plant compound that produces soaplike foam when mixed in water.

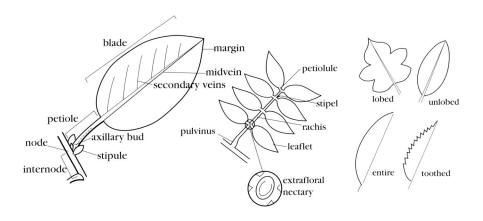

Figure 5. Parts of a Leaf and Leaf Margins

scandent. Climbing.

scurfy. Covered with tiny, flaky scales or scalelike particles.

skototropic. Growing or moving toward dark objects.

second growth. The type of vegetation that grows in an area where the original vegetation was destroyed by humans or by a natural disturbance.

sepal. One of the lobes of the calyx when these are distinct from each other; the sepals are usually green and leaflike. See fig. 1.

sessile. Without a stalk.

sheath. Any long or more or less tubular structure surrounding an organ or part.

shingle leaves. Overlapping leaves that cling to tree trunks. Seen in the climbing, juvenile phases of some Araceae and a few other plant families, they often differ in shape and size from leaves produced in adult plants.

simple leaf. A leaf that is not divided into leaflets (see **compound leaf** and fig. 3).

sleep movements. When leaflets fold up and the stem they are on collapses downward; caused by water moving in and out of certain cells, resulting in a decrease in the turgidity of cells in the pulvinus, the thickening at the base of a leaf stem.

sorus (PL. **sori**). In ferns, a cluster of **sporangia**.

spadix. Special term used for a congested spike inflorescence of certain plant families (e.g., Araceae, Cyclanthaceae). It is usually associated with a bract called a spathe.

spathe. 1) A bract that surrounds, or is found at the base of, a spadix. 2) Sometimes used to describe a large bract of a palm inflorescence.

spikelet. A small spike. In grasses, such a spike has reduced flowers and tiny, specialized bracts.

sporangia. Sacs that contain or produce spores.

sporophyll. A leaf or leaflike structure that is spore-bearing.

stamen. The male or staminate part of a flower, consisting of the filament or stalk and the anther or pollen-producing sac. See fig. 1.

staminate. Flowers that have stamens but lack functional pistils.

staminode (PL. **staminodia**). Sterile stamen (i.e., not producing pollen); those that resemble petals are called petaloid staminodia.

stellate. Star-shaped; stellate hairs appear on some species of Malvaceae and Tiliaceae. See fig. 7.

sterile frond. In ferns, leaves that lack spore-producing structures.

stigma. The receptive part of a style, usually located at the tip, where pollen is received during pollination. See fig. 1.

stipel. The stipule of a leaflet. See fig. 5.

stipule. A small leaflike or bractlike structure found at the base of some leaf petioles; most easily seen on new growth because stipules are often deciduous. See fig. 5.

stolon. A trailing stem producing roots at nodes and sometimes sending up new shoots.

strangler. A form in some species of figs (*Ficus* spp.) that begin growing as epiphytes but later produce adventitious roots that coalesce to form a woody trunk around the host tree. The constriction of the host's trunk, along with shading of its crown, eventually kills the host, leaving the fig as a self-supporting tree.

strobilus (PL. **strobili**). A conelike cluster of sporophylls; found in *Equisetum*, *Lycopodium*, *Selaginella*, cycads, and conifers.

subopposite. More or less opposite; when one of two leaves at a node is attached slightly above or below its facing leaf (can also apply to leaflets).

substrate. The surface upon which an organism attaches or grows.

style. A more or less slender stalk (sometimes branched), attached at the top of the ovary, bearing the stigma (pollen-collecting tissue) at its tip. See fig. 1.

tendril. An elongated, coiled extension of a stem, leaf, or inflorescence that aids in climbing and support.

tepal. The term for a petal or sepal of a flower when these two parts are similar in appearance.

terrestrial. Growing on the ground.

tomentose. Covered with a wool-like pubescence.

toothed. In this book, the term used to describe a leaf margin that has any sort of toothlike projections (includes botanical terms such as crenate, dentate, serrate, biserrate, and serrulate).

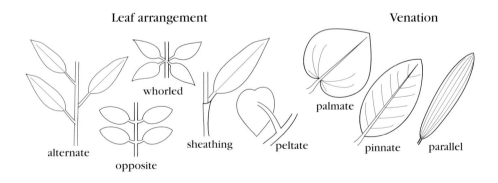

Figure 6. Leaf Arrangement and Venation

tubercle. A small, rounded projection.

umbel. An inflorescence in which flower stems all radiate from a central point.

urticating. Causing irritation to the skin.

vascular bundle. A cluster of vessels or transport tissue.

vascular plant. One that possesses a well-developed system of vessels to transport water and nutrients. Ferns, gymnosperms, and angiosperms are vascular plants; mosses and other bryophytes are not.

whorled. In groups of three or more at a node, usually referring to leaves. See fig. 6.

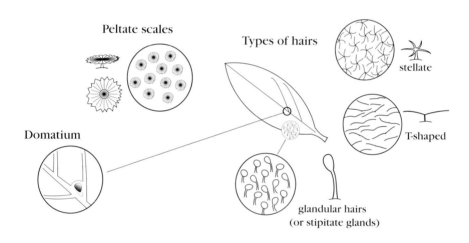

Figure 7. Leaf Surface Structures

List of Species by Family

All species featured in species accounts are included in this list, in addition to those species that are represented solely by a photograph or an illustration.

Bibliography

Acevedo-Rodriguez, P. and R.O. Woodbury. 1985. *Los bejucos de Puerto Rico. Volumen I.* Gen. Tech. Rep. SO-58.: U.S.D.A. Forest Service, Southern Forest Experiment Station, New Orleans, LA.

Ackery, P.R. and R.I. Vane-Wright. 1984. *Milkweed Butterflies: Their Cladistics and Biology.* Cornell Univ. Press, Ithaca, NY.

Ågren, J. and D.W. Schemske. 1991. Pollination by deceit in a neotropical monoecious herb, *Begonia involucrata. Biotropica* 23: 235-241.

Agriculture & Resource Management Council of Australia & New Zealand, Australian & New Zealand Environment & Conservation Council and Forestry Ministers. 2000. *Weeds of National Significance Parkinsonia (Parkinsonia aculeata) Strategic Plan* [Electronic version]. National Weeds Strategy Executive Committee, Launceston, Tasmania, AU. http:// www.weeds.org.au/docs/parstrat.pdf

Aguiar, A.C.F., D. Kass, P. Salvador, J. Mustonen, and E. Gomes de Moura. 2002. Avances en Investigación: Enriquecimiento de la fertilidad del suelo en condiciones de invernadero con especies usadas como abono verde [Electronic version]. *Revista Agroforestería en las Américas* 9 (35-36).

Alfaro, E. and B. Gamboa. 1999. *Plantas comunes del Parque Nacional Chirripó.* Instituto Nacional de Biodiversidad, Heredia, CR.

Allen, P.H. 1977. *The Rain Forests of Golfo Dulce.* Stanford Univ. Press, Stanford.

Almeda Jr., F. 1978. *Systematics of the genus Monochaetum (Melastomataceae) in Mexico and Central America.* (UC Publications in Botany). Univ. of California Press, Berkeley.

Armstrong, W.P. 1998. A marriage between a fern and a cyanobacterium. Wayne's Word Noteworthy Plants: Nov. 1998. http://waynesword.palomar.edu/plnov98.htm

Arroyo, M. T. K. and P. H. Raven. 1975. The evolution of subdioecy in morphologically gynodioecious species of *Fuchsia* sect. Encliandra (Onagraceae) *Evolution* 29: 500-511.

Arvigo, R. and M. Balick. 1998. *Rainforest Remedies.* 2nd revised and expanded ed. Lotus Press, Twin Lakes, WI.

Association of Societies for Growing Australian Plants. 2004. http://farrer.csu.edu.au/ASGAP/index.html

Atwood, J.T. 1988. The vascular flora of La Selva Biological Station, Costa Rica: Orchidaceae. *Selbyana* 10: 76-145.

Atwood, J.T. 1989. *Orchids of Costa Rica, Part 1.* Icones Plantarum Tropicarum. The Marie Selby Botanical Gardens, Sarasota, FL.

Atwood, J.T. and R.L. Dressler. 1998. Clarifications and new combinations in the *Phragmipedium caudatum* complex from Central America. *Selbyana* 19: 245-248.

Badilla, B., T. Miranda, G. Mora, and K. Vargas. 1998. Actividad gastrointestinal del extracto acuoso bruto de *Quassia amara* (Simaroubaceae) [Electronic version]. *Rev. Biol. Trop.*: 46(2).

Balick, M. 1990. Production of coyol wine from *Acrocomia mexicana* (Arecaceae) in Honduras. *Econ. Bot.* 44: 84-93.

Barquero, H. 1982. *José María Orozco.* Ministerio de Cultura, Juventud y Deportes, Dirección de Publicaciones, San José, CR.

Barrios, C., J. Beer, and M. Ibrahim. 1999. Pastoreo regulado y bostas del ganado para la protección de plántulas de *Pithecolobium saman* en potreros [Electronic version]. *Revista Agroforestería en las Américas* 6 (23).

Bar-zvi, D. 1996. *Tropical Gardening*. Random House, NY & Toronto.

Bawa, K.S. 1985. Reproductive biology of tropical lowland rain forest trees. II. Pollination systems. *Amer. J. Bot.* 72(3): 346-356.

Beath, D. no date. Pollination ecology of the Araceae. http://www.aroid.org/pollination/beath/index.html

Belt, T. 1874. *A Naturalist in Nicaragua*. E.P. Dutton, NY.

Bennett, B.C., R. Alarcón, and C. Cerón. 1992. The ethnobotany of *Carludovica palmata* Ruíz & Pavón (Cyclanthaceae) in Amazonian Ecuador. *Econ. Bot.* 46: 233-240.

Bentley, B.L. 1977. The protective function of ants visiting the extrafloral nectaries of *Bixa orellana* (Bixaceae). *J. Ecol.* 65: 27-38.

Benzing, D.H. 1970. An investigation of two bromeliad myrmecophytes: *Tillandsia butzii* Mez, *T. caput-medusae* E. Morren, and their ants. *Bull. Torrey Bot. Club* 97: 109-115.

Benzing, D.H. 1990. *Vascular Epiphytes*. Cambridge Univ. Press, Cambridge.

Bernhardt, E. 1995. *The Costa Rican Organic Home Gardening Guide*. New Dawn Center, San Isidro de El General, CR.

Berry, F. and W. J. Kress. 1991. *Heliconia: An Identification Guide*. Smithsonian Institution Press, Wash., DC.

Bierzychudek, P. 1981. *Asclepias*, *Lantana*, and *Epidendrum*: A floral mimicry complex? *Biotropica* 13(2), Suppl. 54-58.

Bjartiya, H.P. and P.C. Gupta. 1981. A chalcone glycoside from the seeds of *Bauhinia purpurea*. *Phytochemistry* 20: 2051.

Bonaccorso, F.J., W.E. Glanz, and C.M. Sandford. 1980. Feeding assemblages of mammals at fruiting *Dipteryx panamensis* (Papilionaceae) trees in Panama: seed predation, dispersal, and parasitism. *Rev. Biol. Trop.* 28(1): 61-72.

Boza, M.A. 1988. *Costa Rica National Parks*. Editorial Heliconia, San José, CR.

Britton, N.L. and J.N. Rose. 1963. *The Cactaceae*. Dover Publications, NY.

Brokaw, N.V.L. 1987. Gap-phase regeneration of three pioneer tree species in a tropical forest. *J. Ecol.* 75: 9-19.

Bronstein, J. L. 1986. The origin of bract liquid in a neotropical *Heliconia* species. *Biotropica* 18(2): 111-114.

Brown, D. 2000. *Aroids: Plants of the Arum Family*. 2nd ed. Timber Press, Portland, OR.

Brücher, H. 1989. *Useful Plants of Neotropical Origin and their Wild Relatives*. Springer-Verlag, Germany & NY.

Bruggeman, L. 1957. *Tropical Plants and their Cultivation*. Thames & Hudson, London.

Buckley, R. and H. Harries. 1984. Self-sown wild-type coconuts from Australia. *Biotropica* 16: 148-151.

Burger, W. and collaborators. 1971 to present. *Flora Costaricensis*, Fieldiana Bot. Vol. 35 & 40 plus New series 4, 13, 18, 23, 28, 33, 36, 40.

Burnham, R.M. 1999. Threat of bioterrorism in America. Statement before the U.S. House of Representatives. http://www.fbi.gov/congress/congress99/bioleg3.htm

Butterfield, R.P. 1996. Early species selection for tropical reforestation: A consideration of stability. *Forest Ecol. Manag.* 81: 161-168.

Butterfield, R.P. and M. Espinoza. 1995. Screening trial of 14 tropical hardwoods with an emphasis on species native to Costa Rica: Fourth year results. *New Forests* 9: 135-45.

CABI-*Biocontrol News and Information* 20 (3) September 1999 News-Biorational, Plant Tricks. http://pest.cabweb.org/Journals/BNI/BNI20-3/BIORAT.HTM

Calvert, G. March 2001. Association of Societies for Growing Australian Plants, Australian plants online: Rotten cheesefruit?...or Great morinda?? Reprinted from "The Native Gardener" newsletter of the SGAP Townsville Branch April 1998 and November 1998. http://farrer.csu.edu.au/ASGAP/APOL21/mar01-5.html

Carpio Malavassi, I.M. 1992. *Maderas de Costa Rica: 150 especies forestales.* Editorial de la Universidad de Costa Rica, San José, CR.

Castner, J.L., S.L. Timme, and J.A. Duke. 1998. *A Field Guide to Medicinal and Useful Plants of the Upper Amazon.* Feline Press, Gainesville, FL.

Center for Aquatic and Invasive Plants. 2004. http://aquat1.ifas.ufl.edu/

Chavarría, U., J. González, and N. Zamora. 2001. *Árboles comunes del Parque Nacional Palo Verde/Common Trees of Palo Verde National Park.* Editorial INBio, Heredia, CR.

Chazdon, R.L. 1992. Patterns of growth and reproduction of *Geonoma congesta,* a clustered understory palm. *Biotropica* 24: 43-51.

Chickering, C. 1973. *Flowers of Guatemala.* Univ. of Oklahoma Press, Norman.

Churchill, S.P., H. Balslev, E. Forero, and J. L. Luteyn (eds.). 1995. *Biodiversity and Conservation of Neotropical Montane Forests.* The New York Botanical Garden, Bronx.

Cintron, B.B. 1990. *Cedrela odorata* L. – Cedro hembra, Spanish-cedar. In: *Silvics of North America.* Hardwoods. Agric. Handb. 654(2): 250-257. USDA, Washington, DC.

Clark, D.B., and D.A. 1987. Population ecology and microhabitat distribution of *Dipteryx panamensis,* a neotropical rain forest emergent tree. *Biotropica* 19(3): 236-244.

Clement, C.R., contributor. 1995. New Crop FactSHEET: Pejibaye. http://www.hort.purdue.edu/newcrop/cropfactsheets/pejibaye.html

College of Tropical Agriculture and Human Resources, Univ. of Hawaii at Manoa. 2002. Cover Crop-Perennial Peanut (*Arachis pintoi*). http://www.ctahr.hawaii.edu/sustainag/CoverCrops/perennial_peanut.asp

Collins, J.P., R.C. Berkelhamer, and M. Mesler. 1977. Notes on the natural history of the mangrove *Pelliciera rhizophorae* Tr. & Pl., (Theaceae). *Brenesia* 10/11: 17-29.

Constantine Jr., A. 1987. *Know your Woods.* Macmillan Publishing Co., NY.

Courtright, G. 1988. *Tropicals.* Timber Press, Portland, OR.

Crane, E. and P. Walker. 1984. *Pollination Directory for World Crops.* International Bee Research Association, London.

Croat, T.B. 1978. *Flora of Barro Colorado Island.* Stanford Univ. Press, Stanford.

Croat, T.B. 1983. A revision of the genus *Anthurium* (Araceae) of Mexico and Central America. Part I: Mexico and Middle America. *Ann. Missouri Bot. Gard.* 70: 211-420.

Crouch, J.H., D. Vuylsteke, and R. Ortiz. 1998. Perspectives on the application of biotechnology to assist the genetic enhancement of plantain and banana (*Musa* spp.). *EJB: Electronic Journal of Biotechnology* 1 (1). http://www.ejb.org/content/vol1/issue1/full/2/

Dalling, J.W., M.D. Swaine, and N.C. Garwood. 1998. Dispersal patterns and seed bank dynamics of pioneer trees in moist tropical forest. *Ecology* 79: 564-578. http://www.findarticles.com/cf_0/m2120/n2_v79/20574299/print.jhtml

Daniels, G.S. and F.G. Stiles. 1979. The *Heliconia* taxa of Costa Rica: Keys and descriptions. *Brenesia* 15 (Supl.): 1-150.

D'Arcy, W.G. and M.D. Correa A. (eds.). 1985. *The Botany and Natural History of Panama.* Missouri Botanical Garden, St. Louis.

Davidse, G., M. Sousa, and A.O. Chater (eds.). 1994. *Flora Mesoamericana*. Vol. 6. Alismataceae a Cyperaceae. Missouri Botanical Garden Press, Universidad Nacional Autonoma de Mexico (UNAM), and the Natural History Museum (London).

de Laubenfels, D. J. 1990. The Podocarpaceae of Costa Rica. *Brenesia* 33: 119-121.

Deuth, D. 1977. The function of extra-floral nectaries in *Aphelandra deppeana* Schl. & Cahm. (Acanthaceae). *Brenesia* 10/11: 135-145.

Devall, M. and R. Kiester. 1987. Notes on *Raphia* at Corcovado. *Brenesia* 28: 89-96.

Dodson, C.H., A.H. Gentry, and F.M. Valverde. 1985. *La Flora de Jauneche, Los Ríos, Ecuador*. Banco Central de Ecuador, Quito.

Downum, K.R., J.T. Romeo, and H.A. Stafford (eds.). 1993. *Phytochemical Potential of Tropical Plants*. Plenum Press, NY.

Dressler, R.L. 1993. *Field Guide to the Orchids of Costa Rica and Panama*. Cornell Univ. Press, Ithaca, NY.

Duke, J.A. 1989. *CRC Handbook of Nuts*. CRC Press, Boca Raton, FL.

Duke, J.A. 1989. Third International Poisonous Plant Symposium. *HerbalGram* 21: 39.

Duke, J.A. updated 1998. Agricultural Research Service. 2004. Dr. Duke's Phytochemical and Ethnobotanical Databases. http://www.ars-grin.gov/duke/plants.html

Duke, J.A. and R. Vasquez. 1994. *Amazonian Ethnobotanical Dictionary*. CRC Press, Boca Raton, FL.

Earle, C. (ed.). Gymnosperm Database. *Cupressus lusitanica*. University of Bonn. http://www.botanik.uni-bonn.de/conifers/cu/cup/lusitanica.htm

Emboden, W. 1979. *Narcotic Plants*. Revised and enlarged. Macmillan Publishing Co., NY.

Endress, P.K. 1994. *Diversity and Evolutionary Biology of Tropical Flowers*. Cambridge Univ. Press, Cambridge.

Enquist, B.J. and J.J. Sullivan. 2001. *Vegetative key and descriptions of tree species of the tropical dry forest of upland Sector Santa Rosa, Área de Conservación Guanacaste, Costa Rica* [Electronic version].

Faegri, K. and L. van der Pijl. 1971. *The Principles of Pollination Ecology, ed. 2*. Pergamon, Oxford.

Flores, E.M. 1992. *Árboles y semillas del trópico*. Museo Nacional de CR. Vol. 1: 1.

Flowerdew, B. 1995. *The Complete Book of Fruit*. Penguin Books, NY.

Food and Agriculture Organization Forestry Dept. 1986. *Food and Fruit-bearing Forest Species 3: Examples from Latin America*. Forestry Paper 44/3. FAO, United Nations, Rome.

Food and Agriculture Organization. 1989. *Utilization of Tropical Foods: Tropical Oil Seeds*. Food and Nutrition Paper 47/5. FAO, United Nations, Rome.

Foster, S. and R.A. Caras. 1994. *A Field Guide to Venomous Animals and Poisonous Plants of North America North of Mexico* (The Peterson Field Guide Series: 46). Houghton Mifflin, NY.

Fournier, L.A. and E.G. García. 1998. *Nombres vernaculares y científicos de los árboles de Costa Rica*. Editorial Guayacán, San José, CR.

Franz, N.M. and C.W. O'Brien. 2001. Revision and phylogeny of *Perelleschus* (Coleoptera: Curculionidae), with notes on its association with *Carludovica* (Cyclanthaceae). *Transactions of the American Entomological Society* 127: 255-287.

Fryxell, P.A. 1997. The American genera of Malvaceae – II. *Brittonia* 49: 204-269.

García-Franco, J.G., V. Rico-Gray, and O. Zayas. 1991. Seed and seedling predation of *Bromelia pinguin* L. by the red land crab *Gecarcinus lateralis* Frem. in Veracruz, Mexico. *Biotropica* 23: 96-97.

Gentry, A.H. 1993. *A Field Guide to the Families and Genera of Woody Plants of Northwest South America (Colombia, Ecuador, Peru)*. Conservation International, Wash., DC.

George, A.S. (ed.) 1989. *Flora of Australia Vol. 3, Hamamelidales to Casuarinales*, Australian Government Publishing Service, Canberra.

Gerhardt, K. and D. Fredriksson. 1995. Biomass allocation by broad-leaf mahogany seedlings, *Swietenia macrophylla* (King), in abandoned pasture and secondary dry forest in Guanacaste, Costa Rica. *Biotropica* 27: 174-182.

Gerhardt, K. and Håkan Hytteborn. 1992. Natural dynamics and regeneration methods in tropical dry forests – an introduction. *J. Veg. Sci.* 3: 361-364.

Gómez, L.D. *Las plantas acuáticas y anfibias de Costa Rica y Centroamérica. 1. Liliopsida*. Editorial Universidad Estatal a Distancia, San José, CR.

González, J.E. and R.F. Fisher. 1994. Growth of native forest species planted on abandoned pasture land in Costa Rica. *Forest Ecol. Manag.* 70: 159-167.

Grayum, M.H. and H.W. Churchill. 1989. The vascular flora of La Selva Biological Station, Costa Rica: Polypodiophyta. *Selbyana* 11: 66-118.

Grayum, M.H. and B.E. Hammel. 1996. The genus *Tetranema* (Scrophulariaceae) in Costa Rica, with two new species. *Phytologia* 79: 269-280.

Grayum, M.H., B.E. Hammel, and N. Zamora *The Cutting Edge* Vols. III (2) 1996, IV (4) 1997, V (3) 1998, VIII (1) 2001, IX (4) 2002, XI (1, 2) 2004, XII (1) 2005, XIII (1, 2, 3) 2006.

Griffiths, M. 1994. *The New R.H.S. Dictionary Index of Garden Plants*. Timber Press, Portland, OR.

Gruezo, W.S. and H.C. Harries. 1984. Self-sown, wild-type coconuts in the Philippines. *Biotropica* 16: 140-147.

Guanacaste Conservation Area/Area de Conservación de Guanacaste (ACG). División de plantas. Plants of ACG. http://www.acguanacaste.ac.cr/paginas_especie/plantae_online/division.html

Guariguata, M.R., R. Rheingans, and F. Montagini. 1995. Early woody invasion under tree plantations in Costa Rica: Implications for forest restoration. *Restoration Ecology* 3(4): 252-260.

Gunn, C.R. and J.V. Dennis. 1976. *World Guide to Tropical Drift Seeds and Fruits*. Quadrangle, NY.

Gutteridge, R.C. and H.M. Shelton (eds.). 1998. Forage Tree Legumes in Tropical Agriculture [Electronic version]. Tropical Grassland Society of Australia Inc., Queensland, AU.

Haber, W.A. 1984. Pollination by deceit in a mass-flowering tropical tree *Plumeria rubra* L. (Apocynaceae). *Biotropica* 16: 269-275.

Haber, W.A. and G.W. Frankie. 1982. Pollination of *Luehea* (Tiliaceae) in Costa Rican deciduous forest. *Ecology* 63: 1740-1750.

Haber, W. A., W. Zuchowski, and E. Bello. 2000. *An Introduction to Cloud Forest Trees: Monteverde, Costa Rica*. 2nd ed. Mountain Gem Publications, Monteverde, CR.

Hall, P., L.C. Orrell, and K.S. Bawa. 1994. Genetic diversity and mating system in a tropical tree, *Carapa guianensis* (Meliaceae). *Amer. J. Bot.* 81: 1104-1111.

Hammel, B.E. 1999. *Plantas ornamentales nativas de Costa Rica*. Instituto Nacional de Biodiversidad, Heredia, CR.

Hammel, B.E., M.H. Grayum, C. Herrera, y N. Zamora (eds.). 2003. *Manual de plantas de Costa Rica. Volumen II. Gimnospermas y monocotiledóneas (Agavaceae-Musaceae)* [and Electronic version of draft treatments: http://www.mobot.org/MOBOT/research/treat/]. Missouri Botanical Garden Press, St. Louis, MO.

Hammel, B.E., M.H. Grayum, C. Herrera, y N. Zamora (eds.). 2003. *Manual de plantas de Costa Rica. Volumen III. Monocotiledóneas (Orchidaceae-Zingiberaceae)* [and Electronic version of draft treatments: http://www.mobot.org/MOBOT/research/treat/]. Missouri Botanical Garden Press, St. Louis, MO.

Hargreaves, D. and B. 1960. *Tropical Blossoms of the Caribbean*. Ross-Hargreaves, Lahaina, Hawaii.

Hargreaves, D. and B. 1965. *Tropical Trees*. Hargreaves, Kailua, Hawaii.

Hargreaves, R.T., R.D. Johnson, D.S. Millington, M.H. Mondal, W. Beavers, L. Becker, C. Young, and K.L. Rinehart, Jr. 1974. Alkaloids of American species of *Erythrina. Lloydia* 37: 569-580.

Hay, R., F.R. McQuown, G. Beckett, and K. Beckett. 1974. *The Dictionary of Indoor Plants in Color*. Exeter Books, NY.

Henderson, A. 1986. A review of pollination studies in the Palmae. *Bot. Rev.* 52: 221-259.

Henderson, A., G. Galeano, and R. Bernal. 1995. *Field Guide to Palms of the Americas*. Princeton Univ. Press, Princeton, NJ.

HerbalGram. 1989. Mangoes for herpes? 21: 10.

Hernández, D. 1993. *La flora acuática del humedal de Palo Verde*. Editorial de la Universidad Nacional, Heredia, CR.

Hilje, L. 1984. Fenología y ecología floral de *Aristolochia grandiflora* Swartz (Aristolochiacee) en Costa Rica. *Brenesia* 22: 1-44.

Hinton, H.L., Jr. 1999. Combating terrorism: Observations on the threat of chemical and biological terrorism. Testimony before the U.S. House of Representatives committee on government reform. http://www.homelanddefense.org/GAO/T-NSIAD-00-50.pdf

Holdridge, L.R. 1967. *Life Zone Ecology* (revised ed.). Tropical Science Center, San José, CR.

Holdridge, L.R. and L.J. Poveda (2da. ed. actualizada por Q. Jiménez). 1997. *Arboles de Costa Rica: Vol 1*. Centro Científico Tropical, San José, CR.

Honychurch, P.N. 1986. *Caribbean Wild Plants and their Uses: An Illustrated Guide to some Medicinal and Wild Ornamental Plants of the West Indies*. Macmillan Publ. Ltd., London.

Horvitz, C.C. 1991. Light environments, stage structure, and dispersal syndromes of Costa Rican Marantaceae. In: *Ant-Plant Interactions*, C.R. Huxley & D.F. Cutler (eds.), 463-485. Oxford Univ. Press, Oxford, UK.

Horvitz, C.C., M. A. Pizo, B. Bello y Bello, J. LeCorff, and R. Dirzo. 2002. Are plant species that need gaps for recruitment more attractive to seed-dispersing birds and ants than other species? In: *Seed Dispersal and Frugivory: Ecology, Evolution and Conservation*, D.J. Levey, W.R. Silva and M. Galetti (eds.), 145-159. CAB International Press, Oxon, UK.

Hoshizaki, J.B. and R.C. Moran. 2001. *Fern Grower's Manual, Revised and Expanded Edition*. Timber Press, OR.

Hostettmann, K., A. Marston, M. Maillard, and M. Hamburger (eds.). 1995. *Phytochemistry of Plants Used in Traditional Medicine*. (Proceedings of the Phytochemical Society of Europe). Oxford Univ. Press, Oxford.

Howard, R.A. 1981. Three experiences with the manchineel (*Hippomane* spp., Euphorbiaceae). *Biotropica* 13: 224-227.

Hsiao, S.-C., J. D. Mauseth, and L. D. Gomez. 1994. Growth and anatomy of the vegetative body of the parasitic angiosperm *Langsdorffia hypogaea* (Balanophoraceae). *Bull. Torrey Bot. Club* 121: 24-39.

Hsiao, S.-C., J. D. Mauseth, and C.-I. Peng. 1995. Composite bundles, the host/parasite interface in the holoparasitic angiosperms *Langsdorffia* and *Balanophora* (Balanophoraceae). *Amer. J. Bot.* 82: 81-91.

Instituto Nacional de Biodiversidad (INBio). Familias de plantas-BIMMS. http://www.inbio.ac.cr/bims/PLANTAE.html

Instituto Nacional de Biodiversidad (INBio). UBIs: unidades básicas de información. http://darnis.inbio.ac.cr/ubis/find.html

INBio-Reserva Natural Absoluta Cabo Blanco. 1997. *Árboles de la Reserva Absoluta Cabo Blanco: Especies selectas.* Instituto Nacional de Biodiversidad, Heredia, CR.

Itino, T., M. Kato, and M. Hotta. 1991. Pollination ecology of the two wild bananas, *Musa acuminata* subsp. *halabanensis* and *M. salaccensis*: chiropterophily and ornithophily. *Biotropica* 23: 151-158.

Janzen, D.H., ed. 1983. *Costa Rican Natural History.* Univ. of Chicago Press, Chicago.

Jiménez, G. 1998. Proyecto: Manejo y tratamiento natural de cáscaras de naranja [Electronic version]. *Rothschildia* (revista de ACG) 5 (1).

Jiménez, J.A. 1985. *Rhizophora mangle L. – Red mangrove.* Res. Pap. SO-ITF-SM-2, USDA Forest Service, Institute of Tropical Forestry, Río Piedras, PR.

Jiménez, J.A. 1988. Floral and fruiting phenology of trees in a mangrove forest on the dry Pacific coast of Costa Rica. *Brenesia* 29: 33-50.

Jiménez, J.A. 1994. *Los manglares del Pacífico de Centroamérica.* Editorial Fundación UNA, Heredia, CR.

Jiménez, J.A. and A.E. Lugo. 1985. *Avicennia germinans (L) L. – Black mangrove.* Res. Pap. SO-ITF-SM-4, USDA Forest Service, Institute of Tropical Forestry, Río Piedras, PR.

Jiménez, Q. 1999. *Árboles maderables en peligro de extinción en Costa Rica.* 2da. edición revisada y ampliada. Instituto Nacional de Biodiversidad, Heredia, CR.

Jiménez, Q. and L.J. Poveda. no date. *Lista actualizada de los árboles maderables de Costa Rica.* Aportes al desarrollo sostenible. Universidad Nacional, Heredia, CR.

Jones, C. E. and R. J. Little (eds.). 1983. *Handbook of Experimental Pollination Biology.* Van Nostrand Reinhold, NY.

Jones, D.L. 1987. *Encyclopaedia of Ferns.* Timber Press, Portland, OR.

Jones, D.L. 1993. *Cycads of the World.* Reed New Holland, Sydney, Australia.

Judd, W.S., C.S. Campbell, E.A. Kellogg, and P.F. Stevens. 1999. *Plant Systematics: A Phylogenetic Approach.* Sinauer Press, Sunderland, MA.

Kaplan, E.H. 1988. *A Field Guide to Southeastern and Caribbean Seashores* (The Peterson Field Guide Series: 36). Houghton Mifflin, Boston.

Kappelle, M. 1995. *Ecology of Mature and Recovering Talamancan Montane Quercus Forests, Costa Rica.* Dissertation. University of Amsterdam.

Kappelle, M. 1996. *Los bosques de roble (Quercus) de la Cordillera de Talamanca, Costa Rica.* Instituto Nacional de Biodiversidad (Heredia, CR) & Universidad de Amsterdam.

Kass, D. 1999 Noticias: Proyecto *Tithonia diversifolia* [Electronic version]. *Revista Agroforestería en las Américas* 6 (23).

Katiyar, S.K., R. Agarwal, and H. Mukhtar. 1996. Inhibition of tumor promotion in SENCAR mouse skin by ethanol extract of *Zingiber officinale* rhizome. *Cancer Research* 56: 1023-1030.

Kay, M.A. 1996. *Healing with Plants.* The University of Arizona Press, Tucson.

Keller, R. 1996. *Identification of Tropical Woody Plants in the Absence of Flowers or Fruits: A Field Guide.* Birkhaüser Verlag, Basel, Switzerland.

Kingsbury, J.M. 1988. *200 Conspicuous, Unusual, or Economically Important Tropical Plants of the Caribbean.* Bullbrier Press, Ithaca, NY.

Koptur, S., E.N. Dávila, D.R. Gordon, B.J. Davis McPhail, C.G. Murphy, and J.B. Slowinski. 1990. The effect of pollen removal on the duration of the staminate phase of *Centropogon talamancensis*. *Brenesia* 33: 15-18.

Kress, W. J. 1985. Bat pollination of an old world *Heliconia*. *Brenesia* 17: 302-308.

Kuck, L.E. and R.C. Tongg. 1960. *Hawaiian Flowers & Flowering Trees: A Guide to Tropical and Semitropical Flora*. Charles E. Tuttle Co., Rutland, VT.

La Selva Biological Station. La flora digital de La Selva. http://www.ots.ac.cr/local/florula/index.htm

Lellinger, D.B. 1985. *A Field Manual of the Ferns and Fern-allies of the United States and Canada*. Smithsonian Institution Press, Washington, DC.

Lellinger, D.B. 1989. The ferns and fern-allies of Costa Rica, Panama, and the Chocó. (Part 1: Psilotaceae through Dicksoniaceae). *Pteridologia* 2A: 5-364.

Lemke, T.O. 1985. Pollen carrying by the nectar-feeding bat *Glossophaga soricina* in a suburban environment. *Biotropica* 17: 107-111.

Lennox, G.W. and S.A. Seddon. 1978. *Flowers of the Caribbean*. Macmillan Education, London.

León, J. and L.J. Poveda. 2000. *Los nombres comunes de las plantas en Costa Rica*. Editorial Guayacán, San José, CR.

Lewis, W.H. and M.P.F. Elvin-Lewis. 1977. *Medical Botany*. John Wiley & Sons, NY.

Liegel, L.H. and J.W. Stead. 1990. *Cordia alliodora* (Ruiz & Pav.) Oken – Laurel, Capá prieto. In: *Silvics of North America*. Hardwoods. Agric. Handb. 654 (2): 270-277. USDA, Washington, DC.

Lötschert, W. and G. Beese. 1983. *Collins Photo Guide: Tropical Plants*. HarperCollins, Hong Kong.

Luteyn, J.L. 1983. *Cavendishia* (Ericaceae: Vaccinieae). Flora Neotropica Monograph 35. The New York Botanical Garden, Bronx.

Lutterodt, G.D. 1989. Inhibition of gastrointestinal release of acetylcholine by quercetin as a possible mode of action of *Psidium guajava* leaf extracts in the treatment of acute diarrhoeal disease (Electronic version of abstract retrieved from: http://www.rain-tree.com/clinic/clinicga.htm#G6). *J. Ethnopharmacol.* 25: 235-47.

Maas, P.J.M. and L.Y.Th. Westra. 1993. *Neotropical Plant Families*. Koeltz Scientific Books, Koenigstein, Germany.

Mabberly, D.J. 1997. *The Plant-Book*. Cambridge Univ. Press, Cambridge.

Macey, A. 1975. The vegetation of Volcán Poás National Park, Costa Rica. *Rev. Biol. Trop.* 23: 239-255.

Madison, M. 1977. A revision of *Monstera* (Araceae). *Contr. Gray Herb.* 207: 1-100.

Martin, F.W. and R.M. Ruberté. 1975. *Edible Leaves of the Tropics*. Mayagüez Institute of Tropical Agriculture (with A.I.D. and U.S.D.A.). Mayagüez, PR.

McCaleb, R. 1989. Guava leaf follow-up. *HerbalGram* 21: 17.

McClatchey, W. 2002. From Polynesian healers to health food stores: Changing perspectives of *Morinda citrifolia* (Rubiaceae). *Integrative Cancer Therapies* 1(2): 110-120.

McDade, L.A. 1985. Breeding systems of Central American *Aphelandra* (Acanthaceae). *Amer. J. Bot.* 72(10): 1515-1521.

McDade, L.A., K.S. Bawa, H.A. Hespenheide, and G.S. Hartshorn (eds.). 1994. *La Selva: Ecology and Natural History of a Neotropical Rainforest*. The University of Chicago Press, Chicago.

McGregor, S.E. 1976. *Insect Pollination of Cultivated Crop Plants*. Agric. Handbook No. 496. U.S.D.A. U.S. Government Printing Office, Washington, DC.

Missouri Botanical Garden. VAST (VAScular Tropicos): Nomenclatural database and associated authority files. http://mobot.mobot.org/W3T/Search/vast.html

Montiel Longhi, M.B. 1991. *Introducción a la flora de Costa Rica*. Editorial de la Universidad de Costa Rica, San José, CR.

Mora de Retana, D.E. and J.T. Atwood. 1992. *Orchids of Costa Rica, Part 2*. Icones Plantarum Tropicarum. The Marie Selby Botanical Gardens, Sarasota, FL.

Mora de Retana, D.E. and J.T. Atwood. 1993. *Orchids of Costa Rica, Part 3*. Icones Plantarum Tropicarum. The Marie Selby Botanical Gardens, Sarasota, FL.

Morales, J.F. 2000. *Bromelias de Costa Rica / Costa Rica Bromeliads*. 2da. ed. Instituto Nacional de Biodiversidad, Heredia, CR.

Morales, J.F. 2001. *Orquídeas, cactus y bromelias del bosque seco de Costa Rica/ Orchids, Cacti and Bromeliads of the Dry Forest*. Editorial INBio, Heredia, CR.

Moran, R.C. 1997. The little nitrogen factories, In: *Fiddlehead Forum – Bulletin of the American Fern Society*, 24(2).

Moran, R.C. 2000. *The Genera of Neotropical Ferns: A Guide for Students* (prepared for Tropical Plant Systematics, OTS). R.C. Moran, The New York Botanical Garden, Bronx.

Moran, R.C., S. Klimas, and M. Carlsen. 2003. Low-trunk epiphytic ferns on tree ferns versus angiosperms in Costa Rica. *Biotropica* 35(1): 48–56.

Moran, R.C. and R. Riba (eds.). 1995. Psilotaceae a Salviniaceae. In: G. Davidse, M. Sousa S., & S. Knapp (eds.), *Flora Mesoamericana*. Vol. 1. [Electronic version]. Missouri Botanical Garden Press, Universidad Nacional Autonoma de Mexico (UNAM), and the Natural History Museum (London).

Moreno, N.P. 1984. *Glosario Botánico Ilustrado*. Compañía Editorial Continental, México, DF.

Mori, S.A., G. Cremers, C. Gracie, J. de Granville, M. Hoff, and J.D. Mitchell. 1997. *Guide to the Vascular Plants of Central French Guiana*. The New York Botanical Garden, Bronx.

Mori, S.A. and G.T. Prance. 1990. Lecythidaceae-Part II. *Flora Neotropica Monograph* 21 (II). The New York Botanical Garden, Bronx.

Morton, J.F. 1977. *Exotic Plants for House and Garden*. A Golden Guide. Golden Press, NY.

Morton, J.F. 1981. *Atlas of Medicinal Plants of Middle America*. Charles C. Thomas Publisher, Springfield, IL.

Morton, J.F. 1985. Indian Almond (*Terminalia catappa*), Salt-tolerant, useful, tropical tree with "nut" worthy of improvement. *Econ. Bot.* 39(2): 101-112.

Morton, J.F. 1992. The ocean-going noni, or Indian mulberry (*Morinda citrifolia*, Rubiaceae) and some of its "colorful" relatives. *Econ. Bot.* 46(3): 241-256.

Morton, J.F. 1994. Lantana, or red sage (*Lantana camara* L., [Verbenaceae]), notorious weed and popular garden flower; Some cases of poisoning in Florida. *Econ. Bot.* 48(3): 259-270.

Mueller, G.M. and R.E. Halling. 2002. Macrofungi of Costa Rica. http://www.nybg.org/bsci/res/hall/sumry2.html

Mulkey, S.S., R.L. Chazdon, and A.P. Smith (eds.). 1996. *Tropical Forest Plant Ecophysiology*. Chapman and Hall, NY.

Murray, K.G. and J.M. García-C. 2002. Contributions of seed dispersal and demography to recruitment limitation in a Costa Rican cloud forest. Pp. 323-338 in: *Seed Dispersal and Frugivory: Ecology, Evolution, and Conservation*. D.J. Levey, W.R. Silva, and M. Galetti (eds.). CAB International.

Murray, K.G., S. Russell, C.M. Picone, K. Winnett-Murray, W. Sherwood, and M.L. Kuhlmann. 1994. Fruit laxatives and seed passage rates in frugivores: Consequences for plant reproductive success. *Ecology* 75(4): 989-994.

Nadkarni, N.M. and N.T. Wheelwright (eds.). 2000. *Monteverde: Ecology and Conservation of a Tropical Cloud Forest.* Oxford Univ. Press, NY.

Niembro, A. 1986. *Árboles y arbustos útiles de México.* Editorial Limusa S.A., México, DF.

Nuñez, E. 1978. *Plantas medicinales de Costa Rica y su folclore.* Editorial de la Universidad de Costa Rica, San José, CR.

Nutri-Tech Solutions of Australia. no date. Perennial Peanut, *Arachis pintoi.* http://www.nutri-tech.com.au/articles/pintoi.htm

O'Hair, S.K., contributor. 1995. New Crop FactSHEET: Cassava [Electronic version retrieved from http://www.hort.purdue.edu/newcrop/cropfactsheets/cassava .html]. Tropical Research and Education Center, Univ. of Florida.

Oliveira, P.S., M. Galetti, F. Pedroni, and L.P.C. Morellato. 1995. Seed cleaning by *Mycocepurus goeldii* ants (Attini) facilitates germination in *Hymenaea courbaril* (Caesalpiniaceae). *Biotropica* 27: 518-522.

Opler, P.A., H.G. Baker, and G.W. Frankie. 1975. Reproductive biology of some Costa Rican *Cordia* species (Boraginaceae). *Biotropica* 7: 234-247.

Ortiz, J. 2000. Helecho macho, leche y cáncer gástrico [Electronic version]. *Ambientico* 80 (mayo).

Padilla, V. 1973. *Bromeliads: A Descriptive Listing of the Various Genera and the Species most often Found in Cultivation.* Crown Publishers, NY.

Pennington, T.D. and J. Sarukhan. 1968. *Árboles tropicales de México.* Instituto Nacional de Investigaciones Forestales Mex. and United Nations, FAO, Rome.

Perkins, K.D. and W.W. Payne. 1978. *Guide to the Poisonous and Irritant Plants of Florida.* Cooperative Extension Service. Univ. of Florida, IFAS, Gainesville.

Perry Jr., J.P. 1991. *The Pines of Mexico and Central America.* Timber Press, Portland, OR.

Perry, F. and R. Hay. 1982. *A Field Guide to Tropical and Subtropical Plants.* Van Nostrand Reinhold Co., NY.

Pittier, H. 1978. *Plantas usuales de Costa Rica.* Editorial Costa Rica, San José, CR.

Plotkin, M.J. 1993. *Tales of a Shaman's Apprentice.* Penguin Books, NY.

Plowman, T.C. 1998. A revision of the South American species of *Brunfelsia* (Solanaceae). *Fieldiana.* New Series, no. 39. Field Museum of Natural History, Chicago.

Poll, D.J., S.C. Snedaker, and A.E. Lugo. 1977. Structure of mangrove forests in Florida, Puerto Rico, Mexico, and Costa Rica. *Biotropica* 9: 195-212.

Poveda, L.J. and P.E. Sánchez-Vindas. 1999. *Árboles y palmas del Pacífico norte de Costa Rica* (claves dendrológicas). Editorial Guayacán, San José, CR.

Proctor, M., P. Yeo, and A. Lack. 1996. *The Natural History of Pollination.* Timber Press, Portland, OR.

Pupulin, F. 1998. *Orchids of Manuel Antonio National Park.* MesoAmerican Press. San José, CR.

Quesada, F. J., Q. Jiménez, N. Zamora, R. Aguilar, and J. González. 1997. *Árboles de la Península de Osa.* Instituto Nacional de Biodiversidad, Heredia, CR.

Quesada, M.A. 1996. *Nuevo diccionario de costarriqueñismos.* Editorial Tecnológica de Costa Rica, Cartago.

Rabinowitz, D. 1978. Dispersal properties of mangrove propagules. *Biotropica* 10: 47-57.

Reichert, R. 1997. Bromelain and musculoskeletal injuries. *HerbalGram* 39: 17.

Rejmánek, M. and S.W. Brewer. 2001. Vegetative identification of tropical woody plants: State of the art and annotated bibliography. *Biotropica* 33: 214-228.

Rickson, F.R. and M.M. Rickson. 1998. The cashew nut, *Anacardium occidentale* (Anacardiaceae), and its perennial association with ants: Extrafloral nectary location and the potential for ant defense. *Amer. J. Bot.* 85: 835-849.

Rico-Gray, V. and L.B. Thien. 1987. Some aspects of the reproductive biology of *Schomburgkia tibicinis* Batem. (Orchidaceae) in Yucatan, Mexico. *Brenesia* 28: 13-24.

Rivas Rossi, M. 1998. *Cactáceas de Costa Rica*. Editorial Universidad Estatal a Distancia, San José, Costa Rica.

Rodale Institute Research Center. 1997. Information Sheet #2 - Tropical Legumes. http://www.forages.css.orst.edu/Organizations/Agriculture/RIRC/Information2.html

Rodríguez, R.L., D.E. Mora, M.E. Barahona, and N.H. Williams. 1986. *Géneros de orquídeas de Costa Rica*. Editorial de la Universidad de Costa Rica, San José, CR.

Rojas, A. 1996. Aportes a la flora pteridophyta neotrópica 1. Notas sobre el género *Niphidium* (Polypodiaceae). *Brenesia* 45/46: 27-32.

Rojas, A. 1999. *Helechos arborescentes de Costa Rica*. Instituto Nacional de Biodiversidad, Heredia, CR.

Ross de Cerdas, M. 1995. *Las frutas del paraíso*. Editorial de la Universidad de Costa Rica, San José, CR.

Rubatzky, V. E., and M. Yamaguchi. 1997. *World Vegetables: Principles, Production, and Nutritive Values*. Chapman and Hall, NY.

Rymer, L. 1976. The history and ethnobotany of bracken. *Bot. J. Linn. Soc.* 73: 151-176.

Sakai, S., M. Kato, and H. Nagamasu. 2000. *Artocarpus* (Moraceae)-gall midge pollination mutualism mediated by male-flower parasitic fungus [Electronic version]. *Amer. J. Bot.* 87: 440-445.

Salas, J.B. 1993. *Árboles de Nicaragua*. Instituto Nicaragüense de Recursos Naturales y del Ambiente, Managua.

Sánchez-Vindas, P.E. 1983. *Flórula del Parque Nacional Cahuita*. Editorial Universidad Estatal a Distancia, San José, CR.

Sánchez-Vindas, P.E. 1989. Flora de Nicaragua: Myrtaceae. *Brenesia* 31: 53-73.

Sánchez-Vindas, P.E. and L.J. Poveda. 1997. *Claves dendrológicas para la identificación de los principales árboles y palmas de la zona norte y Atlántica de Costa Rica*. Overseas Development Administration, San José, CR.

Sancho, E. and M. Baraona. 1997. *Frutas del trópico: Guía Fotográfica*. E. Sancho, San José, CR.

Scariot, A.O., E. Lleras, and J.D. Hay. 1991. Reproductive biology of the palm *Acrocomia aculeata* in central Brazil. *Biotropica* 23: 12-22.

Schmutz, E.M. and L.B. Hamilton. 1979. *Plants that Poison: An Illustrated Guide for the American Southwest*. Northland Press, Flagstaff, AZ.

Schultes, R.E. 1976. *Hallucinogenic Plants*. A Golden Guide. Golden Press, NY.

Schultes, R.E. and A. Hofmann. 1992. *Plants of the Gods*. Healing Arts Press, Rochester VT.

Schultes, R.E. and R.F. Raffauf. 1990. *The Healing Forest: Medicinal and Toxic Plants of the Northwest Amazonia*. Dioscorides Press, Portland, OR.

Seddon, S.A. and G.W. Lennox. 1980. *Trees of the Caribbean*. Macmillan Education, London.

Segelman, A.B., F.P. Segelman, J. Karliner, and R.D. Sofia. 1976. Sassafras and herb tea: Potential health hazards. *JAMA* 236 (No. 5): 477.

Seymour, R.S. 1997. Plants that warm themselves. *Scientific American* March: 90-95.

Sheehan, T. and M. 1985. *Orchid Genera Illustrated.* Cornell Univ. Press, Ithaca.

Shuttleworth, F.S., H.S. Zim, and G.W. Dillon. 1970. *Orchids.* A Golden Guide. Golden Press, NY.

Siegel, R.K. 1976. Herbal intoxication: Psychoactive effects from herbal cigarettes, tea, and capsules. *JAMA* 236: 473-476.

Simpson, B.B. and M.C. Ogorzaly. 1995. *Economic Botany: Plants in our World.* McGraw-Hill, NY.

Smith, N.J.H., J.T. Williams, D.L. Plucknett, and J.P. Talbot. 1992. *Tropical Forests and their Crops.* Cornell Univ. Press, Ithaca.

Smith, R.T. and J.A. Taylor (eds.). 1986. *Bracken: Ecology, Land use and Control Technology.* Parthenon Publishing, Leeds, England.

Soderstrom, T.R. and C.E. Calderón. 1979. A commentary on the bamboos (Poaceae: Bambusoideae). *Biotropica* 11: 161-172.

Solano, J.A. 1992. *Guía de campo de las especies más comunes del Parque Nacional Tortuguero.* PACTo, Ministerio de Recursos Naturales, Energía, y Minas, CR.

Standley, P.C. 1937-8. *Flora of Costa Rica.* Bot. Series 18 (1-4). Field Museum of Natural History, Chicago.

Stevens, W.D., C. Ulloa, A. Pool, and O.M. Montiel (eds.). 2001. *Flora de Nicaragua.* Tomo I, II, y III. Missouri Botanical Garden Press, St. Louis, MO.

Stiles, F.G. 1975. Ecology, flowering phenology, and hummingbird pollination of some Costa Rican *Heliconia* species. *Ecology* 56: 285-301.

Stiles, F.G. 1979. Notes on the natural history of *Heliconia* (Musaceae) in Costa Rica *Brenesia* 15 (supl.): 151-180.

Stiles, F.G. and A.F. Skutch. 1989. *A Guide to the Birds of Costa Rica.* Cornell Univ. Press, Ithaca, NY.

Strong Jr., D.R. and T.S. Ray. 1975. Host tree location behavior of a tropical vine (*Monstera gigantea*) by skototropism. *Science* 90: 804-806.

Tavares, F.C., J. Beer, F. Jimémez, G. Schroth, and C. Fonseca. 1999. Avances de investigación: Experiencia de agricultores de Costa Rica con la introducción de árboles maderables en plantaciones de café [Electronic version]. *Revista Agroforestería en las Américas* 6(23).

Tempel, A.S. 1983. Bracken fern (*Pteridium aquilinum*) and nectar-feeding ants: A nonmutualistic interaction. *Ecology* 64: 1411-1422.

Toledo, V.M. 1977. Pollination of some rain forest plants by non-hovering birds in Veracruz, Mexico. *Biotropica* 9: 262-267.

Tomlinson, P.B. 1980. *The Biology of Trees Native to Tropical Florida.* Harvard Univ. Printing. Allston, MA.

Tryon, R.M. and A.F. Tryon. 1982. *Ferns and Allied Plants with Special Reference to Tropical America.* Springer-Verlag, NY.

Uhl, N. and J. Dransfield. 1987. *Genera Palmarum.* Allen Press, Lawrence, KS.

USDA Forest Service. no date. Index of species Information: *Pteridium aquilinum* botanical and ecological characteristics. http://www.fs.fed.us/database/feis/plants/fern/pteaqu/botanical_and_ecological_characteristics.html

Valerio, C.E. 1984. Insect visitors to the inflorescence of the aroid *Dieffenbachia oerstedii* (Araceae) in Costa Rica. *Brenesia* 22: 139-146.

van der Pijl, L. and C.H. Dodson. 1966. *Orchid Flowers: Their Pollination and Evolution.* Univ. of Miami Press, Coral Gables.

Van Roosmalen, M.G.M. 1985. *Fruits of the Guianan Flora*. Institute of Systematic Botany, Utrecht, The Netherlands.

VanDusen Joyce, K. 1988. *El Consejero Guanacasteco*. Katy VanDusen Joyce, Monteverde, CR.

Vasquez, R. and A.H. Gentry. 1989. Use and misuse of forest-harvested fruits in the Iquitos area. *Conservation Biology* 3: 350-361.

Wainwright, M. 2002. *The Natural History of Costa Rican Mammals*. Distribuidores Zona Tropical, S.A. Miami.

Wallace, J.W. and R.L. Mansell (eds.). 1976. *Biochemical Interactions between Plants and Animals*. Recent Advances in Phytochemistry, Vol. 10. Plenum Press, NY.

Watanabe, I. (revised Dec. 1, 2003). ABC of Azolla. http://www.asahi-net.or.jp/~it6i-wtnb/azolla~E.html

Webb, E.L. 1999. Growth ecology of *Carapa nicaraguensis* Aublet. (Meliaceae): Implications for natural forest management. *Biotropica* 31: 102-110.

Weber, A. (ed.) and collaborators. 2001. *An Introductory Field Guide to the Flowering Plants of the Golfo Dulce Rain Forests, Costa Rica*. Landes Museum, Austria.

Wheelwright, N.T., W.A. Haber, and K.G. Murray. 1984. Tropical fruit-eating birds and their food plants: A survey of a Costa Rican lower montane forest. *Biotropica* 16: 173-192.

Wilbur, R. L. 1976. A synopsis of the Costa Rican species of the genus *Centropogon* Presl (Campanulaceae, Lobelioideae). *Brenesia* 8: 59-84.

Wilson, L. 1977. *Bromeliads for Modern Living*. Merchants Publishing Co. Kalamazoo, MI.

Woodson Jr., R.E., R.W. Schery, and collaborators. 1943-1981. *Flora of Panama* (41 issues). Annals of the Missouri Botanical Garden, St. Louis.

Wootton, J. T. and I-F. Sun. 1990. Bract liquid as a herbivore defense mechanism for *Heliconia wagneriana* inflorescences. *Biotropica* 22: 155-159.

Young, A.M. 1983. Nectar and pollen robbing of *Thunbergia grandiflora* by *Trigona* bees in Costa Rica. *Biotropica* 15: 78-80.

Young, A.M. 1994. *The Chocolate Tree: A Natural History of Cacao*. Smithsonian Institution Press. Washington, DC.

Zamora, N. 1989. *Flora arborescente de Costa Rica: I. Especies de hojas simples*. Editorial Tecnológica de Costa Rica, Cartago.

Zamora, N., Q. Jiménez, and L. Poveda. 2000. *Árboles de Costa Rica, Vol. II*. Editorial INBio, Heredia, CR.

Zamora, N., Q. Jiménez, and L. Poveda. 2004. *Árboles de Costa Rica, Vol. III*. Editorial INBio, Heredia, CR.

Zamora, N. and T. D. Pennington. 2001. *Guabas y cuajiniquiles de Costa Rica (Inga spp.)*. Editorial INBio, Heredia, CR.

Zheng, M.S. and Z.Y. Lu. 1989. Antiviral effect of mangiferin and isomangiferin on herpes simplex virus (Electronic version of abstract retrieved from National Library of Medicine PubMed. http://www.ncbi.nlm.nih.gov/entrez/query.fcgi?cmd=Retrieve&db=PubMed&list_uids=2554669&dopt=Citation). *Acta Pharmacologica Sinica* 10(1): 85-90

Zomlefer, W.B. 1994. *Guide to Flowering Plant Families*. The Univ. of North Carolina Press, NY & London.

Index

Scientific names are in *italics*; common names appear in regular type. The scientific names of all featured species are in **bold** type.

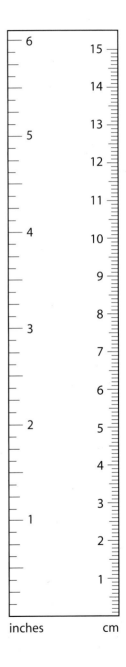

inches   cm